STUDENT'S SOLUTIONS MANUAL

TRIGONOMETRY

SIXTH EDITION

STUDENT'S SOLUTIONS MANUAL

TRIGONOMETRY

SIXTH EDITION

Lial • Hornsby • Schneider

Prepared with the assistance of

Sandra Morris

August Zarcone
John Sullivan
Gerald Krusinski

College of DuPage

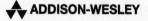
ADDISON-WESLEY

An imprint of Addison Wesley Longman, Inc.

Reading, Massachusetts • Menlo Park, California • New York • Harlow, England
Don Mills, Ontario • Sydney • Mexico City • Madrid • Amsterdam

Reproduced by Addison-Wesley Educational Publishers Inc. from camera-ready copy supplied by the author.

Copyright © 1997 Addison-Wesley Educational Publishers Inc.

ISBN   0-673-98337-4

3 4 5 6 7 8 9 10 CRS 9998

PREFACE

This book provides complete solutions for many of the exercises in *Trigonometry*, Sixth Edition, by Margaret L. Lial, E. John Hornsby, Jr., and David I. Schneider. Solutions are provided for the odd–numbered section exercises and chapter review exercises as well as all "Discovering Connections" exercises and some writing exercises.

In addition, Chapter Tests with answers and Cumulative Review Exercises with solutions are included. You may find these exercises helpful in preparing for examinations.

This book should be used as an aid as you work to master your coursework. Try to solve the exercises that your instructor assigns before you refer to the solutions in the book. Then, if you have difficulty, read these solutions to guide you in solving the exercises. The solutions have been written so they are consistent with the methods used in the textbook.

Solutions to textbook exercises that require graphs will refer to the answer section in the textbook. These graphs are not included in this book.

In addition to solutions, you will find a list of suggestions on how to be successful in mathematics. A careful reading will be helpful to many students.

The following people have made valuable contributions to the production of this *Student's Solution Manual*: Terry McGinnis, editor; Judy Martinez, typist; and Carmen Eldersveld, proofreader. Artwork has been provided by Precision Graphics, Therese Brown, Charles Sullivan, and August Zarcone.

We also want to thank Tommy Thompson of Cedar Valley Community College for his suggestions for the essay "To the Student: Success in Mathematics."

CONTENTS

TO THE STUDENT: SUCCESS IN MATHEMATICS

The main reason students have difficulty with mathematics is that they don't know how to study it. Studying mathematics *is* different from studying subjects like English or history. The key to success is regular practice.

This should not be surprising. After all, can you learn to play the piano or to ski well without a lot of regular practice? The same thing is true for learning mathematics. Working problems nearly every day is the key to becoming successful. Here is a list of things you can do to help you succeed in studying trigonometry.

1. *Attend class regularly.* Pay attention in class to what your instructor says and does, and make careful notes. In particular, note the problems the instructor works on the board and copy the complete solutions. Keep these notes separate from your homework to avoid confusion when you read them over later.

2. Don't hesitate to ask questions in class. It is not a sign of weakness, but of strength. There are always other students with the same question who are too shy to ask.

3. *Read your text carefully.* Many students read only enough to get by, usually only the examples. Reading the complete section will help you to be successful with the homework problems. Most exercises are keyed to specific examples or objectives that will explain the procedures for working them.

4. Before you start on your homework assignment, rework the problems the instructor worked in class. This will reinforce what you have learned. Many students say, "I understand it perfectly when you do it, but I get stuck when I try to work the problem myself."

5. Do your homework assignment only *after* reading the text and reviewing your notes from class. Check your work with the answers in the back of the book. If you get a problem wrong and are unable to see why, mark that problem and ask your instructor about it. Then practice working additional problems of the same type to reinforce what you have learned.

6. Work as neatly as you can. Write your symbols clearly, and make sure the problems are clearly separated from each other. Working neatly will help you to think clearly and also make it easier to review the homework before a test.

7. After you have completed a homework assignment, look over the text again. Try to decide what the main ideas are in the lesson. Often they are clearly highlighted or boxed in the text.

8. Keep any quizzes and tests that are returned to you and use them when you study for future tests and the final exam. These quizzes and tests indicate what your instructor considers most important. Be sure to correct any problems on these tests that you missed, so you will have the corrected work to study.

9. Don't worry if you do not understand a new topic right away. As you read more about it and work through the problems, you will gain understanding. Each time you look back at a topic you will understand it a little better. No one understands each topic completely right from the start.

STUDENT'S SOLUTIONS MANUAL

TRIGONOMETRY

SIXTH EDITION

CHAPTER 1 THE TRIGONOMETRIC FUNCTIONS

Section 1.1

For Exercises 1–7, see the answer graph in the back of the textbook.

1. To graph (3, 2), go three units from the origin to the right along the x-axis, and then go two units up parallel to the y-axis. The point (3, 2) is in quadrant I since the x- and y- coordinates are both positive.

3. To graph (−7, −4), go seven units from the origin to the left along the x-axis, and then go four units down parallel to the y-axis. The point (−7, −4) is in quadrant III since the x- and y-coordinates are both negative.

5. To graph (0, 5), do not move along the x-axis at all since the x-coordinate is 0. Move five units along the y-axis. The point (0, 5) is not in any quadrant since it is on the y-axis.

7. To graph (4.5, 7), go about four and one half units from the origin to the right along the x-axis, then go seven units up parallel to the y-axis. The point (4.5, 7) is in quadrant I since the x- and y-coordinates are both positive.

9. (−5, π), lies in quadrant II since the x-coordinate is negative and the y-coordinate ($\pi \approx 3.14$) is positive.

11. (−$\sqrt{2}$, −2$\sqrt{2}$) lies in quadrant III since the x- and y-coordinates are both negative.

15. Since xy = 1, and 1 is a positive number attained by the product of two variables x and y, then either x and y are both positive (quadrant I), or both negative (quadrant III). The graph of xy = 1 will lie in quadrants I and III.

In Exercises 17–23, the formula
$$d = \sqrt{(x_2 - x_1)^2 + (y_2 - y_1)^2}$$
is used to find the distance between (x_1, y_1) and (x_2, y_2).

17. (x_1, y_1) = (2, −1) and (x_2, y_2) = (−3, −4)
$$d = \sqrt{(-3 - 2)^2 + [-4 - (-1)]^2}$$
$$= \sqrt{(-5)^2 + (-3)^2}$$
$$= \sqrt{25 + 9}$$
$$= \sqrt{34}$$

19. (x_1, y_1) = (−1, 0) and (x_2, y_2) = (−4, −5)
$$d = \sqrt{[-4 - (-1)]^2 + (-5 - 0)^2}$$
$$= \sqrt{(-3)^2 + (-5)^2}$$
$$= \sqrt{9 + 25}$$
$$= \sqrt{34}$$

21. (x_1, y_1) = ($\sqrt{2}$, −$\sqrt{5}$) and (x_2, y_2) = (3$\sqrt{2}$, 4$\sqrt{5}$)
$$d = \sqrt{(3\sqrt{2} - \sqrt{2})^2 + [4\sqrt{5} - (-\sqrt{5})]^2}$$
$$= \sqrt{(2\sqrt{2})^2 + (5\sqrt{5})^2}$$
$$= \sqrt{8 + 125}$$
$$= \sqrt{133}$$

23. $(x_1, y_1) = (5, -6)$. Let $(x_2, y_2) =$ $(5, 0)$ since we want the point on the x–axis with x–value 5.

$$d = \sqrt{(5 - 5)^2 + [0 - (-6)]^2}$$
$$= \sqrt{0^2 + 6^2}$$
$$= \sqrt{36}$$
$$= 6$$

25. In the distance formula, we find the square root of $(x_2 - x_1)^2 + (y_2 - y_1)^2$. From the display, $x_2 = -2$, $x_1 = 3$, $y_2 = 1$, and $y_1 = -\pi$. Therefore, the two points (x_1, y_1) and (x_2, y_2) are $(3, -\pi)$ and $(-2, 1)$.

29. $(a, b, c) = (9, 12, 5)$

$$a^2 + b^2 = 9^2 + 12^2$$
$$= 81 + 144$$
$$= 225$$
$$= 15^2 = c^2$$

This triple is a Pythagorean triple since $a^2 + b^2 = c^2$.

31. $(a, b, c) = (5, 10, 15)$

$$a^2 + b^2 = 5^2 + 10^2$$
$$= 25 + 100$$
$$= 125$$
$$\neq 15^2 = c^2$$

This triple is not a Pythagorean triple since $a^2 + b^2 \neq c^2$.

33. If, for positive integers a, b, and c, $a^2 + b^2 = c^2$, then it is not necessarily true that $a + b = c$. For example,

let $a = 3$, $b = 4$, and $c = 5$. Then $a^2 = 9$, $b^2 = 16$, and $c^2 = 25$.

$$a^2 + b^2 = 9 + 16$$
$$= 25 = c^2.$$

Note that $a + b = 3 + 4 = 7 \neq 5 = c$.

35. Are $(-2, 5)$, $(1, 5)$, and $(1, 9)$ the vertices of a right triangle? Use the distance formula to find the lengths of the sides connecting the points.

$$a = \sqrt{[1 - (-2)]^2 + (5 - 5)^2}$$
$$= \sqrt{3^2 + 0^2} = \sqrt{9} = 3$$
$$b = \sqrt{(1 - 1)^2 + (9 - 5)^2}$$
$$= \sqrt{0^2 + 4^2} = \sqrt{16} = 4$$
$$c = \sqrt{[1 - (-2)]^2 + (9 - 5)^2}$$
$$= \sqrt{3^2 + 4^2} = \sqrt{9 + 16} = \sqrt{25} = 5$$
$$a^2 + b^2 = 3^2 + 4^2$$
$$= 9 + 16 = 25 = c^2$$

Yes, the given points are vertices of a right triangle.

37. $(\sqrt{3}, 2\sqrt{3} + 3)$, $(\sqrt{3} + 4, -\sqrt{3} + 3)$, $(2\sqrt{3}, 2\sqrt{3} + 4)$

$$a = \sqrt{(\sqrt{3} + 4 - \sqrt{3})^2 + [-\sqrt{3} + 3 - (2\sqrt{3} + 3)]^2}$$
$$= \sqrt{4^2 + (-3\sqrt{3})^2}$$
$$= \sqrt{16 + 27} = \sqrt{43}$$

$$b = \sqrt{(2\sqrt{3} - \sqrt{3})^2 + [2\sqrt{3} + 4 - (2\sqrt{3} + 3)]^2}$$
$$= \sqrt{(\sqrt{3})^2 + 1^2}$$
$$= \sqrt{3 + 1} = \sqrt{4} = 2$$

$$c = \sqrt{[2\sqrt{3} - (\sqrt{3} + 4)]^2 + [2\sqrt{3} + 4 - (-\sqrt{3} + 3)]^2}$$

$$= \sqrt{(\sqrt{3} - 4)^2 + (3\sqrt{3} + 1)^2}$$

$$= \sqrt{19 - 8\sqrt{3} + 28 + 6\sqrt{3}}$$

$$= \sqrt{47 - 2\sqrt{3}}$$

$$a^2 + b^2 = (\sqrt{43})^2 + 2^2$$

$$= 43 + 4 \neq c^2,$$

or $\qquad 47 - 2\sqrt{3}$

No, the triangle is not a right triangle.

39. Find x such that the distance between (x, 7) and (2, 3) is 5. Use the distance formula, and solve for x.

$$5 = \sqrt{(2 - x)^2 + (3 - 7)^2}$$

$$= \sqrt{(2 - x)^2 + (-4)^2}$$

$$= \sqrt{(2 - x)^2 + 16}$$

Square both sides.

$$25 = (2 - x)^2 + 16$$

$$9 = (2 - x)^2$$

Take the square root of both sides.

$$3 = 2 - x \quad \text{or} \quad -3 = 2 - x$$

$$x = -1 \qquad \text{or} \quad x = 5$$

41. Find y such that the distance between (3, y) and (-2, 9) is 12.

$$12 = \sqrt{(-2 - 3)^2 + (9 - y)^2}$$

$$= \sqrt{(-5)^2 + (9 - y)^2}$$

$$= \sqrt{25 + (9 - y)^2}$$

$$144 = 25 + (9 - y)^2$$

$$119 = (9 - y)^2$$

$$\sqrt{119} = 9 - y \qquad \text{or} \quad -\sqrt{119} = 9 - y$$

$$y = 9 - \sqrt{119} \quad \text{or} \qquad y = 9 + \sqrt{119}$$

43. Let P = (x, y) be a point 5 units from (0, 0). Then, by the distance formula,

$$5 = \sqrt{(x - 0)^2 + (y - 0)^2}$$

$$= \sqrt{x^2 + y^2}$$

$$25 = x^2 + y^2 \quad \text{or} \quad x^2 + y^2 = 25.$$

This is the required equation. The graph is a circle with center (0, 0) and radius 5. See the answer graph in the back of the textbook.

45. $AC^2 + BC^2 = AB^2$

$$(1000)^2 + BC^2 = (1000.5)^2$$

$$1{,}000{,}000 + BC^2 = 1{,}001{,}000.25$$

$$BC^2 = 1{,}001{,}000.25 - 1{,}000{,}000$$

$$= 1000.25$$

$$BC = \sqrt{1000.25} \approx 31.6 \text{ ft}$$

47. Let b = the leg of the right triangle with leg h and hypotenuse a.

$$b^2 + h^2 = a^2$$

$$b^2 = a^2 - h^2$$

$$= (184.7)^2 - (144)^2$$

$$= 13{,}378$$

$$b \approx 115.66$$

The length ℓ of a side is 2b, or $2(115.66) \approx 231.3$ m.

51. $(x_1, y_1) = (4, -3)$ and $(x_2, y_2) = (-1, 2)$

The x-coordinate of the midpoint is

$$x = \frac{x_1 + x_2}{2} = \frac{4 + (-1)}{2} = \frac{3}{2} \text{ or } 1.5.$$

The y-coordinate of the midpoint is

$$y = \frac{y_1 + y_2}{2} = \frac{-3 + 2}{2} = \frac{-1}{2} \text{ or } -.5.$$

The midpoint is $(1.5, -.5)$.

53. $(x_1, y_1) = (\pi, 3.5)$ and $(x_2, y_2) = (4 - \pi, -5.5)$

The x-coordinate of the midpoint is

$$x = \frac{x_1 + x_2}{2} = \frac{\pi + (4 - \pi)}{2}$$
$$= \frac{4}{2} = 2.$$

The y-coordinate of the midpoint is

$$y = \frac{y_1 + y_2}{2} = \frac{3.5 + (-5.5)}{2}$$
$$= \frac{-2}{2} = -1.$$

The midpoint is $(2, -1)$.

55. Let $(x_1, y_1) = (-4, 1)$. Since the x-coordinate of the midpoint $(5, 6)$ is 5,

$$\frac{x_1 + x_2}{2} = 5$$

$$\frac{-4 + x_2}{2} = 5$$

$$-4 + x_2 = 10$$

$$x_2 = 14.$$

Since the y-coordinate of the midpoint is 6,

$$\frac{y_1 + y_2}{2} = 6$$

$$\frac{1 + y_2}{2} = 6$$

$$1 + y_2 = 12$$

$$y_2 = 11.$$

The other endpoint (x_2, y_2) is $(14, 11)$.

57. $\{x \mid x > 6\}$ This is an open interval with interval notation $(6, \infty)$.

59. $\{p \mid p \geq 10\}$ This is a half-open interval with interval notation $[10, \infty)$.

61. $\{y \mid 8 \leq y \leq 13\}$ This is a closed interval with interval notation $[8, 13]$.

63. The graph shows an open interval. The interval does not include 2, but consists of all real numbers less than 2, written in interval notation as $(-\infty, 2)$.

65. $|y| \leq 1$ is the same as $y \leq 1$ and $y \geq 1$. In other words, those numbers with absolute values that are less than or equal to 1 are those numbers that are between -1 and 1, including -1 and 1. The interval notation for these numbers is $[-1, 1]$.

In Exercises 67-71, $f(x) = -2x^2 + 4x + 6$.

67. $f(0) = -2(0)^2 + 4(0) + 6$
$$= 0 + 0 + 6$$
$$= 6$$

69. $f(-1) = -2(-1)^2 + 4(-1) + 6$
$$= -2(1) + 4(-1) + 6$$
$$= -2 + (-4) + 6$$
$$= 0$$

71. $f(1 + a)$

    $= -2(1 + a)^2 + 4(1 + a) + 6$

    $= -2(1 + 2a + a^2) + 4(1 + a) + 6$

    $= -2 - 4a - 2a^2 + 4 + 4a + 6$

    $= -2a^2 + 8$

73. **(a)** We want the value of f (or Y_1) when X is 2. When X = 2 in the table, Y_1 = 1.2, so f(2) = 1.2.

    (b) Now we want the value of X for which f (or Y_1) equals −2.4. When $f(x) = Y_1 = -2.4$, X = 5 in the table.

    (c) The graph of y = f(x) crosses the y−axis at the y−intercept (0, 3.6).

    (d) The graph of y = f(x) crosses the x−axis at the x−intercept (3, 0).

75. $y = 4x - 3$

    Domain: x can take on any value, so the domain is (−∞, ∞).
    Range: y can take on any value, so the range is (−∞, ∞).
    $y = 4x - 3$ is a function because each x−value gives exactly one y−value.

77. $y = x^2 + 4$

    Domain: (−∞, ∞)
    Range: The smallest value of x^2 is 0, so the smallest value of $y = x^2 + 4 = 0 + 4 = 4$. So the range is [4, ∞).

$y = x^2 + 4$ is a function because each x−value gives exactly one y−value.

79. $y = -2(x - 3)^2 + 4$

    Domain: (−∞, ∞)
    Range: Since $(x - 3)^2 \geq 0$

    $-2(x - 3)^2 \leq 0$

    $-2(x - 3)^2 + 4 \leq 4$.

    The range is (−∞, 4].
    $y = -2(x - 3)^2 + 4$ is a function because each x−value gives exactly one y−value.

81. $x = y^2$

    Domain: $y^2 \geq 0$, so the domain is [0, ∞).
    Range: (−∞, ∞)
    $x = y^2$ is not a function because all x−values (except x = 0) give two y−values. For example, if x = 4, y = 2 or y = −2.

83. $y = \sqrt{x^2 + 1}$

    Domain: (−∞, ∞)
    Range: For all x−values, $x^2 \geq 0$, $x^2 + 1 \geq 1$, and $\sqrt{x^2 + 1} \geq 1$. The range is [1, ∞).
    $y = \sqrt{x^2 + 1}$ is a function because each x−value gives only one y−value.

85. Domain: [−5, 4]
    Range: [−2, 6]

    This is a function because any vertical line intersects the graph in at most one point.

87. Domain: $(-\infty, -3] \cup [3, \infty)$
     Range: $(-\infty, \infty)$

     This is not a function because some vertical lines, for example, $x = 4$, intersect the graph in two points.

In Exercises 89 and 91, the domain includes all real numbers for which the denominator does not equal zero.

89. $y = \dfrac{1}{x}$

     The denominator cannot be zero, so $x \neq 0$.
     Domain: $(-\infty, 0) \cup (0, \infty)$

91. $y = \dfrac{-1}{\sqrt{x^2 + 25}}$

     Since the numerator is a constant, the domain is determined by the requirements that the denominator must not be 0 and the expression under the radical sign must be nonnegative.
     Since $x^2 \geq 0$ for all real numbers x,

$$x^2 + 25 \geq 25$$
$$\sqrt{x^2 + 25} \geq 5 > 0.$$

     Thus, the denominator is not 0 for any real number x and the expression $x^2 + 25$ is nonnegative for any x.
     The domain is $(-\infty, \infty)$.

93. From the figure, the length of a side of the large square is $a + b$, so its area is $(a + b)^2$ or $a^2 + 2ab + b^2$.

94. The area of each right triangle is $(1/2)ab$, so the sum of the areas of the four right triangles is $4(1/2)ab$ or $2ab$. The smaller square has side of length c, so its area is $c \cdot c$ or c^2.

95. The sum of the areas of the four right triangles and the smaller square from part (b) is $2ab + c^2$.

96. Setting the expressions from parts (a) and (c) equal to each other we obtain $a^2 + 2ab + b^2 = 2ab + b^2$.

97. Simplifying the expression in part (d) results in the Pythagorean theorem $a^2 + b^2 = c^2$.

Section 1.2

3. Let x = the measure of the angle. If the angle is its own complement,

$$x + x = 90$$
$$2x = 90$$
$$x = 45.$$

    A 45° angle is its own complement.

5. A merry-go-round turns in a counter-clockwise direction.

7. $7x + 11x = 180$

$18x = 180$

$x = 10$

The measures of the two angles are

$(7x)° = 7(10°)$

$= 70°$

and $(11x)° = 11(10°)$

$= 110°.$

9. $(5k + 5) + (3k + 5) = 90$

$8k + 10 = 90$

$8k = 80$

$k = 10$

The measures of the two angles are

$(5k + 5)° = 5(10°) + 5°$

$= 55°$

and $(3k + 5)° = 3(10°) + 5°$

$= 35°.$

11. $6x - 4 + 8x - 12 = 180$

$14x - 16 = 180$

$14x = 196$

$x = 14$

The measures of the two angles are

$(6x - 4)° = 6(14°) - 4°$

$= 80°$

and $(8x - 12)° = 8(14°) - 12°$

$= 100°.$

13. If an angle measures x degrees, and two angles are complementary if their sum is 90°, then the complement of an angle of x degrees is 90 − x degrees.

15. The first negative angle coterminal with x between 0° and 60° is x − 360 degrees.

17. $62° \ 18' + 21° \ 41' = 83° \ 59'$

19. $71° \ 58' + 47° \ 29'$

$= 118° \ 87'$

$= 118° + 60' + 27'$

$= 118° + 1° + 27'$

$= 119° \ 27'$

21. $90° - 51° \ 28' = 89° \ 60' - 51° \ 28'$

$= 38° \ 32'$

23. $90° - 72° \ 58' \ 11''$

$= 89° \ 59' \ 60'' - 72° \ 58' \ 11''$

$= 17° \ 1' \ 49''$

Use a calculator in Exercises 25–35. Step–by–step solutions (without using calculators) are given here.

25. $20° \ 54' = 20° + \dfrac{54°}{60}$

$= 20° + .900°$

$= 20.900°$

27. $91° \ 35' \ 54''$

$= 91° + \dfrac{35°}{60} + \dfrac{54}{3600}°$

$= 91.598°$

29. $274° \ 18' \ 59''$

$= 274° + \dfrac{18°}{60} + \dfrac{59}{3600}°$

$= 274.316°$

31. $31.4296° = 31° + (.4296)60'$

$= 31° + 25.776$

$= 31° + 25' + (.776)(60'')$

$= 31° \ 25' \ 47''$

33. $89.9004°$

$= 89° + (.9004)(60')$

$= 89° + 54.024'$

$= 89° + 54' + (.024)(60'')$

$= 89° \ 54' \ 1''$

35. $178.5994°$

$= 178° + (.5994)(60')$

$= 178° + 35.964'$

$= 178° + 35' + (.964)(60'')$

$= 178° \ 35' \ 58''$

39. $-40°$ is coterminal with

$-40° + 360° = 320°.$

41. $-125°$ is coterminal with

$-125° + 360° = 235°.$

43. $539°$ is coterminal with

$539° - 360° = 179°.$

45. $850°$ is coterminal with

$850° - 2(360°) = 850° - 720°$

$= 130°$

47. $30°$

A coterminal angle can be obtained by adding an integer multiple of $360°$.

$30° + n \cdot 360°$

49. $60°$

A coterminal angle can be obtained by adding an integer multiple of $360°$.

$60° + n \cdot 360°$

51. $135°$

A coterminal angle can be obtained by adding an integer multiple of $360°$.

$135° + n \cdot 360°$

53. $-90°$

A coterminal angle can be obtained by adding an integer multiple of $360°$.

$-90° + n \cdot 360°$

55. The answers to Exercises 52 and 53 give the same set of angles since $270°$ and $-90°$ are coterminal. Notice that $-90° + 360° = 270°.$

57. $Y_1 = 360((X/360) - \text{int}(X/360))$

$X = -40°$

$Y_1 = 360((-40/360) - \text{int}(-40/360))$

$= 320°$

For Exercises 59–69, angles other than those given are possible. See the answer graphs in the back of the textbook.

59. A positive angle coterminal with $75°$ is

$75° + 360° = 435°.$

A negative angle coterminal with 75° is

$$75° - 360° = -285°.$$

These angles are in quadrant I.

61. A positive angle coterminal with 122° is

$$122° + 360° = 482°.$$

A negative angle coterminal with 122° is

$$122° - 360° = -238°.$$

These angles are in quadrant II.

63. A positive angle coterminal with 234° is

$$234° + 360° = 594°.$$

A negative angle coterminal with 234° is

$$234° - 360° = -126°.$$

These angles are in quadrant III.

65. A positive angle coterminal with 300° is

$$300° + 360° = 660°.$$

A negative angle coterminal with 300° is

$$300° - 360° = -60°.$$

These angles are in quadrant IV.

67. 624° is coterminal with 624° − 360° = 264° and 624° − 2(360°) = −96°; these angles are in quadrant III.

69. −61° is coterminal with 299° and −421°; these angles are in quadrant IV.

For Exercises 71–75, see the answer graphs in the back of the textbook.

71. Points: (−3, −3), (0, 0)
$$r = \sqrt{(-3-0)^2 + (-3-0)^2}$$
$$= \sqrt{9+9}$$
$$= \sqrt{18}$$
$$= \sqrt{9 \cdot 2}$$
$$= 3\sqrt{2}$$

73. Points: (−3, −5), (0, 0)
$$r = \sqrt{(-3-0)^2 + (-5-0)^2}$$
$$= \sqrt{9+25}$$
$$= \sqrt{34}$$

75. Points: (−2, 2√3), (0, 0)
$$r = \sqrt{(-2-0)^2 + (2\sqrt{3}-0)^2}$$
$$= \sqrt{4+12}$$
$$= \sqrt{16}$$
$$= 4$$

77. 90 revolutions per minute
$$= \frac{90}{60} \text{ revolutions per second}$$
$$= 1.5 \text{ revolutions per second}$$

79. 600 rotations per minute
= 600/60 rotations per second
= 10 rotations per second
= 5 rotations per 1/2 second
= 5(360°) = 1800° per 1/2 second

81. 75° per minute

    = 75(60°) per hour

    = 4500° per hour

    $= \dfrac{4500}{360}$ rotations per hour

    = 12.5 rotations per hour

83. Earth rotates at a speed of 360° per 24 hours. This is equivalent to 15° per hour = 1° every 4 minutes = 1′ every 4 seconds. The motor should rotate the telescope through an angle of 1′ every 4 seconds.

Section 1.3

1. In geometry, opposite angles are called vertical angles.

3. $5x - 129 = 2x - 21$

since vertical angles are equal.

$3x - 129 = -21$

$\qquad 3x = 108$

$\qquad\ \ x = 36$

$5x - 129 = 5(36) - 129$

$\qquad\qquad = 180 - 129 = 51$

$2x - 21 = 2(36) - 21$

$\qquad\qquad = 72 - 21 = 51$

Both angles are 51°.

5. $x + (x + 20) + (210 - 3x) = 180$

since the sum of the angles of a triangle is 180°.

$-x + 230 = 180$

$\qquad -x = -50$

$\qquad\ \ x = 50$

$x + 20 = 50 + 20 = 70$

$210 - 3x = 210 - 3(50)$

$\qquad\qquad = 210 - 150 = 60$

The three angles are 50°, 70°, and 60°.

7. $(x - 30) + (2x - 120) + \left(\frac{1}{2}x + 15\right) = 180$

since the sum of the angles of a triangle is 180°.

$\frac{7}{2}x - 135 = 180$

$\qquad \frac{7}{2}x = 315$

$\qquad\ \ x = (315)\frac{2}{7} = 90$

$x - 30 = 90 - 30 = 60$

$\frac{1}{2}x + 15 = \frac{1}{2}(90) + 15$

$\qquad\qquad = 45 + 15 = 60$

$2x - 120 = 2(90) - 120$

$\qquad\qquad = 180 - 120 = 60$

All three angles are 60°.

9. $(3x + 5) + (5x + 15) = 180$

since these two angles are supplementary.

$8x + 20 = 180$

$\qquad 8x = 160$

$\qquad\ \ x = 20$

$3x + 5 = 3(20) + 5$

$\qquad\quad = 60 + 5 = 65$

$5x + 15 = 5(20) + 15$

$\qquad\qquad = 100 + 15 = 115$

The two angles are 65° and 115°.

11. $2x - 5 = x + 22$

since alternate interior angles are equal.

$x - 5 = 22$

$x = 27$

$2x - 5 = 2(27) - 5$

$= 54 - 5 = 49$

$x + 22 = 27 + 22 = 49$

Both angles have a measure of 49°.

13. $(x + 1) + (4x - 56) = 180$

since the sum of interior angles on the same side of the transversal is 180°.

$5x - 55 = 180$

$5x = 235$

$x = 47$

$x + 1 = 47 + 1 = 48$

$4x - 56 = 4(47) - 56$

$= 188 - 56 = 132$

The angles are 48° and 132°.

15. Let x = the measure of the third angle.

$37° + 52° + x = 180°$

$89° + x = 180°$

$x = 91°$

17. Let x = the measure of the third angle.

$147° \, 12' + 30° \, 19' + x = 180°$

$177° \, 31' + x = 180°$

$x = 179° \, 60' - 177° \, 31'$

$x = 2° \, 29'$

19. Let x = the measure of the third angle.

$74.2° + 80.4° + x = 180°$

$154.6° + x = 180°$

$x = 25.4°$

23. angle 1 = 55° (vertical angle with 55°)

angle 5 = 180° - 120° = 60° (interior angles on same side of transversal)

angle 4 = 180° - 60° - 55° = 65° (supplemental angle with angle 5 and 55°)

angle 2 = 65° (vertical angle with angle 4)

angle 3 = 60° (vertical angle with angle 5)

angle 6 = 120° (vertical angles)

angle 7 = 180° - 120° = 60° (supplemental angle with angle 6)

angle 8 = 60° (vertical angle with angle 7)

angle 10 = 180° - 60° - 65° = 55° (sum of angles of triangle is 180°)

angle 9 = 55° (vertical angle with angle 10)

25. The triangle has a right angle, but each side has a different measure. The triangle is a right triangle and a scalene triangle.

27. The triangle has three acute angles and three equal sides, so it is acute and equilateral.

29. The triangle has a right angle and three unequal sides, so it is right and scalene.

31. The triangle has a right angle and two equal sides, so it is right and isosceles.

33. The triangle has one obtuse angle and three unequal sides, so it is obtuse and scalene.

35. The triangle has three acute angles and two equal sides, so it is acute and isosceles.

37. An isosceles right triangle is a triangle with two sides perpendicular (form a right angle) and the same length.

39. All equilateral triangles are similar because all angles have a measure of 60°, and since all sides have equal measure, the ratio of any side of triangle 1 to the corresponding side of triangle 2 will always be the same (proportional).

41. Corresponding angles are A and P, B and Q, C and R; corresponding sides are AB and PQ, CB and RQ, AC and PR.

43. Corresponding angles:

    A and C, ABE and CBD, E and D

    Corresponding sides:

    AB and CB, EB and DB, AE and CD

45. Q = A so Q = 42°.

    A + B + C = 180° so
    $$B = 180° - 90° - 42° = 48°.$$
    B = R so R = 48°.
    P = C so P = 90°.

47. K = B so B = 106°.

    A + B + C = 180° so
    $$A = 180° - 30° - 106° = 44°.$$
    A = M so M = 44°.

49. X + Y + Z = 180° so
    $$X = 180° - 90° - 38° = 52°.$$
    X = M so M = 52°.

In Exercises 51–55, corresponding sides of similar triangles are proportional.

51. $\dfrac{25}{10} = \dfrac{a}{8}$

    $200 = 10a$

    $20 = a$

    $\dfrac{25}{10} = \dfrac{b}{6}$

    $150 = 10b$

    $15 = b$

53. $\dfrac{6}{12} = \dfrac{a}{12}$

    $72 = 12a$

    $6 = a$

    $\dfrac{b}{15} = \dfrac{6}{12}$

    $\dfrac{b}{15} = \dfrac{1}{2}$

    $2b = 15$

    $b = \dfrac{15}{2}$ or $7\dfrac{1}{2}$

55. $\dfrac{x}{4} = \dfrac{9}{6}$

$\dfrac{x}{4} = \dfrac{3}{2}$

$2x = 12$

$x = 6$

57. Let T = the height of the tree.

The triangle formed by the tree and its shadow is similar to the triangle formed by the stick and its shadow.

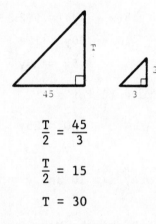

$\dfrac{T}{2} = \dfrac{45}{3}$

$\dfrac{T}{2} = 15$

$T = 30$

The tree is 30 m high.

59. Let x = the middle side of the actual triangle,

and y = the longest side of the actual triangle.

The triangles in the photograph and on the actual piece of land are similar. The shortest side on the land corresponds to the shortest side in the photograph.

$\dfrac{400}{4} = \dfrac{x}{5}$

$100 = \dfrac{x}{5}$

$x = 500$

$\dfrac{400}{4} = \dfrac{y}{7}$

$100 = \dfrac{y}{7}$

$y = 700$

The other two sides are 500 m and 700 m long.

61. Let h = the height of the building.

The triangle formed by the house and its shadow is similar to the triangle formed by the building and its shadow.

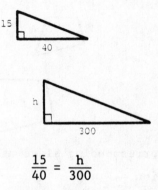

$\dfrac{15}{40} = \dfrac{h}{300}$

$4500 = 40h$

$112.5 = h$

The height of the building is 112.5 ft.

63. $\dfrac{x}{50} = \dfrac{100 + 120}{100}$

$\dfrac{x}{50} = \dfrac{220}{100}$

$\dfrac{x}{50} = \dfrac{11}{5}$

$5x = 550$

$x = 110$

65. $\dfrac{c}{100} = \dfrac{10 + 90}{90}$

$\dfrac{c}{100} = \dfrac{100}{90}$

$\dfrac{c}{100} = \dfrac{10}{9}$

$c = \dfrac{1000}{9} \approx 111.1$

67. Let a = the length of the longest side of the first quadrilateral,

   x = the length of the shortest side of the second quadrilateral,

and y = the other unknown length.

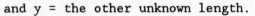
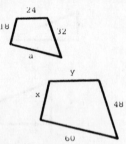

Corresponding sides are in proportion.

Since 32/48 = 2/3,

$\dfrac{a}{60} = \dfrac{2}{3}$

$3a = 120$

$a = 40$

$\dfrac{24}{y} = \dfrac{2}{3}$

$2y = 72$

$y = 36$

$\dfrac{18}{x} = \dfrac{2}{3}$

$2x = 54$

$x = 27.$

The unknown side in the first quadrilateral is 40 cm; the unknown sides in the second quadrilateral are 36 cm and 27 cm.

69. Corresponding ides are in proportion.

$\dfrac{x - 5}{15} = \dfrac{5}{15}$

$15(x - 5) = 75$

$15x - 75 = 75$

$15x = 150$

$x = 10$

$\dfrac{x - 2y}{(x - 2y) + (x + y)} = \dfrac{5}{15}$

$\dfrac{x - 2y}{2x - y} = \dfrac{5}{15}$

$15(x - 2y) = 5(2x - y)$

$15x - 30y = 10x - 5y$

$5x = 25y$

Substitute 10 for x.

$50 = 25y$

$y = 2$

Therefore, x = 10 and y = 2.

71. (a) Let D_s be the Mars–Sun distance, d_s the diameter of the sun, D_m the Mars–Phobes distance, and d_m the diameter of the Phobes. Then, by similar triangles

$\dfrac{D_s}{D_m} = \dfrac{d_s}{d_m}$

$D_m = \dfrac{D_s d_m}{d_s}$

$= \dfrac{142,000,000(17.4)}{865,000}$

≈ 2856 mi.

(b) No. Phobes does not come close enough to the surface of Mars.

Section 1.4

For Exercises 1 and 3, see the answer graphs in the back of the textbook.

5. $(-3, 4)$

$x = -3, \ y = 4$

$r = \sqrt{(-3)^2 + 4^2}$

$ = \sqrt{9 + 16}$

$ = \sqrt{25}$

$ = 5$

$\sin \theta = \dfrac{y}{r} = \dfrac{4}{5}$

$\cos \theta = \dfrac{x}{r} = -\dfrac{3}{5}$

$\tan \theta = \dfrac{y}{x} = -\dfrac{4}{3}$

$\cot \theta = \dfrac{x}{y} = -\dfrac{3}{4}$

$\sec \theta = \dfrac{r}{x} = -\dfrac{5}{3}$

$\csc \theta = \dfrac{r}{y} = \dfrac{5}{4}$

7. $(0, 2)$

$x = 0, \ y = 2$

$r = \sqrt{0^2 + 2^2}$

$ = \sqrt{0 + 4}$

$ = \sqrt{4}$

$ = 2$

$\sin \theta = \dfrac{2}{2} = 1$

$\cos \theta = \dfrac{0}{2} = 0$

$\tan \theta = \dfrac{2}{0} \quad \text{Undefined}$

$\cot \theta = \dfrac{0}{2} = 0$

$\sec \theta = \dfrac{2}{0} \quad \text{Undefined}$

$\csc \theta = \dfrac{2}{2} = 1$

9. $(1, \sqrt{3})$

$x = 1, \ y = \sqrt{3}$

$r = \sqrt{1^2 + (\sqrt{3})^2}$

$ = \sqrt{1 + 3}$

$ = \sqrt{4}$

$ = 2$

$\sin \theta = \dfrac{\sqrt{3}}{2}$

$\cos \theta = \dfrac{1}{2}$

$\tan \theta = \dfrac{\sqrt{3}}{1} = \sqrt{3}$

$\cot \theta = \dfrac{1}{\sqrt{3}} = \dfrac{1}{\sqrt{3}} \cdot \dfrac{\sqrt{3}}{\sqrt{3}} = \dfrac{\sqrt{3}}{3}$

$\sec \theta = \dfrac{2}{1} = 2$

$\csc \theta = \dfrac{2}{\sqrt{3}} = \dfrac{2}{\sqrt{3}} \cdot \dfrac{\sqrt{3}}{\sqrt{3}} = \dfrac{2\sqrt{3}}{3}$

11. $(8.7691, -3.2473)$

$x = 8.7691, \ y = -3.2473$

$r = \sqrt{(8.7691)^2 + (-3.2473)^2}$

$ = \sqrt{87.442072}$

$ = 9.3510466$

$\sin \theta = \dfrac{y}{r} \approx -.34727$

$\cos \theta = \dfrac{x}{r} \approx .93777$

$\tan \theta = \dfrac{y}{x} \approx -.37031$

$\cot \theta = \dfrac{x}{y} \approx -2.7004$

$\sec \theta = \dfrac{r}{x} \approx 1.0664$

$\csc \theta = \dfrac{r}{y} \approx -2.8796$

13. Since angle θ is a nonquadrantal angle, $\sin\theta = \dfrac{y}{r}$ and $\csc\theta = \dfrac{r}{y}$. Since $r = \sqrt{x^2 + y^2}$, r is always positive. Therefore, if y is positive, $\dfrac{y}{r}$ and $\dfrac{r}{y}$ will be positive. If y is negative, $\dfrac{y}{r}$ and $\dfrac{r}{y}$ will be negative. Therefore, $\sin\theta$ and $\csc\theta$ will have the same sign.

15. In the definitions of the sine, cosine, secant, and cosecant functions, the value of r is the hypotenuse of the triangle formed by the x-axis, the line connecting (x, y) and $(0, 0)$ (assuming (x, y) is on the terminal side of the angle), and the vertical line through (x, y), perpendicular to the x-axis.

In Exercises 17–23, $r = \sqrt{x^2 + y^2}$, which is positive.

17. In quadrant II, y is positive so y/r is positive.

19. In quadrant III, y is negative so y/r is negative.

21. In quadrant IV, x is positive so x/r is positive.

23. In quadrant IV, x is positive and y is negative so y/x is negative.

25. Since $x \geq 0$, the graph of the line $2x + y = 0$ is shown to the right of the y-axis. A point on this line is $(1, -2)$ since $2(1) + (-2) = 0$.

$$r = \sqrt{1^2 + (-2)^2} = \sqrt{1 + 4} = \sqrt{5}$$

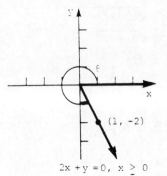

$$2x + y = 0, \; x \geq 0$$

$\sin\theta = -\dfrac{2\sqrt{5}}{5}$

$\cos\theta = \dfrac{\sqrt{5}}{5}$

$\tan\theta = -2$

$\cot\theta = -\dfrac{1}{2}$

$\sec\theta = \sqrt{5}$

$\csc\theta = -\dfrac{\sqrt{5}}{2}$

27. Since $x \leq 0$, the graph of the line $-4x + 7y = 0$ is shown to the left of the y-axis. A point on this line is $(-7, -4)$ since $-4(-7) + 7(-4) = 0$.

$$r = \sqrt{(-7)^2 + (-4)^2} = \sqrt{49 + 16} = \sqrt{65}$$

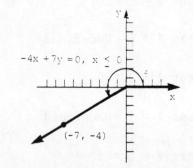

$$-4x + 7y = 0, \; x \leq 0$$

$$\sin \theta = -\frac{4\sqrt{65}}{65}$$

$$\cos \theta = -\frac{7\sqrt{65}}{65}$$

$$\tan \theta = \frac{4}{7}$$

$$\cot \theta = \frac{7}{4}$$

$$\sec \theta = -\frac{\sqrt{65}}{7}$$

$$\csc \theta = -\frac{\sqrt{65}}{4}$$

29. Since $x \leq 0$, the graph of the line
$-5x - 3y = 0$ is shown to the left of
the y-axis. A point on this line is
$(-3, 5)$ since $-5(-3) - 3(5) = 0$.

$$r = \sqrt{(-3)^2 + 5^2} = \sqrt{9 + 25} = \sqrt{34}$$

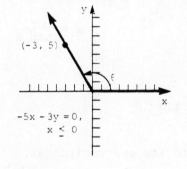

$$\sin \theta = \frac{5\sqrt{34}}{34}$$

$$\cos \theta = -\frac{3\sqrt{34}}{34}$$

$$\tan \theta = -\frac{5}{3}$$

$$\cot \theta = -\frac{3}{5}$$

$$\sec \theta = -\frac{\sqrt{34}}{3}$$

$$\csc \theta = \frac{\sqrt{34}}{5}$$

33. $\cos 90° + 3 \sin 270°$

    $= 0 + 3(-1) = -3$

35. $3 \sec 180° - 5 \tan 360°$

    $= 3(-1) - 5(0) = -3$

37. $\tan 360° + 4 \sin 180° + \cos^2 180°$

    $= 0 + 4(0) + 5(-1)^2 = 5$

39. $\sin^2 180° + \cos^2 180°$

    $= 0^2 + (-1)^2$

    $= 1$

41. $\sec^2 180° - 3 \sin^2 360° + 2 \cos 180°$

    $= (-1)^2 - 3(0)^2 + 2(-1)$

    $= 1 - 0 - 2$

    $= -1$

43. $\sin [n \cdot 180°]$

This angle is a quadrantal angle
whose terminal side lies on either
the positive part of the x-axis or
the negative part of the x-axis.
Any point on these terminal sides
would have the form $(k, 0)$, where k
is any real number $k \neq 0$.

$$\sin [n \cdot 180°] = \frac{y}{r} = \frac{0}{\sqrt{k^2 + 0^2}} = 0$$

45. $\tan [(2n + 1) \cdot 90°]$

This angle is a quadrantal angle
whose terminal side lies on either
the positive part of the y-axis or
the negative part of the y-axis.

Any point on these terminal sides would have the form (0, k), where k is any real number, k ≠ 0.

$$\tan[(2n+1) \cdot 90°] = \frac{y}{x} = \frac{k}{0}$$

$\tan[(2n+1) \cdot 90°]$ is undefined.

47. Using a calculator,

$$\sin 15° = .258819045$$

and $\cos 75° = .258819045$.

We can conjecture that the sines and cosines of complementary angles are equal. Try another pair of complementary angles.

$$\sin 30° = \cos 60°$$
$$.5 = .5$$

Therefore, our conjecture is true.

49. Using a calculator,

$$\sin 10° = .173648178$$
and $\sin(-10°) = .173648178$.

We can conjecture that the sines of an angle and its negative are negatives of each other. Using a circle, an angle θ having the point (x, y) on its terminal side has a corresponding angle $-\theta$ with a point (x, −y) on its terminal side. From the definition of sine,

$$\sin(-\theta) = -\frac{y}{r} \text{ and } \sin\theta = \frac{y}{r}.$$

The sines are negatives of each other.

53. cos 20° is about .940, and sin 20° is about .342.

55. Use the TRACE feature to move around the circle in quadrant I.

$$\sin 35° \approx .574 \text{ so } T = 35°.$$

57. As T increases from 0° to 90°, the cosine decreases and the sine increases.

59. **(a)** From the figure in the text and the definition of tan θ, we can see that

$$\tan\theta = \frac{y}{x}.$$

(b) Solve for x.

$$\tan\theta = \frac{y}{x}$$
$$x\tan\theta = y$$
$$x = \frac{y}{\tan\theta}$$

Section 1.5

1. 1 is its own reciprocal.

$$\sin\theta = \csc\theta = 1$$
$$\sin 90° = \csc 90° = 1$$

Therefore, $\theta = 90°$.

3. $\csc\theta = 3$

$$\sin\theta = \frac{1}{\csc\theta} = \frac{1}{3}$$

5. $\tan\beta = -\frac{1}{5}$

$$\cot\beta = \frac{1}{\tan\beta}$$

$$= \frac{1}{-\frac{1}{5}}$$

$$= -5$$

7. $\sin \alpha = \dfrac{\sqrt{2}}{4}$

   $\csc \alpha = \dfrac{1}{\sin \alpha}$

   $ = \dfrac{1}{\dfrac{\sqrt{2}}{4}}$

   $ = \dfrac{4}{\sqrt{2}} \cdot \dfrac{\sqrt{2}}{\sqrt{2}}$

   $ = \dfrac{4\sqrt{2}}{2} = 2\sqrt{2}$

9. $\cot \theta = -\dfrac{\sqrt{5}}{3}$

   $\tan \theta = \dfrac{1}{\cot \theta}$

   $ = \dfrac{1}{-\dfrac{\sqrt{5}}{3}}$

   $ = -\dfrac{3}{\sqrt{5}} \cdot \dfrac{\sqrt{5}}{\sqrt{5}}$

   $ = -\dfrac{3\sqrt{5}}{5}$

11. $\csc \theta = 1.42716321$

    $\sin \theta = \dfrac{1}{\csc \theta}$

    $ = \dfrac{1}{1.42716327}$

    $ = .70069071$

13. It is impossible for $\sin \gamma > 0$ and $\csc \gamma < 0$. Suppose $\sin \gamma > 0$.

    Since $\csc \gamma = \dfrac{1}{\sin \gamma}$, one divided by a positive number yields a positive number. Therefore, $\csc \gamma$ must be positive.

15. Since $\tan \theta = \dfrac{1}{\cot \theta}$,

    multiply both sides by $\cot \theta$.

    $\tan \theta \cot \theta = 1$

    Now divide both sides by $\tan \theta$.

    $\cot \theta = \dfrac{1}{\tan \theta}$

17. $\cot \gamma = 2$

    $\tan \gamma = \dfrac{1}{\cot \gamma}$

    $ = \dfrac{1}{2}$

19. $\cot \omega = \dfrac{\sqrt{3}}{3}$

    $\tan \omega = \dfrac{1}{\cot \omega}$

    $ = \dfrac{1}{\dfrac{\sqrt{3}}{3}}$

    $ = \dfrac{3}{\sqrt{3}} = \sqrt{3}$

21. $\cot \alpha = -.01$

    $\tan \alpha = \dfrac{1}{\cot \alpha}$

    $ = \dfrac{1}{-.01}$

    $ = -100$

23. $\tan (3B - 4°) = \dfrac{1}{\cot (5B - 8°)}$

    $\tan (3B - 4°) = \tan (5B - 8°)$

    $3B - 4° = 5B - 8°$

    $4° = 2B$

    $B = 2°$

25. $\sec (2\alpha + 6°) \cos (5\alpha + 3°) = 1$

$\cos (5\alpha + 3°) = \dfrac{1}{\sec (2\alpha + 6°)}$

$\cos (5\alpha + 3°) = \cos (2\alpha + 6°)$

$5\alpha + 3° = 2\alpha + 6°$

$3\alpha = 3°$

$\alpha = 1°$

27. $\dfrac{1}{\tan (2k + 1°)} = \cot (4k - 3°)$

$\cot (2k + 1°) = \cot (4k - 3°)$

$2k + 1° = 4k - 3°$

$4° = 2k$

$k = 2°$

29. $\sin \alpha > 0$ implies α is in quadrant I or II. $\cos \alpha < 0$ implies α is in quadrant II or III. α is in quadrant II.

31. $\tan \gamma > 0$ implies γ is in quadrant I or III. $\cot \gamma > 0$ implies γ is in quadrant I or III. γ is in quadrant I or III.

33. $\tan \omega < 0$ implies ω is in quadrant II or IV. $\cot \omega < 0$ implies ω is in quadrant II or IV. ω is in quadrant II or IV.

35. $\cos \beta < 0$ implies β is in quadrant II or III.

In Exercises 37–45, the signs + for positive and − for negative are given in the following order: sine and cosecant, cosine and secant, tangent and cotangent.

37. 74° is in quadrant I.

+; +; +

39. 183° is in quadrant III.

−; −; +

41. 302° is in quadrant IV.

−; +; −

43. 412° is in quadrant I.

+; +; +

45. −14° is in quadrant IV.

−; +; −

47. $\tan 30° = \dfrac{\sin 30°}{\cos 30°}.$

$\sin 30° < \cos 30°,$
so $\sin 30° < \tan 30°.$

Therefore, $\tan 30°$ is greater.

49. $\sec 33° = \dfrac{1}{\cos 33°}$

$\sin 33° < \cos 33°,$
so $\sin 33° < \sec 33°.$

Therefore, $\sec 33°$ is greater.

51. $-1 \le \sin \theta \le 1$ for all θ and 2 is not in this range.

$\sin \theta = 2$ is impossible.

53. tan β can take on all values.

    tan β = .92 is possible.

55. csc $\alpha \geq 1$ or csc $\alpha \leq -1$ for all α
    and $1/2$ is not in this range.

    csc $\alpha = 1/2$ is impossible.

57. tan θ can take on all values.

    tan θ = 1 is possible.

59. sec ω + 1 = 1.3 gives

    sec ω = .3 which

    is impossible since

    sec $\omega \leq -1$ and

    sec $\omega \geq 1$.

61. sin α = 1/2 is possible because
    $-1 \leq \sin \alpha \leq 1$. Furthermore,
    when

    $$\sin \alpha = \frac{1}{2},$$

    $$\csc \alpha = \frac{1}{\sin \alpha}$$

    $$= \frac{1}{1/2} = 2.$$

    sin α = 1/2 and csc α = 2 is

    possible.

63. Find tan α if sec α = 3, with α in
    quadrant IV.
    Start with the identity
    $\tan^2 \alpha + 1 = \sec^2 \alpha$, and replace
    sec α with 3.

    $$\tan^2 \alpha + 1 = 3^2$$

    $$\tan^2 \alpha = 8$$

    $$\tan \alpha = \pm\sqrt{8} = \pm 2\sqrt{2}$$

Since α is in quadrant IV,

tan α < 0, so

$$\tan \alpha = -2\sqrt{2}.$$

65. Find sin α if cos α = -1/4, with
    α in quadrant II.
    Start with the identity
    $\sin^2 \alpha + \cos^2 \alpha = 1$, and replace
    cos α with -1/4.

    $$\sin^2 \alpha + \left(-\frac{1}{4}\right)^2 = 1$$

    $$\sin^2 \alpha + \frac{1}{16} = 1$$

    $$\sin^2 \alpha = \frac{15}{16}$$

    $$\sin \alpha = \pm\sqrt{\frac{15}{16}} = \pm\frac{\sqrt{15}}{4}$$

Since α is in quadrant II,

sin α > 0, so sin $\alpha = \frac{\sqrt{15}}{4}$.

67. Find tan θ, if cos $\theta = \frac{1}{3}$ with θ in

    quadrant IV.
    Start with the identity
    $\sin^2 \theta + \cos^2 \theta = 1$ and replace

    cos θ with $\frac{1}{3}$.

    $$\sin^2 \theta + \left(\frac{1}{3}\right)^2 = 1$$

    $$\sin^2 \theta + \frac{1}{9} = 1$$

    $$\sin^2 \theta = \frac{8}{9}$$

    $$\sin \theta = \pm\sqrt{\frac{8}{9}} = \pm\frac{\sqrt{8}}{3} = \pm\frac{2\sqrt{2}}{3}$$

Since θ is in quadrant IV, $\sin \theta =$ $-\dfrac{2\sqrt{2}}{3}$.

$$\tan \theta = \frac{\sin \theta}{\cos \theta} = \frac{-\dfrac{2\sqrt{2}}{3}}{\dfrac{1}{3}} = -2\sqrt{2}$$

69. Find $\cos \beta$, if $\csc \beta = -4$ with β in quadrant III.

Start with the identity

$$\sin \beta = \frac{1}{\csc \beta}.$$

$$\sin \beta = \frac{1}{-4} = -\frac{1}{4}$$

Now use the identity $\sin^2 \beta + \cos^2 \beta = 1$ and replace $\sin \beta$ with $-1/4$.

$$\left(-\frac{1}{4}\right)^2 + \cos^2 \beta = 1$$

$$\frac{1}{16} + \cos^2 \beta = 1$$

$$\cos^2 \beta = \frac{15}{16}$$

$$\cos \beta = \pm\sqrt{\frac{15}{16}} = \pm\frac{\sqrt{15}}{4}$$

Since β is in quadrant III,

$\cos \beta = -\dfrac{\sqrt{15}}{4}$.

71. Find $\sin \beta$, if $\cot \beta = 2.40129813$ with β in quadrant I. Start with the identity $1 + \cot^2 \beta = \csc^2 \beta$ and replace $\cot \beta$ with 2.40129813.

$$1 + (2.40129813)^2 = \csc^2 \beta$$

$$1 + 5.76623271 = \csc^2 \beta$$

$$6.76623271 = \csc^2 \beta$$

$$\pm2.60119832 = \csc \beta$$

Since β is in quadrant I,

$\csc \beta = 2.60119832$.

Since $\sin \beta = \dfrac{1}{\csc \beta}$,

$$\sin \beta = \frac{1}{2.60119832}$$

$$= .38443820.$$

73. Find $\csc \alpha$, if $\tan \alpha = .98244655$ with α in quadrant III. Start with the identity $\cot \alpha = \dfrac{1}{\tan \alpha}$ and replace $\tan \alpha$ with .98244655.

$$\cot \alpha = \frac{1}{.98244655}$$

$$\cot \alpha = 1.01786708$$

Now use the identity $1 + \cot^2 \alpha = \csc^2 \alpha$ and replace $\cot \alpha$ with 1.01786708.

$$1 + (1.01786708)^2 = \csc^2 \alpha$$

$$1 + 1.03605339 = \csc^2 \alpha$$

$$2.03605339 = \csc^2 \alpha$$

$$\pm\sqrt{2.03605339} = \csc \alpha$$

$$\pm1.4269034 = \csc \alpha$$

Since α in quadrant III,

$\csc \alpha = -1.4269034$.

75. $\sin X = .8$

$$\sin^2 X + \cos^2 X = 1$$

$$\cos^2 X = 1 - \sin^2 X$$

$$\cos^2 X = 1 - (.8)^2$$

$$\cos^2 X = (\cos X)^2 = .36$$

77. $\sin^2 \theta + \cos^2 \theta = 1$

$$(.8)^2 + (-.6)^2 \stackrel{?}{=} 1$$

$$.64 + .36 \stackrel{?}{=} 1$$

$$1 = 1 \quad \textit{True}$$

Yes, there is an angle θ for which $\cos \theta = -.6$ and $\sin \theta = .8$.

For Exercises 79–87, remember that r is always positive.

79. $\tan \alpha = -\dfrac{15}{8}$

$\tan \alpha = \dfrac{y}{x}$ and α is in quadrant II,

so let $y = 15$ and $x = -8$.

$$x^2 + y^2 = r^2$$
$$(-8)^2 + 15^2 = r^2$$
$$64 + 225 = r^2$$
$$289 = r^2$$
$$17 = r$$

$\sin \alpha = \dfrac{y}{r} = \dfrac{15}{17}$

$\cos \alpha = \dfrac{x}{r} = \dfrac{-8}{17} = -\dfrac{8}{17}$

$\tan \alpha = \dfrac{y}{x} = \dfrac{15}{-8} = -\dfrac{15}{8}$

$\cot \alpha = \dfrac{x}{y} = \dfrac{-8}{15} = -\dfrac{8}{15}$

$\sec \alpha = \dfrac{r}{x} = \dfrac{17}{-8} = -\dfrac{17}{8}$

$\csc \alpha = \dfrac{r}{y} = \dfrac{17}{15}$

81. $\cot \gamma = \dfrac{3}{4}$

$\cot \gamma = \dfrac{x}{y}$ and γ is in quadrant III,

so let $x = -3$ and $y = -4$.

$$x^2 + y^2 = r^2$$
$$(-3)^2 + (-4)^2 = r^2$$
$$9 + 16 = r^2$$
$$25 = r^2$$
$$5 = r$$

$\sin \gamma = \dfrac{y}{r} = \dfrac{-4}{5} = -\dfrac{4}{5}$

$\cos \gamma = \dfrac{x}{r} = \dfrac{-3}{5} = -\dfrac{3}{5}$

$\tan \gamma = \dfrac{y}{x} = \dfrac{-4}{-3} = \dfrac{4}{3}$

$\cot \gamma = \dfrac{x}{y} = \dfrac{-3}{-4} = \dfrac{3}{4}$

$\sec \gamma = \dfrac{r}{x} = \dfrac{5}{-3} = -\dfrac{5}{3}$

$\csc \gamma = \dfrac{r}{y} = \dfrac{5}{-4} = -\dfrac{5}{4}$

83. $\tan \beta = \sqrt{3}$

$\tan \beta = \dfrac{y}{x}$ and β is in quadrant III,

so let $y = -\sqrt{3}$ and $x = -1$.

$$x^2 + y^2 = r^2$$
$$(-1)^2 + (-\sqrt{3})^2 = r^2$$
$$1 + 3 = r^2$$
$$4 = r^2$$
$$2 = r$$

$\sin \beta = \dfrac{y}{r} = \dfrac{-\sqrt{3}}{2} = -\dfrac{\sqrt{3}}{2}$

$\cos \beta = \dfrac{x}{r} = \dfrac{-1}{2} = -\dfrac{1}{2}$

$\tan \beta = \dfrac{y}{x} = \dfrac{-\sqrt{3}}{-1} = \sqrt{3}$

$\cot \beta = \dfrac{x}{y} = \dfrac{-1}{-\sqrt{3}} = \dfrac{1}{\sqrt{3}} = \dfrac{\sqrt{3}}{3}$

$\sec \beta = \dfrac{r}{x} = \dfrac{2}{-1} = -2$

$\csc \beta = \dfrac{r}{y} = \dfrac{2}{-\sqrt{3}} = -\dfrac{2\sqrt{3}}{3}$

85. $\sin \beta = \dfrac{\sqrt{5}}{7}$, with $\tan \beta > 0$

Since $\sin \beta$ is positive, β is in quadrant I or II.
Since $\tan \beta > 0$, β is in quadrant I or III.
β is in quadrant I.

Since $\sin \beta = \frac{y}{r}$, let $y = \sqrt{5}$ and $r = 7$.

$$x^2 + y^2 = r^2$$
$$x^2 + (\sqrt{5})^2 = 7^2$$
$$x^2 + 5 = 49$$
$$x^2 = 44$$
$$x = \pm\sqrt{44} = \pm2\sqrt{11}$$

Since β is in quadrant I, $x = 2\sqrt{11}$.

$$\sin \beta = \frac{y}{r} = \frac{\sqrt{5}}{7}$$

$$\cos \beta = \frac{x}{r} = \frac{2\sqrt{11}}{7}$$

$$\tan \beta = \frac{y}{x} = \frac{\sqrt{5}}{2\sqrt{11}} = \frac{\sqrt{55}}{22}$$

$$\cot \beta = \frac{x}{y} = \frac{2\sqrt{11}}{\sqrt{5}} = \frac{2\sqrt{55}}{5}$$

$$\sec \beta = \frac{r}{x} = \frac{7}{2\sqrt{11}} = \frac{7\sqrt{11}}{22}$$

$$\csc \beta = \frac{r}{y} = \frac{7}{\sqrt{5}} = \frac{7\sqrt{5}}{5}$$

87. $\cot \theta = -1.49586$ with θ in quadrant IV.

Since $\cot \theta = \frac{x}{y}$ and θ is in quadrant IV, let $x = 1.49586$ and $y = -1$.

$$x^2 + y^2 = r^2$$
$$(1.49586)^2 + (-1)^2 = r^2$$
$$2.23760 + 1 = r^2$$
$$3.23760 = r^2$$
$$1.79933 = r$$

$$\sin \theta = \frac{y}{r} = \frac{-1}{1.79933} = -.555762$$

$$\cos \theta = \frac{x}{r} = \frac{1.49586}{1.79933} = .831342$$

$$\tan \theta = \frac{y}{x} = \frac{-1}{1.49586} = -.668512$$

$$\cot \theta = \frac{x}{y} = \frac{1.49586}{-1} = -1.49586$$

$$\sec \theta = \frac{r}{x} = \frac{1.79933}{1.49586} = 1.20287$$

$$\csc \theta = \frac{r}{y} = \frac{1.79933}{-1} = -1.79933$$

89. $x^2 + y^2 = r^2$

Divide by y^2.

$$\frac{x^2}{y^2} + \frac{y^2}{y^2} = \frac{r^2}{y^2}$$

$$\frac{x^2}{y^2} + 1 = \frac{r^2}{y^2}$$

$$\left(\frac{x}{y}\right)^2 + 1 = \left(\frac{r}{y}\right)^2$$

$$\cot^2 \theta + 1 = \csc^2 \theta$$

91. The statement is false.

$$\sin 30° + \cos 30° \overset{?}{=} 1$$
$$.5 + .8660 \overset{?}{=} 1$$
$$1.3660 \neq 1$$

Since $1.3660 \neq 1$, the given statement is false.

Chapter 1   Review Exercises

3. $\{x \mid x \leq -4\}$ This is a half-open interval with interval notation $(-\infty, -4]$.

5. $(4, -2)$ and $(1, -6)$

$$d = \sqrt{(1 - 4)^2 + [-6 - (-2)]^2}$$
$$= \sqrt{(-3)^2 + (-4)^2}$$
$$= \sqrt{9 + 16}$$
$$= \sqrt{25} = 5$$

7. Let A = (-2, -2), B = (8, 4) and
   C = (2, 14).

   Distance from A to B is

   $$\sqrt{[8 - (-2)]^2 + [4 - (-2)]^2}$$
   $$= \sqrt{100 + 36}$$
   $$= \sqrt{136} = 2\sqrt{34}.$$

   Distance from A to C is

   $$\sqrt{(-2 - 2)^2 + (-2 - 14)^2} = \sqrt{16 + 256}$$
   $$= \sqrt{272}$$
   $$= 4\sqrt{17}.$$

   Distance from B to C is

   $$\sqrt{(8 - 2)^2 + (4 - 14)^2} = \sqrt{36 + 100}$$
   $$= \sqrt{136} = 2\sqrt{34}.$$

   Does $(2\sqrt{34})^2 + (2\sqrt{34})^2 = (4\sqrt{17})^2$?
   Yes, since 136 + 136 = 272.
   Therefore, these points are vertices
   of a right triangle.

In Exercises 9 and 11, $f(x) = -x^2 + 3x + 2$.

9. $f(0) = -(0)^2 + 3(0) + 2$
    $$= -0 + 0 + 2$$
    $$= 2$$

11. $f(x) = -x^2 + 3x + 2$
     $f(x + 1) = -(x + 1)^2 + 3(x + 1) + 2$
     $$= -(x^2 + 2x + 1) + 3x + 3 + 2$$
     $$= -x^2 - 2x - 1 + 3x + 3 + 2$$
     $$= -x^2 + x + 4$$

13. $y = 9x + 2$

   Domain: x can take on any value so
            the domain is $(-\infty, \infty)$.

   Range: y can take on any value, so
           the range is $(-\infty, \infty)$.

Each x-value gives one y-value so
$y = 9x + 2$ is a function.

15. $y = |x|$

   Domain: x can take on any value, so
            the domain is $(-\infty, \infty)$.

   Range: y will be any nonnegative
           number since the absolute
           value of a number is non-
           negative. The range is
           $[0, \infty)$.

Each x-value gives one y-value, so
$y = |x|$ is a function.

17. $x + 1 = y^2$

   Domain: y^2 is always ≥ 0, so
            $x + 1 \geq 0$ and $x \geq -1$.
            The domain is $[-1, \infty)$.

   Range: y can take on any value,
           so the range is $(-\infty, \infty)$.

Some x-values give more than one
y-value. For example, if x = 3,

   $$y = 2 \quad \text{or} \quad y = -2.$$

So $x + 1 = y^2$ is not a function.

19. Domain: Since every point on this
              graph has an x-value
              between -3 and 3 inclusive,
              the domain is $[-3, 3]$.

   Range: Since every point on this
           graph has a y-value between
           -5 and 5 inclusive, the
           range is $[-5, 5]$.

Notice that the points (0, 5) and (0, −5) lie on the graph. Therefore, one x-value gives more than one y-value. So this is not the graph of a function.

21. −51° is coterminal with

   −51° + 360° = 309°.

23. 792° is coterminal with

   792° − 2(360°) = 72°.

25. 320 rotations per minute

   = 320(360°) per minute

   = 115,200° per minute

   = $\dfrac{115,200°}{60}$ per second

   = 1920° per second

   = 1920° $\left(\dfrac{2}{3}\right)$ per $\dfrac{2}{3}$ second

   = 1280° per $\dfrac{2}{3}$ second

27. 47° 25′ 11″

   = $47° + \dfrac{25°}{60} + \dfrac{11}{3600}°$

   = 47° + .417° + .003°

   = 47.420°

29. 74.2983°

   = 74° + (.2983)(60′)

   = 74° + 17.898′

   = 74° + 17′ + (.898)(60″)

   = 74° + 17′ + 53.88″

   ≈ 74° 17′ 54″

31. 183.0972°

   = 183° + (.0972)(60′)

   = 183° + 5.832′

   = 183° + 5′ + (.832)(60″)

   = 183° + + 5′ + 49.92″

   = 183° 5′ 50″

33. The angles (9x + 4)° and (12x − 14)° are vertical angles, which have equal measure.

$$9x + 4 = 12x - 4$$
$$4 = 3x - 14$$
$$18 = 3x$$
$$6 = x$$

   The angles would be 58° and 58°.

Hint for Exercises 35–39: In similar triangles, corresponding sides are in proportion and corresponding angles are equal.

35. V = X = 41°

   Z = T = 32°

   The sum of the angles in a triangle is 180° so

   Y = U = 180° − (41° + 32°)

         = 107°.

37. $\dfrac{75}{50} = \dfrac{3}{2}$

   $\dfrac{m}{30} = \dfrac{3}{2}$

   2m = 90

   m = 45

   $\dfrac{n}{40} = \dfrac{3}{2}$

   2n = 120

   n = 60

39. $\dfrac{6}{r} = \dfrac{7}{11 + 7}$

$\dfrac{6}{r} = \dfrac{7}{18}$

$7r = 108$

$r = \dfrac{108}{7}$

41. It two triangles are similar, their corresponding sides are *proportional* and their corresponding angles are *equal*.

43. $x = -3, \ y = -3$

$r = \sqrt{x^2 + y^2}$

$ = \sqrt{(-3)^2 + (-3)^2}$

$ = \sqrt{9 + 9}$

$ = \sqrt{18}$

$ = 3\sqrt{2}$

$\sin \theta = \dfrac{y}{r} = -\dfrac{3}{3\sqrt{2}} = -\dfrac{1}{\sqrt{2}} = -\dfrac{\sqrt{2}}{2}$

$\cos \theta = \dfrac{x}{r} = -\dfrac{3}{3\sqrt{2}} = -\dfrac{1}{\sqrt{2}} = -\dfrac{\sqrt{2}}{2}$

$\tan \theta = \dfrac{y}{x} = \dfrac{-3}{-3} = 1$

$\cot \theta = \dfrac{x}{y} = \dfrac{-3}{-3} = 1$

$\sec \theta = \dfrac{r}{x} = \dfrac{3\sqrt{2}}{-3} = -\sqrt{2}$

$\csc \theta = \dfrac{r}{y} = \dfrac{3\sqrt{2}}{-3} = -\sqrt{2}$

45. $\sin 180° = 0$

$\cos 180° = -1$

$\tan 180° = 0$

$\cot 180°$ is undefined.

$\sec 180° = -1$

$\csc 180°$ is undefined.

47. $(-8, 15), \ x = -8, \ y = 15,$

$r = \sqrt{(-8)^2 + 15^2}$

$ = \sqrt{64 + 225}$

$ = \sqrt{289} = 17$

$\sin \theta = \dfrac{y}{r} = \dfrac{15}{17}$

$\cos \theta = \dfrac{x}{r} = -\dfrac{8}{17}$

$\tan \theta = \dfrac{y}{x} = \dfrac{15}{-8} = -\dfrac{15}{8}$

$\cot \theta = \dfrac{x}{y} = -\dfrac{8}{15}$

$\sec \theta = \dfrac{r}{x} = \dfrac{17}{-8} = -\dfrac{17}{8}$

$\csc \theta = \dfrac{r}{y} = \dfrac{17}{15}$

49. $(1, -5), \ x = 1, \ y = -5,$

$r = \sqrt{1^2 + (-5)^2}$

$ = \sqrt{1 + 25}$

$ = \sqrt{26}$

$\sin \theta = \dfrac{y}{r} = -\dfrac{5}{\sqrt{26}} = -\dfrac{5\sqrt{26}}{26}$

$\cos \theta = \dfrac{x}{r} = \dfrac{1}{\sqrt{26}} = \dfrac{\sqrt{26}}{26}$

$\tan \theta = \dfrac{y}{x} = -\dfrac{5}{1} = -5$

$\cot \theta = \dfrac{x}{y} = \dfrac{1}{-5} = -\dfrac{1}{5}$

$\sec \theta = \dfrac{r}{x} = \dfrac{\sqrt{26}}{1} = \sqrt{26}$

$\csc \theta = \dfrac{r}{y} = \dfrac{\sqrt{26}}{-5} = -\dfrac{\sqrt{26}}{5}$

51. $(6\sqrt{3}, -6), \ x = 6\sqrt{3}, \ y = -6$

$r = \sqrt{(6\sqrt{3})^2 + (-6)^2}$

$ = \sqrt{108 + 36}$

$ = \sqrt{144} = 12$

$\sin \theta = \dfrac{y}{r} = -\dfrac{6}{12} = -\dfrac{1}{2}$

$\cos \theta = \dfrac{x}{r} = \dfrac{6\sqrt{3}}{12} = \dfrac{\sqrt{3}}{2}$

$\tan \theta = \dfrac{y}{x} = -\dfrac{6}{6\sqrt{3}} = -\dfrac{1}{\sqrt{3}} = -\dfrac{\sqrt{3}}{3}$

$\cot \theta = \dfrac{x}{y} = \dfrac{6\sqrt{3}}{-6} = -\sqrt{3}$

$\sec \theta = \dfrac{r}{x} = \dfrac{12}{6\sqrt{3}} = \dfrac{2}{\sqrt{3}} = \dfrac{2\sqrt{3}}{3}$

$\csc \theta = \dfrac{r}{y} = \dfrac{12}{-6} = -2$

53. Since the terminal side of the angle is defined by $5x - 3y = 0$, $x \geq 0$, a point on this terminal side would be $(3, 5)$ since $5(3) - 3(5) = 0$.

$r = \sqrt{3^2 + 5^2} = \sqrt{9 + 25} = \sqrt{34}$

$\sin \theta = \dfrac{y}{r} = \dfrac{5}{\sqrt{34}} = \dfrac{5\sqrt{34}}{34}$

$\cos \theta = \dfrac{x}{r} = \dfrac{3}{\sqrt{34}} = \dfrac{3\sqrt{34}}{34}$

$\tan \theta = \dfrac{y}{x} = \dfrac{5}{3}$

$\cot \theta = \dfrac{x}{y} = \dfrac{3}{5}$

$\sec \theta = \dfrac{r}{x} = \dfrac{\sqrt{34}}{3}$

$\csc \theta = \dfrac{r}{y} = \dfrac{\sqrt{34}}{5}$

55. See the answer graph in the back of the textbook.

57. $\tan \theta = -5$

59. $-\cot^2 90° + 4 \sin 270° - 3 \tan 180°$

$= -(0)^2 + 4(-1) - 3(0)$

$= 0 - 4 - 0$

$= -4$

61. From the screen, $\cos X = .1$.

$\sin^2 X + \cos^2 X = 1$

$\qquad \sin^2 X = 1 - (.1)^2$

$\qquad \sin^2 X = .99$

$\tan^2 X = (\tan X)^2 = \dfrac{\sin^2 X}{\cos^2 X}$

$\qquad (\tan X)^2 = \dfrac{.99}{.01} = 99$

63. For any angle θ, $\sec \theta \leq -1$ or $\sec \theta \geq 1$. Therefore, $\sec \theta = -2/3$ is impossible.

65. If $\cos \theta = .25$, $\sec \theta = \dfrac{1}{.25} = 4$, so $\cos \theta = .25$ and $\sec \theta = -4$ is impossible.

67. $\cos \gamma = -\dfrac{5}{8}$, with γ in quadrant III.

$\cos \gamma = \dfrac{x}{r}$, so $x = -5$ and $r = 8$.

$x^2 + y^2 = r^2$

$(-5)^2 + y^2 = (8)^2$

$25 + y^2 = 64$

$y^2 = 39$

$y = \pm\sqrt{39}$

Since γ is in quadrant III, $y = -\sqrt{39}$.

$\sin \gamma = \dfrac{y}{r} = \dfrac{-\sqrt{39}}{8} = -\dfrac{\sqrt{39}}{8}$

$\cos \gamma = \dfrac{x}{r} = -\dfrac{5}{8}$

$$\tan \gamma = \frac{y}{x} = \frac{-\sqrt{39}}{-5} = \frac{\sqrt{39}}{5}$$

$$\cot \gamma = \frac{x}{y} = \frac{-5}{-\sqrt{39}} = \frac{5\sqrt{39}}{39}$$

$$\sec \gamma = \frac{r}{x} = \frac{8}{-5} = -\frac{8}{5}$$

$$\csc \gamma = \frac{r}{y} = \frac{8}{-\sqrt{39}} = -\frac{8\sqrt{39}}{39}$$

69. $\sec \beta = -\sqrt{5}$, with β in quadrant II.

$\sec \beta = \frac{r}{x}$, so $r = \sqrt{5}$ and $x = -1$.

$$x^2 + y^2 = r^2$$
$$(-1)^2 + y^2 = (\sqrt{5})^2$$
$$1 + y^2 = 5$$
$$y^2 = 4$$
$$y = \pm 2$$

Since β is in quadrant II, $y = 2$.

$$\sin \beta = \frac{y}{r} = \frac{2}{\sqrt{5}} = \frac{2\sqrt{5}}{5}$$

$$\cos \beta = \frac{x}{r} = \frac{-1}{\sqrt{5}} = -\frac{\sqrt{5}}{5}$$

$$\tan \beta = \frac{y}{x} = \frac{2}{-1} = -2$$

$$\cot \beta = \frac{x}{y} = \frac{-1}{2} = -\frac{1}{2}$$

$$\sec \beta = \frac{r}{x} = \frac{\sqrt{5}}{-1} = -\sqrt{5}$$

$$\csc \beta = \frac{r}{y} = \frac{\sqrt{5}}{2}$$

71. $\sec \alpha = \frac{5}{4}$, with α in quadrant IV.

$\sec \alpha = \frac{r}{x}$, so $r = 5$ and $x = 4$.

$$x^2 + y^2 = r^2$$
$$4^2 + y^2 = 5^2$$
$$16 + y^2 = 25$$
$$y^2 = 9$$
$$y = \pm 3$$

Since α is in quadrant IV, $y = -3$.

$$\sin \alpha = \frac{y}{r} = \frac{-3}{5} = -\frac{3}{5}$$

$$\cos \alpha = \frac{x}{r} = \frac{4}{5}$$

$$\tan \alpha = \frac{y}{x} = \frac{-3}{4} = -\frac{3}{4}$$

$$\cot \alpha = \frac{x}{y} = \frac{4}{-3} = -\frac{4}{3}$$

$$\sec \alpha = \frac{r}{x} = \frac{5}{4}$$

$$\csc \alpha = \frac{r}{y} = \frac{5}{-3} = -\frac{5}{3}$$

73. Keystrokes may vary based on the type and/or model of calculator used.
If $\tan \theta = 1.6778490$, first input the number 1.6778490. Since $\cot \theta = \frac{1}{\tan \theta}$, then press the [1/x] key or the [x⁻¹] key to get the reciprocal of 1.6778490. This would yield .5960012, the $\cot \theta$.

75. Let x be the depth of the crater Autolycus. Then,

$$\frac{x}{11,000} = \frac{1.3}{1.5}$$
$$1.5x = 1.3(11,000)$$
$$x = \frac{14,300}{1.5}$$
$$x \approx 9500.$$

The depth of the crater Autolycus is about 9500 ft.

CHAPTER 1   TEST

[1.1] Find the distance between each pair of points.

    1. (−3, −1) and (5, −3)   **2.** $(-6\sqrt{3},\ 3\sqrt{6})$ and $(4\sqrt{3},\ -\sqrt{6})$

[1.1] Write each set using interval notation.

    3. $\{x \mid x \geq -4\}$ **4.** $\{t \mid -2 < t \leq 3\}$

[1.1] Let $f(x) = -2x^2 + x - 3$. Find the following.

    5. $f(-2)$ **6.** $f(-b)$

[1.1] Find the domain and the range.

    7. $2x - 3y + 1 = 0$ **8.** $y = \sqrt{x^2 - 1}$

[1.2] Perform each calculation.

    9. 58° 43′ + 121° 31′   **10.** 90° − 36° 15′ 38″

[1.2] **11.** Convert 307.2104° to degrees, minutes, seconds. Round to the nearest second.

[1.2] **12.** Convert 94° 15′ 27″ to decimal degrees. Round to the nearest thousandth of a degree.

[1.2] Find the angles of smallest possible positive measure coterminal with each angle.

    13. 980°   **14.** −47°

[1.3] **15.** The measures of two angles of a triangle are 38° 40′ and 51° 30′. Find the measure of the third angle.

[1.3] **16.** Find the unknown side lengths in the following pair of similar triangles.

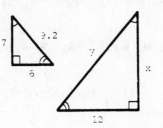

1.4] Find the values of the sin, cosine, and tangent functions for angle a in standard position having the given point on its terminal side.

17. (−8, 6)   **18.** (−3, −2)

[1.4] Evaluate each statement.

19. 2 tan 180° + 5 sin 90°   **20.** sec² 180° + 2 cot² 90°

[1.5] **21.** Find cot θ and csc θ, given sin θ = 3/5 and cos θ < 0.

[1.5] **22.** Find the values of all the trigonometric functions, given cos $\theta = -\frac{1}{2}$, where θ is in quadrant III.

[1.5] Decide whether each statement is *possible* or *impossible*.

23. tan $\alpha = \dfrac{3\sqrt{8}}{8}$ **24.** cos s = $-\sqrt{3}$

[1.5] **25.** If for some particular angle θ, cot θ is less in value than −1, what is the range in value of tan θ?

32   Chapter 1   The Trigonometric Functions

CHAPTER 1 TEST ANSWERS

1. $2\sqrt{17}$ **2.** $6\sqrt{11}$ **3.** $[-4, \infty)$ **4.** $(-2, 3]$ **5.** -13

6. $-2b^2 - b - 3$ **7.** $(-\infty, \infty)$; $(-\infty, \infty)$ **8.** $(-\infty, -1] \cup [1, \infty)$; $[0, \infty)$

9. $180°\ 14'$ **10.** $53°\ 44'\ 22''$ **11.** $307°\ 12'\ 37''$ **12.** $94.258°$

13. $260°$ **14.** $313°$ **15.** $89°\ 50'$ **16.** $x = 14$, $y = 18.4$

17. $\sin \alpha = 3/5$; $\cos \alpha = -4/5$; $\tan \alpha = -3/4$ **18.** $\sin \alpha = -2\sqrt{13}/13$;

$\cos \alpha = -3\sqrt{13}/13$; $\tan \alpha = 2/3$ **19.** 5 **20.** 1

21. $\cot \theta = -4/3$; $\csc \theta = 5/3$ **22.** $\sin \theta = -\sqrt{3}/2$; $\cos \theta = -1/2$;

$\tan \theta = \sqrt{3}$; $\cot \theta = \sqrt{3}/3$; $\sec \theta = -2$; $\csc \theta = -2\sqrt{3}/3$ **23.** Possible

24. Impossible   **25.** $(-1, 0)$

CHAPTER 2 ACUTE ANGLES AND RIGHT
TRIANGLES

Section 2.1

1. $\sin A = \dfrac{\text{side opposite}}{\text{hypotenuse}} = \dfrac{3}{5}$

$\cos A = \dfrac{\text{side adjacent}}{\text{hypotenuse}} = \dfrac{4}{5}$

$\tan A = \dfrac{\text{side opposite}}{\text{side adjacent}} = \dfrac{3}{4}$

$\cot A = \dfrac{\text{side adjacent}}{\text{side opposite}} = \dfrac{4}{3}$

$\sec A = \dfrac{\text{hypotenuse}}{\text{side adjacent}} = \dfrac{5}{4}$

$\csc A = \dfrac{\text{hypotenuse}}{\text{side opposite}} = \dfrac{5}{3}$

3. $\sin A = \dfrac{\text{side opposite}}{\text{hypotenuse}} = \dfrac{21}{29}$

$\cos A = \dfrac{\text{side adjacent}}{\text{hypotenuse}} = \dfrac{20}{29}$

$\tan A = \dfrac{\text{side opposite}}{\text{side adjacent}} = \dfrac{21}{20}$

$\cot A = \dfrac{\text{side adjacent}}{\text{side opposite}} = \dfrac{20}{21}$

$\sec A = \dfrac{\text{hypotenuse}}{\text{side adjacent}} = \dfrac{29}{20}$

$\csc A = \dfrac{\text{hypotenuse}}{\text{side opposite}} = \dfrac{29}{21}$

5. $\sin A = \dfrac{\text{side opposite}}{\text{hypotenuse}} = \dfrac{n}{p}$

$\cos A = \dfrac{\text{side adjacent}}{\text{hypotenuse}} = \dfrac{m}{p}$

$\tan A = \dfrac{\text{side opposite}}{\text{side adjacent}} = \dfrac{n}{m}$

$\cot A = \dfrac{\text{side adjacent}}{\text{side opposite}} = \dfrac{m}{n}$

$\sec A = \dfrac{\text{hypotenuse}}{\text{side adjacent}} = \dfrac{p}{m}$

$\csc A = \dfrac{\text{hypotenuse}}{\text{side opposite}} = \dfrac{p}{n}$

7. $a = 5, \ b = 12$

$c^2 = a^2 + b^2$

$c^2 = 5^2 + 12^2$

$c^2 = 25 + 144$

$c^2 = 169$

$c = 13$

$\sin B = \dfrac{\text{side opposite}}{\text{hypotenuse}} = \dfrac{12}{13}$

$\cos B = \dfrac{\text{side adjacent}}{\text{hypotenuse}} = \dfrac{5}{13}$

$\tan B = \dfrac{\text{side opposite}}{\text{side adjacent}} = \dfrac{12}{5}$

$\cot B = \dfrac{\text{side adjacent}}{\text{side opposite}} = \dfrac{5}{12}$

$\sec B = \dfrac{\text{hypotenuse}}{\text{side adjacent}} = \dfrac{13}{5}$

$\csc B = \dfrac{\text{hypotenuse}}{\text{side opposite}} = \dfrac{13}{12}$

9. $a = 6, \ c = 7$

$c^2 = a^2 + b^2$

$7^2 = 6^2 + b^2$

$49 = 36 + b^2$

$13 = b^2$

$\sqrt{13} = b$

$\sin B = \dfrac{\text{side opposite}}{\text{hypotenuse}} = \dfrac{\sqrt{13}}{7}$

$\cos B = \dfrac{\text{side adjacent}}{\text{hypotenuse}} = \dfrac{6}{7}$

$\tan B = \dfrac{\text{side opposite}}{\text{side adjacent}} = \dfrac{\sqrt{13}}{6}$

$\cot B = \dfrac{\text{side adjacent}}{\text{side opposite}} = \dfrac{6}{\sqrt{13}} = \dfrac{6\sqrt{13}}{13}$

$\sec B = \dfrac{\text{hypotenuse}}{\text{side adjacent}} = \dfrac{7}{6}$

$\csc B = \dfrac{\text{hypotenuse}}{\text{side opposite}} = \dfrac{7}{\sqrt{13}} = \dfrac{7\sqrt{13}}{13}$

13. cot 73° = tan (90° – 73°)
            = tan 17°

15. sec 39° = csc (90° – 39°)
            = csc 51°

17. cot (β – 10°)
     = tan [90° – (β – 10°)]
     = tan (100° – β)

19. sin 38.7° = cos (90° – 38.7°)
              = cos 51.3°

21. tan α = cot (α + 10°)

Since tangent and cotangent are co-functions, this equation is true if the sum of the angles is 90°.

     α + (α + 10°) = 90°
         2α + 10° = 90°
              2α = 80°
               α = 40°

23. sin (2γ + 10°) = cos (3γ – 20°)

Since sine and cosine are cofunctions,

     (2γ + 10°) + (3γ – 20°) = 90°
                  5γ – 10° = 90°
                      5γ = 100°
                       γ = 20°.

25. tan (3B + 4°) = cot (5B – 10°)

Since tangent and cotangent are cofunctions,

     (3B + 4°) + (5B – 10°) = 90°
                  8B – 6° = 90°
                      8B = 96°
                       B = 12°.

27. tan 28° ≤ tan 40°

tan θ increases as θ increases from 0° to 90°. Since

           40° > 28°
        tan 40° > tan 28°.

Therefore, the given statement is true.

29. sin 46° < cos 46°

sin θ increases as θ increases from 0° to 90°. Since

           46° > 44°
        sin 46° > sin 44°.
But
        sin 44° = cos 46°,
so
        sin 46° > cos 46°.
False

31. tan 41° < cot 41°

tan θ increases as θ increases from 0° to 90°. Since

           49° > 41°
        tan 49° > tan 41°.
But
        tan 49° = cot 41°,
so
        cot 41° > tan 41°.
True

33. sin 45° = cos 45°

sin A increases and cos A decreases as A increases from 0° to 90°. Therefore, 0° ≤ A < 45°.

35.

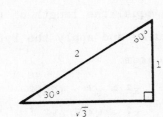

$$\tan 30° = \frac{\text{side opposite}}{\text{side adjacent}}$$

$$= \frac{1}{\sqrt{3}}$$

$$= \frac{\sqrt{3}}{3}$$

37. See figure in Exercise 35.

$$\sin 30° = \frac{\text{side opposite}}{\text{hypotenuse}}$$

$$= \frac{1}{2}$$

39.

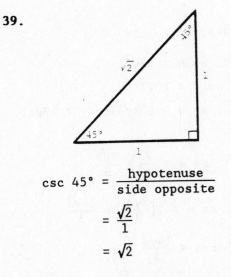

$$\csc 45° = \frac{\text{hypotenuse}}{\text{side opposite}}$$

$$= \frac{\sqrt{2}}{1}$$

$$= \sqrt{2}$$

41. See figure in Exercise 39.

$$\cos 45° = \frac{\text{side adjacent}}{\text{hypotenuse}}$$

$$= \frac{1}{\sqrt{2}}$$

$$= \frac{\sqrt{2}}{2}$$

43. See figure in Exercise 35.

$$\sin 60° = \frac{\text{side opposite}}{\text{hypotenuse}}$$

$$= \frac{\sqrt{3}}{2}$$

45. See figure in Exercise 35.

$$\tan 60° = \frac{\text{side opposite}}{\text{side adjacent}}$$

$$= \frac{\sqrt{3}}{1}$$

$$= \sqrt{3}$$

47. $\sin 0° = 0$

$\sin 45° = .70711$

$\sin 90° = 1$

$\tan 0° = 0$

$\tan 45° = 1$

$\tan 90°$ is undefined.

Y_1 is sin x and Y_2 is tan x.

49. $\sin 60° = .8660$ for A between 0° and 90°. Therefore, A = 60°.

51. Graph $y_1 = x$ and $y_2 = \sqrt{1 - x^2}$ in a window such as $-2 \le x \le 2$, $-2 \le y \le 2$.
Use TRACE or the intersection function to see that the point of intersection is (.70710678, .70710678). (This is the point $(\sqrt{2}/2, \sqrt{2}/2)$.) These coordinates are the sine and cosine of 45°.

53.

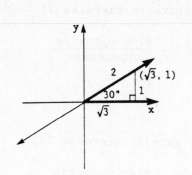

The line passes through $(0, 0)$ and $(\sqrt{3}, 1)$. The slope is $\sqrt{3}/3$. The equation of the line is

$$y = \frac{\sqrt{3}}{3}x.$$

55. The slope of a line is the change in y over the change in x, or $\tan \theta$. Since $y = \sqrt{3}x$,

$$m = \sqrt{3} = \tan \theta$$

and $\tan 60° = \sqrt{3}$.

The line $y = \sqrt{3}x$ makes a 60° angle with the positive x-axis. (See Exercise 52.)

57.

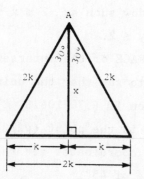

(a) Each angle of every equilateral triangle has a measure of 60°.

(b) The perpendicular bisects the opposite side, so the length of each side opposite each 30° angle is k.

(c) Let x equal the length of the perpendicular and apply the Pythagorean theorem.

$$x^2 + k^2 = (2k)^2$$
$$x^2 + k^2 = 4k^2$$
$$x^2 = 3k^2$$
$$x = k\sqrt{3}$$

The length of the perpendicular is $k\sqrt{3}$.

(d) In a 30°–60° right triangle, the the hypotenuse is always *2* times as long as the shorter leg, and the longer leg has a length that is $\sqrt{3}$ times as long as that of the shorter leg. Also, the shorter leg is opposite the *30°* angle, and the longer leg is opposite the *60°* angle.

59. $a = \frac{1}{2}(24)$, so $a = 12$.

$b = a\sqrt{3}$, so $b = 12\sqrt{3}$.

$d = b$, so $d = 12\sqrt{3}$.

$c = d\sqrt{2}$, so $c = (12\sqrt{3})(\sqrt{2}) = 12\sqrt{6}$.

61. $7 = m\sqrt{3}$, so $m = \dfrac{7}{\sqrt{3}} = \dfrac{7\sqrt{3}}{3}$.

$a = 2m$, so $a = 2\left(\dfrac{7\sqrt{3}}{3}\right) = \dfrac{14\sqrt{3}}{3}$.

$n = a$, so $n = \dfrac{14\sqrt{3}}{3}$.

$q = n\sqrt{2}$, so $q = \left(\dfrac{14\sqrt{3}}{3}\right)\sqrt{2} = \dfrac{14\sqrt{6}}{3}$.

63. Let h be the height of the equilateral triangle. By geometry, h bisects the base, s, and forms two 30°–60° right triangles.

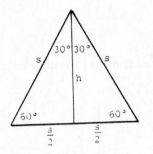

The formula for the area of a triangle is

$$\text{Area} = \frac{1}{2}bh.$$

In this triangle, b = s. The height h of the triangle is the side opposite the 60° angle in either 30°-60° right triangle. The side opposite the 30° angle is s/2. The height is $\sqrt{3}/(s/2)$. So the area of the entire triangle is

$$\frac{1}{2}s\left(\sqrt{3}\frac{s}{2}\right) = \frac{s^2\sqrt{3}}{4}.$$

65. Yes, the third angle can be found by subtracting the given acute angle from 90°, and the remaining two sides can be found using a trigonometric function involving the known angle and side.

67. From our previous work, we see that

$$\cos 0° = 1,$$
$$\cos 30° = \frac{\sqrt{3}}{2},$$
$$\cos 45° = \frac{\sqrt{2}}{2},$$
$$\cos 60° = \frac{1}{2}, \text{ and}$$
$$\cos 90° = 0.$$

Therefore, we have the following table.

θ	$\cos \theta$
0°	$\frac{\sqrt{4}}{2} = 1$
30°	$\frac{\sqrt{3}}{2}$
45°	$\frac{\sqrt{2}}{2}$
60°	$\frac{\sqrt{1}}{2} = \frac{1}{2}$
90°	$\frac{\sqrt{0}}{2} = 0$

69. Using the values from Exercise 68, determine V_2 when D = 200.

(Note that 90 mph = 132 ft/sec.)

$$D = \frac{1.05(V_1^2 - V_2^2)}{64.4(K_1 + K_2 + \sin \theta)}$$

$$200 = \frac{1.05(132^2 - V_2^2)}{64.4(.4 + .02 + \sin (-3.5°))}$$

$$200 = \frac{18,295.2 - 1.05V_2^2}{23.12}$$

$$V_2^2 = \frac{23.12(200) - 18,295.2}{-1.05}$$

$$V_2 \approx 114 \text{ ft/sec}$$

$$V_2 \approx 78 \text{ mph}$$

Section 2.2

5. $\theta = 98°$

θ is in quadrant II.

$$\theta' = 180° - 98° = 82°$$

7. $\theta = -135$

Since $\theta < 0°$, add a multiple of 360° to find a coterminal angle between 0° and 360°.

$$\theta + 360° = -135° + 360° = 225°$$

The coterminal angle is in quadrant III. Find θ'. .

$$\theta' = 225° - 180° = 45°$$

9. $\theta = 750°$

Since $\theta > 360°$, subtract a multiple of 360° to find a coterminal angle.

$$\theta - 2(360°) = 750° - 720° = 30°$$

The coterminal angle is in quadrant I, so it is also the reference angle

$$\theta' = 30°.$$

11. 120°

Use the method illustrated in Example 2.

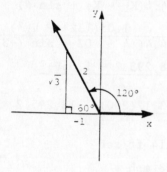

$$x = -1, \ y = \sqrt{3}, \ r = 2$$

$$\sin 120° = \frac{y}{r} = \frac{\sqrt{3}}{2}$$

$$\cos 120° = \frac{x}{r} = \frac{-1}{2} = -\frac{1}{2}$$

$$\tan 120° = \frac{y}{x} = \frac{\sqrt{3}}{-1} = -\sqrt{3}$$

$$\cot 120° = \frac{x}{y} = -\frac{1}{\sqrt{3}} = -\frac{\sqrt{3}}{3}$$

$$\sec 120° = \frac{r}{x} = \frac{2}{-1} = -2$$

$$\csc 120° = \frac{r}{y} = \frac{2}{\sqrt{3}} = \frac{2\sqrt{3}}{3}$$

13. 150°

Use the method illustrated in Example 2.

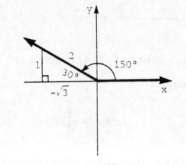

$$x = -\sqrt{3}, \ y = 1, \ r = 2$$

$$\sin 150° = \frac{y}{r} = \frac{1}{2}$$

$$\cos 150° = \frac{x}{r} = \frac{-\sqrt{3}}{2} = -\frac{\sqrt{3}}{2}$$

$$\tan 150° = \frac{y}{x} = \frac{1}{-\sqrt{3}} = -\frac{\sqrt{3}}{3}$$

$$\cot 150° = \frac{x}{y} = \frac{-\sqrt{3}}{1} = -\sqrt{3}$$

$$\sec 150° = \frac{r}{x} = \frac{2}{-\sqrt{3}} = -\frac{2\sqrt{3}}{3}$$

$$\csc 150° = \frac{r}{y} = \frac{2}{1} = 2$$

15. 240°

Use the method illustrated in Example 3.

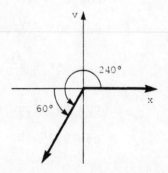

The reference angle is 60°. Since 240° is in quadrant III, the sine, cosine, secant, and cosecant are negative.

$\sin 240° = -\sin 60° = -\dfrac{\sqrt{3}}{2}$

$\cos 240° = -\cos 60° = -\dfrac{1}{2}$

$\tan 240° = \tan 60° = \sqrt{3}$

$\cot 240° = \cot 60° = \dfrac{\sqrt{3}}{3}$

$\sec 240° = -\sec 60° = -2$

$\csc 240° = -\csc 60° = -\dfrac{2\sqrt{3}}{3}$

17. 315°

Use the method illustrated in Example 3.

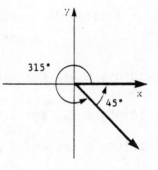

The reference angle is 45°. Since 315° is in quadrant IV, the sine, tangent, cotangent, and cosecant are negative.

$\sin 315° = -\sin 45° = -\dfrac{\sqrt{2}}{2}$

$\cos 315° = \cos 45° = \dfrac{\sqrt{2}}{2}$

$\tan 315° = -\tan 45° = -1$

$\cot 315° = -\cot 45° = -1$

$\sec 315° = \sec 45° = \sqrt{2}$

$\csc 315° = -\csc 45° = -\sqrt{2}$

19. 420°

Use the method illustrated in Example 3.

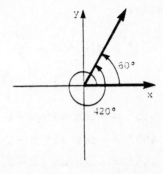

The reference angle is 60°. Since 420° is in quadrant I, all functions are positive.

$\sin 420° = \sin 60° = \dfrac{\sqrt{3}}{2}$

$\cos 420° = \cos 60° = \dfrac{1}{2}$

$\tan 420° = \tan 60° = \sqrt{3}$

$\cot 420° = \cot 60° = \dfrac{\sqrt{3}}{3}$

$\sec 420° = \sec 60° = 2$

$\csc 420° = \csc 60° = \dfrac{2\sqrt{3}}{3}$

21. 495°

Use the method illustrated in Example 3.

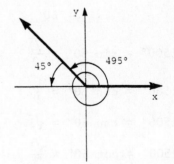

The reference angle is 45°. Since 495° is in quadrant II, cosine, tangent, cotangent, and secant are negative.

$$\sin 495° = \sin 45° = \frac{\sqrt{2}}{2}$$

$$\cos 495° = -\cos 45° = -\frac{\sqrt{2}}{2}$$

$$\tan 495° = -\tan 45° = -1$$

$$\cot 495° = -\cot 45° = -1$$

$$\sec 495° = -\sec 45° = -\sqrt{2}$$

$$\csc 495° = \csc 45° = \sqrt{2}$$

23. 750° is coterminal with

$$750° - 2(360°) = 750° - 720°$$
$$= 30°.$$

$$\sin 750° = \sin 30° = \frac{1}{2}$$

$$\cos 750° = \cos 30° = \frac{\sqrt{3}}{2}$$

$$\tan 750° = \tan 30° = \frac{\sqrt{3}}{3}$$

$$\cot 750° = \cot 30° = \sqrt{3}$$

$$\sec 750° = \sec 30° = \frac{2\sqrt{3}}{3}$$

$$\csc 750° = \csc 30° = 2$$

25. 1500° is coterminal with

$$1500° - 4(360°) = 1500° - 1440°$$
$$= 60°.$$

$$\sin 1500° = \sin 60° = \frac{\sqrt{3}}{2}$$

$$\cos 1500° = \cos 60° = \frac{1}{2}$$

$$\tan 1500° = \tan 60° = \sqrt{3}$$

$$\cot 1500° = \cot 60° = \frac{\sqrt{3}}{3}$$

$$\sec 1500° = \sec 60° = 2$$

$$\csc 1500° = \csc 60° = \frac{2\sqrt{3}}{3}$$

27. -390° is coterminal with

$$-390° - (-2)(360°) = 330°.$$

The reference angle is 30°. Since -390° is in quadrant IV, the sine, tangent, cotangent, and cosecant are negative.

$$\sin (-390°) = -\sin 30° = -\frac{1}{2}$$

$$\cos (-390°) = \cos 30° = \frac{\sqrt{3}}{2}$$

$$\tan (-390°) = -\tan 30° = -\frac{\sqrt{3}}{3}$$

$$\cot (-390°) = -\cot 30° = -\sqrt{3}$$

$$\sec (-390°) = \sec 30° = \frac{2\sqrt{3}}{3}$$

$$\csc (-390°) = -\csc 30° = -2$$

29. -1020° is coterminal with

$$-1020° - (-3)(360°) = -1020° + 1080°$$
$$= 60°.$$

$$\sin (-1020°) = \sin 60° = \frac{\sqrt{3}}{2}$$

$$\cos (-1020°) = \cos 60° = \frac{1}{2}$$

$$\tan (-1020°) = \tan 60° = \sqrt{3}$$

$$\cot (-1020°) = \cot 60° = \frac{\sqrt{3}}{3}$$

$$\sec (-1020°) = \sec 60° = 2$$

$$\csc (-1020°) = \csc 60° = \frac{2\sqrt{3}}{3}$$

31. 30°:

$$\tan 30° = \frac{\sqrt{3}}{3}$$

$$\cot 30° = \sqrt{3}$$

33. 60°:

$$\sin 60° = \frac{\sqrt{3}}{2}$$

$$\cot 60° = \frac{\sqrt{3}}{3}$$

$$\csc 60° = \frac{2\sqrt{3}}{3}$$

35. 135°:

$$\tan 135° = -\tan 45° = -1$$

$$\cot 135° = -\cot 45° = -1$$

37. 210°:

$$\cos 210° = -\cos 30° = -\frac{\sqrt{3}}{2}$$

$$\sec 210° = -\sec 30° = -\frac{2\sqrt{3}}{3}$$

39. For every angle θ, $\sin \theta = \frac{y}{r}$ and

$\cos \theta = \frac{x}{r}$. Thus, for every angle θ,

$$(\sin \theta)^2 + (\cos \theta)^2 = \left(\frac{y}{r}\right)^2 + \left(\frac{x}{r}\right)^2$$

$$= \frac{y^2 + x^2}{r^2}$$

$$= \frac{r^2}{r^2} = 1.$$

Since $(.6)^2 + (-.8)^2 = .36 + .64 = 1$, there is an angle θ for which $\cos \theta = .6$ and $\sin \theta = -.8$. Since $\cos \theta > 0$ and $\sin \theta < 0$, it is an angle in quadrant IV.

41. If θ is in the interval (90°, 180°),

$$90° < \theta < 180°.$$

Thus, dividing by 2, we get

$$45° < \frac{\theta}{2} < 90°.$$

Thus, $\frac{\theta}{2}$ is a quadrant I angle and $\sin \frac{\theta}{2}$ is positive.

43. If θ is in the interval (90°, 180°),

$$90° < \theta < 180°.$$

Adding 180° gives

$$270° < \theta + 180° < 360°.$$

Thus, $\theta + 180°$ is a quadrant IV angle and $\cot (\theta + 180°)$ is negative.

45. If θ is in the interval (90°, 180°),

$$90° < \theta < 180°.$$

Multiplying by −1 gives

$$-90° > -\theta > -180°.$$

Thus, $-\theta$ is a quadrant III angle and $\cos (-\theta)$ is negative.

49. The reference angle is 65°. Since 115° is in quadrant II, the cosine is negative. $\cos \theta$ decreases on the interval [90°, 180°] from 0 to −1. Therefore, $\cos 115°$ is closest to −.4.

51. $\sin \theta = \cos \theta$ when $\theta = 45°$. Sine and cosine are negatives in quadrants II and IV. Thus

$$180° - \theta = 180° - 45° = 135°$$

in quadrant II.

$$360° - \theta = 360° - 45° = 315°$$

in quadrant IV.

53. $\sin 30° + \sin 60° = \sin (30° + 60°)$

Evaluate each side to determine whether this statement is true or false.

$\sin 30° + \sin 60° = \dfrac{1}{2} + \dfrac{\sqrt{3}}{2}$

$\qquad\qquad\qquad = \dfrac{1 + \sqrt{3}}{2}$

$\sin (30° + 60°) = \sin 90°$

$\qquad\qquad\qquad = 1$

Since $\dfrac{1 + \sqrt{3}}{2} \neq 1$, the given statement is false.

55. $\cos 60° = 2 \cos^2 30° - 1$

$\cos 60° = \dfrac{1}{2}$

$2 \cos^2 30° - 1 = 2\left(\dfrac{\sqrt{3}}{2}\right)^2 - 1$

$\qquad\qquad\qquad = 2\left(\dfrac{3}{4}\right) - 1$

$\qquad\qquad\qquad = \dfrac{3}{2} - 1$

$\qquad\qquad\qquad = \dfrac{1}{2}$

Thus, $\dfrac{1}{2} = \dfrac{1}{2}$.

True

57. $\sin 120° = \sin 150° - \sin 30°$

$\sin 120° = \dfrac{\sqrt{3}}{2}$

$\sin 150° - \sin 30° = \dfrac{1}{2} - \dfrac{1}{2} = 0$

Thus, $\dfrac{\sqrt{3}}{2} \neq 0$.

False

59. $\sin 120° = \sin 180° \cdot \cos 60°$

$\qquad\qquad\quad - \sin 60° \cdot \cos 180°$

$\sin 120° = \dfrac{\sqrt{3}}{2}$

$\sin 180° \cdot \cos 60° - \sin 60° \cdot \cos 180°$

$\qquad = 0\left(\dfrac{1}{2}\right) - \left(\dfrac{\sqrt{3}}{2}\right)(-1)$

$\qquad = 0 - \left(-\dfrac{\sqrt{3}}{2}\right)$

$\qquad = \dfrac{\sqrt{3}}{2}$

Thus, $\dfrac{\sqrt{3}}{2} = \dfrac{\sqrt{3}}{2}$.

True

61. (a) $L = \dfrac{(\theta_2 - \theta_1)S^2}{200(h + S \tan \alpha)}$

Let $h = 1.9$, $\alpha = .9°$, $\theta_1 = -3°$, $\theta_2 = 4°$, and $S = 336$.

$L = \dfrac{[4 - (-3)]336^2}{200(1.9 + 336 \tan .9°)}$

$\quad \approx 550$ ft

(b) $L = \dfrac{(\theta_2 - \theta_1)S^2}{200(h + S) \tan \alpha}$

Now let $\alpha = 1.5°$.

$\quad = \dfrac{[4 - (-3)]336^2}{200(1.9 + 336 \tan 1.5°)}$

$\quad \approx 369$ ft

(c) The alignment of the headlights has a large effect on the value of L. A small increase in α allows the driver to see further along the sag curve.

Section 2.3

For the following exercises, be sure your calculator is in degree mode. If your calculator accepts angles in degrees, minutes, and seconds, it is not necessary to change angles to decimal degrees. Keystroke sequences may vary based on the type and/or model of calculator being used.

1. tan 29° 30′

$$29° \ 30′ = \left(29 + \frac{30}{60}\right)°$$
$$= 29.5°$$

Enter: 29.5 　[TAN]

or 　　[TAN] 29.5 [ENTER]

Display: .5657728

3. cot 41° 24′

$$41° \ 24′ = \left(41 + \frac{24}{60}\right)°$$
$$= 41.4°$$

Enter: 41.4 [TAN][1/x]

or 　　[(] [TAN] 41.4 [)] [x⁻¹] [ENTER]

Display: 1.1342773

5. sec 13° 15′

$$13° \ 15′ = \left(13 + \frac{15}{60}\right)°$$
$$= 13.25°$$

Enter: 13.25 [COS][1/x]

or 　　[(] [COS] 13.25 [)] [x⁻¹]
　　　　[ENTER]

Display: 1.0273488

7. sin 39° 40′

$$39° \ 40′ = \left(39 + \frac{40}{60}\right)°$$

Enter: [(] 39 [+] 40 [÷] 60 [)] [SIN]

or 　　[SIN] 39′40′ [ENTER]

Display: .6383201

9. csc 145° 45′

$$145° \ 45′ = \left(145 + \frac{45}{60}\right)°$$
$$= 145.75°$$

Enter: 145.75 [SIN][1/x]

or 　　[(] [SIN] 145.75 [)] [x⁻¹]
　　[ENTER]

Display: 1.7768146

11. cos 421° 30′

$$421° \ 30′ = \left(421 + \frac{30}{60}\right)°$$
$$= 421.5°$$

Enter: 421.5 [COS]

or 　　[COS] 421.5 [ENTER]

Display: .4771588

13. tan (−80° 6′)

$$-80° \ 6′ = -\left(80 + \frac{6}{60}\right)°$$
$$= -80.1°$$

Enter: 80.1 [+/−] [TAN]

or 　　[TAN] [(−)] 80.1 [ENTER]

Display: −5.7297416

15. cot (-512° 20′)

$$-512° 20′ = -\left(512 + \frac{20}{60}\right)°$$

Enter:

(512 + 20 ÷ 60) +/- TAN 1/x

or (TAN (-) 512′20′) x⁻¹ ENTER

Display: 1.9074147

17. $\dfrac{1}{\sec 14.8°} = \cos 14.8°$

Enter: 14.8 COS

or COS 14.8 ENTER

Display: .9668234

19. $\dfrac{1}{\cot 23.4°} = \tan 23.4°$

Enter: 23.4 TAN

or TAN 23.4 ENTER

Display: .4327386

21. $\dfrac{\cos 77°}{\sin 77°} = \cot 77°$

Enter: 77 TAN 1/x

or (TAN 77) x⁻¹ ENTER

Display: .2308682

23. cot (90° - 4.72°) = tan 4.72°

Enter: 4.72 TAN

or TAN 4.72 ENTER

Display: .0825664

25. $\sec^2 47.8° - 1 = \tan^2 47.8°$

Enter: 47.8 TAN x²

or (TAN 47.8) x² ENTER

Display: 1.2162701

27. The student should check the mode of the calculator. The calculator was not set in degree mode.

29. sin θ = .84802194

Enter: .84802194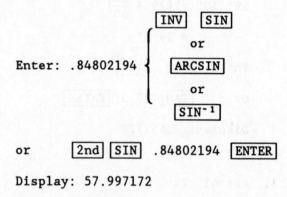

or 2nd SIN .84802194 ENTER

Display: 57.997172

θ = 57.997172°

31. sec θ = 1.1606249

Enter: 1.1606249 1/x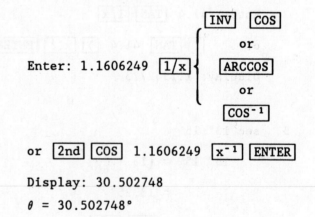

or 2nd COS 1.1606249 x⁻¹ ENTER

Display: 30.502748

θ = 30.502748°

33. sin θ = .72144101

Enter: .72144101 $\begin{cases} \boxed{\text{INV}}\ \boxed{\text{SIN}} \\ \text{or} \\ \boxed{\text{ARCSIN}} \\ \text{or} \\ \boxed{\text{SIN}^{-1}} \end{cases}$

or $\boxed{\text{2nd}}\ \boxed{\text{SIN}}$ .72144101 $\boxed{\text{ENTER}}$

Display: 46.173581

θ = 46.173581°

35. tan θ = 6.4358841

Enter: 6.4358841 $\begin{cases} \boxed{\text{INV}}\ \boxed{\text{TAN}} \\ \text{or} \\ \boxed{\text{ARCTAN}} \\ \text{or} \\ \boxed{\text{TAN}^{-1}} \end{cases}$

or $\boxed{\text{2nd}}\ \boxed{\text{TAN}}$ 6.4358841 $\boxed{\text{ENTER}}$

Display: = 81.168073

θ = 81.168073°

37. tan A = 1.482560969

Enter: 1.482560969 $\begin{cases} \boxed{\text{INV}}\ \boxed{\text{TAN}} \\ \text{or} \\ \boxed{\text{ARCTAN}} \\ \text{or} \\ \boxed{\text{TAN}^{-1}} \end{cases}$

or $\boxed{\text{2nd}}\ \boxed{\text{TAN}}$ 1.482560969 $\boxed{\text{ENTER}}$

Display: 56

A = 56°

39. sin 35° cos 55° + cos 35° sin 55°

Enter: $\boxed{(}$ 35 $\boxed{\text{SIN}}$ $\boxed{)}$ $\boxed{\times}$ $\boxed{(}$ 55 $\boxed{\text{COS}}$ $\boxed{)}$ $\boxed{+}$ $\boxed{(}$ 35 $\boxed{\text{COS}}$ $\boxed{)}$ $\boxed{\times}$ $\boxed{(}$ 55 $\boxed{\text{SIN}}$ $\boxed{)}$ $\boxed{=}$

or $\boxed{\text{SIN}}$ 35 $\boxed{\text{COS}}$ 55 $\boxed{+}$ $\boxed{\text{COS}}$ 35 $\boxed{\text{SIN}}$ 55 $\boxed{\text{ENTER}}$

Display: 1

41. cos 75° 29′ cos 14° 31′
− sin 75° 29′ sin 14° 31′

Enter: $\boxed{(}$ 75 $\boxed{+}$ 29 $\boxed{\div}$ 60 $\boxed{)}$ $\boxed{\text{COS}}$ $\boxed{\times}$ $\boxed{(}$ 14 $\boxed{+}$ 31 $\boxed{\div}$ 60 $\boxed{)}$ $\boxed{\text{COS}}$ $\boxed{-}$ $\boxed{(}$ 75 $\boxed{+}$ 29 $\boxed{\div}$ 60 $\boxed{)}$ $\boxed{\text{SIN}}$ $\boxed{\times}$ $\boxed{(}$ 14 $\boxed{+}$ 31 $\boxed{\div}$ 60 $\boxed{)}$ $\boxed{\text{SIN}}$ $\boxed{=}$

or $\boxed{\text{COS}}$ 75′29′ $\boxed{\text{COS}}$ 14′31′ $\boxed{-}$ $\boxed{\text{SIN}}$ 75′29′ $\boxed{\text{SIN}}$ 14′31′ $\boxed{\text{ENTER}}$

Display: 0

43. $\theta_1 = 46°$, $\theta_2 = 31°$, $c_1 = 3 \times 10^8$ m per sec

$$\frac{c_1}{c_2} = \frac{\sin \theta_1}{\sin \theta_2}$$

$$c_2 = \frac{c_1 \sin \theta_2}{\sin \theta_1}$$

$$c_2 = \frac{(3 \times 10^8)(\sin 31°)}{\sin 46°}$$

$$c_2 = 2 \times 10^8 \text{ m per sec}$$

45. $\theta_1 = 40°$, $c_2 = 1.5 \times 10^8$ m per sec,
$c_1 = 3 \times 10^8$ m per sec

$$\frac{c_1}{c_2} = \frac{\sin \theta_1}{\sin \theta_2}$$

$$\sin \theta_2 = \frac{c_2 \sin \theta_1}{c_1}$$

$$\sin \theta_2 = \frac{1.5 \times 10^8 (\sin 40°)}{3 \times 10^8}$$

$$\theta_2 = 19°$$

47. $\theta_1 = 90°$, $c_1 = 3 \times 10^8$ m per sec,
$c_2 = 2.254 \times 10^8$

$$\frac{c_1}{c_2} = \frac{\sin \theta_1}{\sin \theta_2}$$

$$\sin \theta_2 = \frac{c_2 \sin \theta_1}{c_1}$$

$$\sin \theta_2 = \frac{2.254 \times 10^8 (\sin 90°)}{3 \times 10^8}$$

$$= \frac{2.254 \times 10^8 (1)}{3 \times 10^8}$$

$$= \frac{2.254}{3} = .7513$$

$$\theta_2 = 48.7°$$

49. $\cos 40° = 2 \cos 20°$

Using a calculator gives

$\cos 40° = .766044$ and
$2 \cos 20° = 1.8793852$.

False

51. $\cos 70° = 2 \cos^2 35° - 1$

Using a calculator gives

$\cos 70° = .34202014$ and
$2 \cos^2 35° - 1 = .34202014$.

True

53. $2 \cos 38° 22' = \cos 76° 44'$

Using a calculator gives

$2 \cos 38° 22' = 1.5681094$ and
$\cos 76° 44' = .22948353$.

False

55. $\frac{1}{2} \sin 40° = \sin \frac{1}{2} (40°)$

Using a calculator gives

$\frac{1}{2} \sin 40° = .32139380$ and

$\sin \frac{1}{2}(40°) = .34202014$.

False

57. $F = W \sin \theta$
$F = 2400 \sin (-2.4°)$
≈ -100.5 lb

F is negative because the car is traveling downhill.

59. $F = W \sin \theta$
$150 = 3000 \sin \theta$
$\frac{150}{3000} = \sin \theta$
$.05 = \sin \theta$
$\theta \approx 2.866°$

61. $F = W \sin \theta$
$-145 = W \sin (-3°)$
$\frac{-145}{\sin (-3°)} = W$
$W \approx 2771$ lb

63. F = W sin θ

F = 2200 sin 2°

F = 76.77889275 lb

F = W sin θ

F = 2000 sin 2.2°

F = 76.77561818 lb

The 2200-lb car on a 2° uphill grade has the greater grade resistance.

65. For parts (a) and (b), α = 3°, g = 32.2, and f = .14.

(a) Since 45 mi/hr = 66 ft/sec,

$$R = \frac{v^2}{g(f + \tan \alpha)}$$

$$\approx \frac{66^2}{32.2(.14 + .052)}$$

$$\approx 703 \text{ ft.}$$

(b) Since 70 mi/hr = 102 2/3 ft/sec,

$$R = \frac{v^2}{g(f + \tan \alpha)}$$

$$\approx \frac{102.67^2}{32.2(.14 + .052)}$$

$$\approx 1701 \text{ ft.}$$

(c) Intuitively, increasing α would make it easier to negotiate the curve at a higher speed much like is done at a race track. Mathematically, a larger value of α (acute) will lead to a larger value for tan α. If tan α increases, then the ratio determining R will *decrease*. Thus, the radius can be smaller and the curve sharper if α is increased.

$$R = \frac{v^2}{g(f + \tan \alpha)}$$

$$\approx \frac{66^2}{32.2(.14 + .070)}$$

$$\approx 644 \text{ ft}$$

$$R = \frac{v^2}{g(f + \tan \alpha)}$$

$$\approx \frac{102.67^2}{32.2(.14 + .070)}$$

$$\approx 1559 \text{ ft}$$

As predicted, both values are less.

Section 2.4

3. If h is the actual height of a building and the height is measured as 58.6 ft, then $|h - 58.6| \le .05$.

5. 29,000 ft

This measurement is to the nearest foot. It represents a range of 28,999.5 ft to 29,000.5 ft.

7. If 1650 ft and 160 ft represent accuracy to the nearest foot, then the ranges represented by these two numbers are 1649.5 to 1650.5 and 159.5 to 160.5, respectively.

9. The error is that both 2 and 65 are counting numbers, not measurements. They are exact numbers so all the digits in their product are significant.

11. A = 36° 20′, c = 964 m

A + B = 90°

B = 90° − A

B = 90° − 36° 20′

= 89° 60′ − 36° 20′

B = 53° 40′

$$\sin A = \frac{a}{c}$$

$$a = c \sin A$$

$$a = 964 \sin 36° \ 20'$$

Use a calculator and round answer to three significant digits.

$$a = 571 \text{ m}$$

$$\cos A = \frac{b}{c}$$

$$b = c \cos A$$

$$b = 964 \cos 36° \ 20'$$

Use a calculator and round answer to three significant digits.

$$b = 777 \text{ m}$$

13. N = 51.2°, m = 124 m

$$M + N = 90°$$

$$M = 90° - N$$

$$M = 90° - 51.2°$$

$$M = 38.8°$$

$$\tan N = \frac{n}{m}$$

$$n = m \tan N$$

$$n = 124 \cdot \tan 51.2°$$

$$n = 154 \text{ m}$$

$$\cos N = \frac{m}{p}$$

$$p = \frac{m}{\cos N}$$

$$p = \frac{124}{\cos 51.2°}$$

$$p = 198 \text{ m}$$

15. B = 42.0892°, b = 56.851 cm

$$A + B = 90°$$

$$A = 90° - B$$

$$A = 90° - 42.0892°$$

$$= 47.9108°$$

$$\sin B = \frac{b}{c}$$

$$c = \frac{b}{\sin B}$$

$$c = \frac{56.851}{\sin 42.0892°}$$

$$c = 84.816 \text{ cm}$$

$$\tan B = \frac{b}{a}$$

$$a = \frac{b}{\tan B}$$

$$a = \frac{56.851}{\tan 42.0892°}$$

$$a = 62.942 \text{ cm}$$

23. A = 28.00°, c = 17.4 ft

$$A + B = 90°$$

$$B = 90° - A$$

$$B = 90° - 28.00°$$

$$B = 62.00°$$

$$\sin A = \frac{a}{c}$$

$$a = c \sin A$$

$$a = 17.4 \sin 28.00°$$

$$a = 8.17 \text{ ft}$$

$$\cos A = \frac{b}{c}$$

$$b = c \cos A$$

$$b = 17.4 \cos 28.00°$$

$$b = 15.4 \text{ ft}$$

25. B = 73.00°, b = 128 in

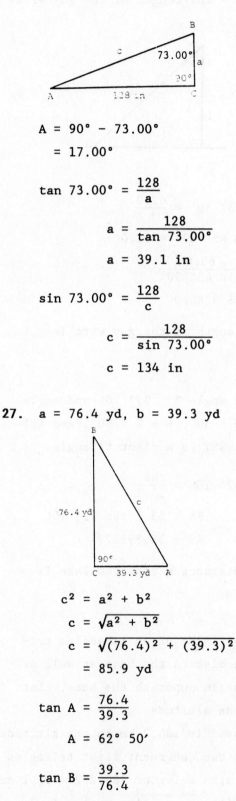

A = 90° − 73.00°

= 17.00°

$\tan 73.00° = \dfrac{128}{a}$

$a = \dfrac{128}{\tan 73.00°}$

a = 39.1 in

$\sin 73.00° = \dfrac{128}{c}$

$c = \dfrac{128}{\sin 73.00°}$

c = 134 in

27. a = 76.4 yd, b = 39.3 yd

$c^2 = a^2 + b^2$

$c = \sqrt{a^2 + b^2}$

$c = \sqrt{(76.4)^2 + (39.3)^2}$

= 85.9 yd

$\tan A = \dfrac{76.4}{39.3}$

A = 62° 50′

$\tan B = \dfrac{39.3}{76.4}$

B = 27° 10′

29. a = 18.9 cm, c = 46.3 cm

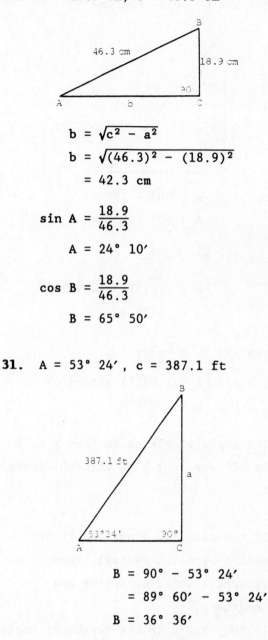

$b = \sqrt{c^2 - a^2}$

$b = \sqrt{(46.3)^2 - (18.9)^2}$

= 42.3 cm

$\sin A = \dfrac{18.9}{46.3}$

A = 24° 10′

$\cos B = \dfrac{18.9}{46.3}$

B = 65° 50′

31. A = 53° 24′, c = 387.1 ft

B = 90° − 53° 24′

= 89° 60′ − 53° 24′

B = 36° 36′

$\sin 53° 24′ = \dfrac{a}{387.1}$

a = 387.1 sin 53° 24′

a = 310.8 ft

$\cos 53° 24′ = \dfrac{b}{387.1}$

b = 387.1 cos 53° 24′

b = 230.8 ft

33. B = 39° 9′, c = .6231 m

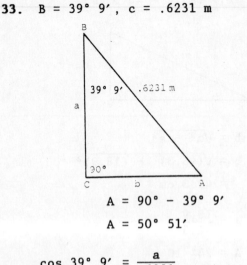

$$A = 90° - 39° 9'$$
$$A = 50° 51'$$

$$\cos 39° 9' = \frac{a}{.6231}$$
$$a = (.6231) \cos 39° 9'$$
$$a = .4832 \text{ m}$$

$$\sin 39° 9' = \frac{b}{.6231}$$
$$b = (.6231) \sin 39° 9'$$
$$b = .3934 \text{ m}$$

35. The angle of elevation from X to Y is 90° whenever Y is directly above X.

37. If two parallel lines are intersected by a transversal, then alternate interior angles are congruent.

In the figure in the textbook, angle DAB and angle ABC are alternate interior angles and are congruent. Therefore, they have the same measure.

39. Let h = the distance the ladder goes up the wall.

$$\sin 43° 50' = \frac{h}{13.5}$$
$$h = 13.5 \sin 43° 50'$$
$$h \approx 9.3496000$$

The ladder goes up the wall 9.35 m.

41. Let x = the length of the guy wire.

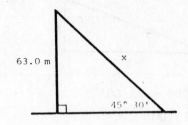

$$\sin 45° 30' = \frac{63.0}{x}$$
$$x \sin 45° 30' = 63.0$$
$$x = \frac{63.0}{\sin 45° 30'}$$
$$x \approx 88.328020$$

The length of the guy wire is 88.3 m.

43. Since angle T = 32° 10′ and angle S = 57° 50′, S + T = 90°, and triangle RST is a right triangle.

$$\tan 32° 10' = \frac{RS}{53.1}$$
$$RS = 53.1 \tan 32° 10'$$
$$RS \approx 33.395727$$

The distance across the lake is 33.4 m.

45. The altitude of an isosceles triangle bisects the base as well as the angle opposite the base. Let h = the altitude.

The base is 184.2 cm so the altitude forms two congruent right triangles each with a leg of 184.2/2 = 92.1 cm and with an opposite angle of

$$\frac{1}{2}(68° 44') = 34° 22'.$$

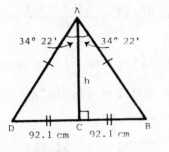

In triangle ABC,

$$\tan 34° \ 22' = \frac{92.1}{h}$$

$$h \tan 34° \ 22' = 92.1$$

$$h = \frac{92.1}{\tan 34° \ 22'}$$

$$h \approx 134.67667.$$

The altitude of the triangle is 134.7 cm.

47. Let h = height of the tower.

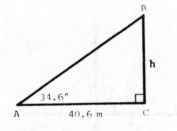

In triangle ABC,

$$\tan 34.6° = \frac{h}{40.6}$$

$$h = 40.6 \tan 34.6°$$

$$h \approx 28.0.$$

The height of the tower is 28.0 m.

49. Let d = the distance from the top B of the building to the point on the ground A.

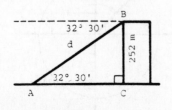

In triangle ABC,

$$\sin 32° \ 30' = \frac{252}{d}$$

$$d = \frac{252}{\sin 32° \ 30'}$$

$$d \approx 469.$$

The distance from the top of the building to the point on the ground is 469 m.

51. Let x = the height of the taller building,

  h = the difference in height between the shorter and taller buildings,

  d = the distance between the buildings along the ground.

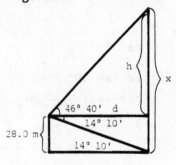

$$\frac{d}{28.0} = \cot 14° \ 10'$$

$$d = 28.0 \cot 14° \ 10'$$

$$= 110.93 \text{ m}$$

$$\frac{h}{d} = \tan 46° \ 40'$$

$$h = d \tan 46° \ 40'$$

$$= 110.93 \tan 46° \ 40'$$

$$= 117.57$$

$$x = h + 28.0$$

$$= 117.57 + 28.0$$

$$\approx 146$$

The height of the taller building is 146 m.

53. Let θ = the angle of depression.

$$\tan \theta = \frac{39.82}{51.74}$$

$$\theta = 37.58° = 37° \, 35'$$

55. The triangle with 60° as vertex angle and x as a base is equilateral. d is the perpendicular bisector of x and forms a 30°-60° right triangle.

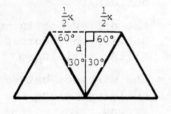

$$\frac{\frac{1}{2}x}{d} = \tan 30°$$

$$x = 2d \tan 30°$$

$$= 2(2.894)(\tan 30°)$$

$$\approx 3.342 \text{ mm}$$

57. **(a)** The height h of the peak above 14,545 would be

$$\sin 5.82° = \frac{h}{27.0134(5280)}$$

$$h \approx 14,463 \text{ ft.}$$

The total height would be about

$$14,545 + 14,463 = 29,008 \text{ ft.}$$

(b) The curvature of the earth would make the peak appear *shorter* than it actually is. Initially the surveyors did not think Mt. Everest was the tallest peak in the Himalayas. It did not look like the tallest peak because it was farther away than the other large peaks.

59. **(a)** $\beta \approx \dfrac{57.3S}{R} = \dfrac{57.3(336)}{600} \approx 32°$

$$d = R\left(1 - \cos \frac{\beta}{2}\right)$$

$$\approx 600(1 - \cos 16°)$$

$$\approx 23.4 \text{ ft}$$

(b) $\beta \approx \dfrac{57.3S}{R} = \dfrac{57.3(485)}{600} \approx 46.3°$

$$d = R\left(1 - \cos \frac{\beta}{2}\right)$$

$$\approx 600\left(1 - \cos \frac{46.3°}{2}\right)$$

$$\approx 48.3 \text{ ft}$$

(c) The faster the speed, the more land needs to be cleared on the inside of the curve.

Section 2.5

1.

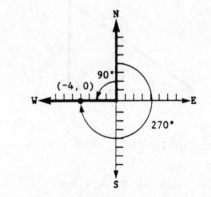

The bearing of the airplane measured in a clockwise direction from due north is 270°. The bearing can also be expressed as N 90° W.

3.

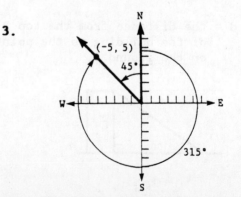

The bearing of the airplane measured in a clockwise direction from due north is 315°. The bearing can also be expressed as N 45° W.

5.

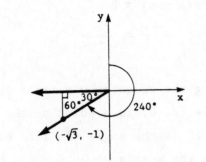

The reference angle is 30°. A point on the ray is $(-\sqrt{3}, -1)$. Using the equation $y = mx + b$, the equation of the ray is

$$y = \frac{\sqrt{3}}{3}x, \ x \le 0$$

since the ray lies in quadrant III.

7. Let x = the distance the plane is from its starting point.

In the figure, the measure of angle ACB is 40° + (180° − 130°) or 90°. Therefore, triangle ACB is a right triangle.

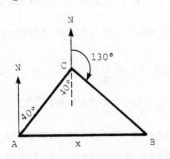

Distance traveled in 1.5 hr:

    (1.5 hr)(110 mph) = 165 mi.

Distance traveled in 1.3 hr:

    (1.3 hr)(110 mph) = 143 mi.

Using the Pythagorean theorem, we have

$$x^2 = 165^2 + 143^2$$
$$x^2 = 47,674$$
$$x \approx 220.$$

The plane is 220 mi from its starting point.

9. Let x = distance the ships are apart.

(1.5 hr)(18 knots) = 27 nautical mi

(1.5 hr)(26 knots) = 39 nautical mi

Since 130° − 40° = 90°, we have a right triangle. Applying the Pythagorean theorem,

$$x^2 = 27^2 + 39^2$$
$$x^2 = 2250$$
$$x \approx 47.$$

The ships are 47 nautical mi apart.

11. Let x = distance between the two ships.

The angle between the bearings of the ships is 180° − (28° 10′ + 61° 50′) or 90°. The triangle formed is a right triangle.

Distance traveled at 24.0 mph:

(4 hr)(24.0 mph) = 96 mi.

Distance traveled at 28.0 mph:

(4 hr)(28.0 mph) = 112 mi.

Applying the Pythagorean theorem gives

$$x^2 = 96^2 + 112^2$$
$$x^2 = 21,760$$
$$x \approx 148.$$

The ships are 148 mi apart.

13. Draw triangle WDG with W representing Winston–Salem, D representing Danville, and G representing Goldsboro. Name any point X on the line due south from D.

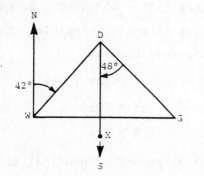

Since the bearing from W to D is 42°, angle WDX = 42°.
Angle XDG = 48°
So angle D = 42° + 48° = 90°, and triangle WDG is a right triangle.
Using d = rt and the Pythagorean theorem, we have

$$WG = \sqrt{(WD)^2 + (DG)^2}$$
$$= \sqrt{[60(1)]^2 + [60(1.05)]^2}$$
$$= \sqrt{15,300}$$
$$\approx 120.$$

The distance from Winston–Salem to Goldsboro is 120 mi.

15. Solve the equation ax = b + cx for x in terms of a, b, and c.

$$ax = b + cx$$
$$ax - cx = b$$
$$x(a - c) = b$$
$$x = \frac{b}{a - c}$$

17. Using the equation $y = \tan \theta \, (x - a)$ where (a, 0) is a point on the line and $\theta°$ is the angle the line makes with the x–axis,

$$y = (\tan 35°)(x - 25).$$

19. Let x = the side adjacent to 49.2° in the smaller triangle.

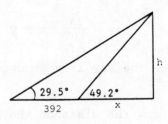

In the larger right triangle:

$$\tan 29.5° = \frac{h}{392 + x}$$
$$h = (392 + x) \tan 29.5°.$$

In the smaller right triangle:

$$\tan 49.2° = \frac{h}{x}$$
$$h = x \tan 49.2°.$$

Substitute the first expression for h in this equation, and solve for x.

$$(392 + x) \tan 29.5°$$
$$= x \tan 49.2°$$
$$392 \tan 29.5° + x \tan 29.5°$$
$$= x \tan 49.2°$$

392 tan 29.5°

  = x tan 49.2° - x tan 29.5°

392 tan 29.5°

  = x(tan 49.2° - tan 29.5°)

$$\frac{392 \tan 29.5°}{\tan 49.2° - \tan 29.5°} = x$$

Then substitute this value for x in the equation for the smaller triangle.

h = x tan 49.2°

$$h = \frac{392 \tan 29.5° (\tan 49.2°)}{\tan 49.2° - \tan 29.5°}$$

h ≈ 433

The height of the triangle is 433 ft.

21. Let x = the distance from the closer point on the ground to the base of height h of the pyramid.

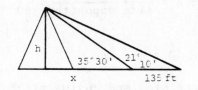

In the larger right triangle:

$$\tan 21° 10' = \frac{h}{135 + x}$$

      h = (135 + x) tan 21° 10'.

In the smaller right triangle:

$$\tan 35° 30' = \frac{h}{x}$$

      h = x tan 35° 30'.

Substitute for h in this equation, and solve for x.

(135 + x) tan 21° 10'

  = x tan 35° 30'

135 tan 21° 10' + x tan 21° 10'

  = x tan 35° 30'

135 tan 21° 10'

  = x tan 35° 30' - x tan 21° 10'

135 tan 21° 10'

  = x(tan 35° 30' - tan 21° 10')

$$\frac{135 \tan 21° 10'}{(\tan 35° 30' - \tan 21° 10')} = x$$

Then substitute for x in the equation for the smaller triangle.

$$h = \frac{135 \tan 21° 10' (\tan 35° 30')}{(\tan 35° 30' - \tan 21° 10')}$$

h ≈ 114

The height of the pyramid is 114 ft.

23. Let x = the height of the antenna and h = the height of the house.

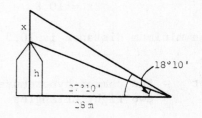

In the smaller right triangle:

$$\tan 18° 10' = \frac{h}{28}$$

      h = 28 (tan 18° 10').

In the larger right triangle:

$$\tan 27° 10' = \frac{x + h}{28}$$

    x + h = 28 (tan 27° 10')

      x = 28(tan 27° 10')

        - 28(tan 18° 10')

      x ≈ 5.18.

The height of the antenna is 5.18 m.

25. (a) From the figure in the text,

$$d = \frac{b}{2} \cot \frac{\alpha}{2} + \frac{b}{2} \cot \frac{\beta}{2}$$

$$d = \frac{b}{2}\left(\cot \frac{\alpha}{2} + \cot \frac{\beta}{2}\right).$$

(b) $d = \frac{b}{2}\left(\cot \frac{\alpha}{2} + \cot \frac{\beta}{2}\right)$

Let $\alpha = 37'\ 48''$, $\beta = 42'\ 3''$, and
$b = 2$.

$$d = \frac{2}{2}\left(\cot \frac{37'\ 48''}{2} + \cot \frac{42'\ 3''}{2}\right)$$

$$d \approx 345.3951\ m$$

27. Let x = the minimum distance that a plant needing full sun can be placed from the fence.

$$\tan 23°\ 20' = \frac{4.65}{x}$$

$$x \tan 23°\ 20' = 4.65$$

$$x = \frac{4.65}{\tan 23°\ 20'}$$

$$x \approx 10.8$$

The minimum distance is 10.8 ft.

29. Let y = the common hypotenuse of the two right triangles.

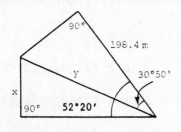

$$\cos 30°\ 50' = \frac{198.4}{y}$$

$$y = \frac{198.4}{\cos 30°\ 50'}$$

$$y = 231.06$$

To find x, first find the angle opposite x in the right triangle by subtracting.

$$52°\ 20' - 30°\ 50' = 51°\ 80' - 30°\ 50'$$
$$= 21°\ 30'$$

$$\sin 21°\ 30' = \frac{x}{y} = \frac{x}{231.06}$$

$$x = 231.06(\sin 21°\ 30')$$

$$\approx 84.7\ m$$

Chapter 2 Review Exercises

1. $\sin A = \dfrac{\text{side opposite}}{\text{hypotenuse}} = \dfrac{60}{61}$

$\cos A = \dfrac{\text{side adjacent}}{\text{hypotenuse}} = \dfrac{11}{61}$

$\tan A = \dfrac{\text{side opposite}}{\text{side adjacent}} = \dfrac{60}{11}$

$\cot A = \dfrac{\text{side adjacent}}{\text{side opposite}} = \dfrac{11}{60}$

$\sec A = \dfrac{\text{hypotenuse}}{\text{side adjacent}} = \dfrac{61}{11}$

$\csc A = \dfrac{\text{hypotenuse}}{\text{side opposite}} = \dfrac{61}{60}$

3. $\sin 4\beta = \cos 5\beta$

Since sine and cosine are co-functions,

$$4\beta + 5\beta = 90°$$
$$9\beta = 90°$$
$$\beta = 10°.$$

5. $\tan (5x + 11°) = \cot (6x + 2°)$

Since tangent and cotangent are co-functions,

$$5x + 11° + 6x + 2° = 90°$$
$$11x = 90° - 11° - 2°$$
$$11x = 77°$$
$$x = 7°.$$

7. sin 46° < sin 58°

    sin θ increases as θ increases from
    0° to 90°. Since

    $$58° > 46°$$
    $$\sin 58° > \sin 46°$$

    True

9. sec 48° ≥ cos 42°

    sec θ ≥ 1 for all θ and cos θ ≤ 1
    for all θ.
    Therefore,

    $$\cos \theta \leq 1 \leq \sec \theta.$$

    True

11. cos A = $\dfrac{\text{side adjacent}}{\text{hypotenuse}}$ and

    sin B = $\dfrac{\text{side opposite}}{\text{hypotenuse}}$

    The side adjacent to angle A is b,
    and the side opposite angle B is b.
    Thus,

    cos A = $\dfrac{b}{c}$ and sin B = $\dfrac{b}{c}$ so

    cos A = sin B.

13. 225° is an angle in quadrant III
    with a reference angle $\theta' = 45°$.

    sin 225° = −sin 45° = $-\dfrac{\sqrt{2}}{2}$

    cos 225° = −cos 45° = $-\dfrac{\sqrt{2}}{2}$

    tan 225° = tan 45° = 1

    cot 225° = cot 45° = 1

    sec 225° = −sec 45° = $-\sqrt{2}$

    csc 225° = −csc 45° = $-\sqrt{2}$

15. 750° − 2(360°) = 750° − 720° = 30°

    Since 750° is coterminal with 30°,
    the values of the trigonometric
    functions for 750° are the same as
    those for 30°.

    sin 750° = $\dfrac{1}{2}$

    cos 750° = $\dfrac{\sqrt{3}}{2}$

    tan 750° = $\dfrac{\sqrt{3}}{3}$

    cot 750° = $\sqrt{3}$

    sec 750° = $\dfrac{2\sqrt{3}}{3}$

    csc 750° = 2

17. −390° is an angle in quadrant IV
    with a reference angle $\theta' = 30°$.

    sin (−390°) = −sin 30° = $-\dfrac{1}{2}$

    cos (−390°) = cos 30° = $\dfrac{\sqrt{3}}{2}$

    tan (−390°) = −tan 30° = $-\dfrac{\sqrt{3}}{3}$

    cot (−390°) = −cot 30° = $-\sqrt{3}$

    sec (−390°) = sec 30° = $\dfrac{2\sqrt{3}}{3}$

    csc (−390°) = −csc 30° = −2

19. cos θ = $-\dfrac{1}{2}$

    Since the cosine is negative, θ is
    in quadrant II or quadrant III.
    Since cos 60° = $\dfrac{1}{2}$, $\theta' = 60°$.

    In quadrant II,

    θ = 180° − θ' = 180° − 60° = 120°.

    In quadrant III,

    θ = 180° + θ' = 180° + 60° = 240°.

21. $\sec \theta = -\dfrac{2\sqrt{3}}{3}$

Since the secant is negative, θ is in quadrant II or quadrant III.

Since $\sec 30° = \dfrac{2\sqrt{3}}{3}$, $\theta' = 30°$.

In quadrant II,

$\theta = 180° - \theta' = 180° - 30° = 150°.$

In quadrant III,

$\theta = 180° + \theta' = 180° + 30° = 210°.$

23. $\tan^2 120° - 2 \cot 240°$

$= (-\sqrt{3})^2 - 2\left(\dfrac{\sqrt{3}}{3}\right)$

$= 3 - \dfrac{2\sqrt{3}}{3}$

25. $\sec^2 300° - 2 \cos^2 150° + \tan 45°$

$= 2^2 - 2\left(\dfrac{\sqrt{3}}{2}\right)^2 + 1$

$= 4 - \dfrac{3}{2} + 1$

$= \dfrac{7}{2}$

In Exercises 27–33, be sure that your calculator is in degree mode. Keystroke sequences may vary based on the type and/ or model of calculator being used.

27. $\sin 72° 30'$

$72° 30' = \left(72 + \dfrac{30}{60}\right)° = 72.5$

Enter: 72.5 [SIN]

or   [SIN] 72.5 [ENTER]

Display: .95371695

29. $\cot 305.6°$

Enter: 305.6 [TAN] [1/x]

or   [(] [TAN] 305.6 [)] [x⁻¹] [ENTER]

Display: -.71592968

31. $\sec 58.9041°$

Enter: 58.9041 [COS] [1/x]

or   [(] [COS] 58.9041 [)]

[x⁻¹] [ENTER]

Display: 1.9362132

33. $\sin 89.0043°$

Enter: 89.0043 [SIN]

or   [SIN] 89.0043 [ENTER]

Display: .99984900

35. The exact answer is $\dfrac{\sqrt{3}}{2}$. The value .8660254038 is a ten-place decimal approximation; it is not exact.

37. $\sin \theta = .82584121$

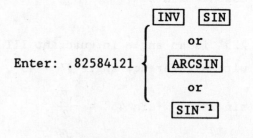

or   [2nd] [SIN] .82584121 [ENTER]

Display: 55.673870

$\theta = 55.673870°$

39. cos θ = .97540415

Enter: .97540415
- $\boxed{\text{INV}}$ $\boxed{\text{COS}}$
- or
- $\boxed{\text{ARCCOS}}$
- or
- $\boxed{\text{COS}^{-1}}$

or $\boxed{\text{2nd}}$ $\boxed{\text{COS}}$ .97540415 $\boxed{\text{ENTER}}$

Display: 12.733938°

θ = 12.733938°

41. tan θ = 1.9633124

Enter: 1.9633124
- $\boxed{\text{INV}}$ $\boxed{\text{TAN}}$
- or
- $\boxed{\text{ARCTAN}}$
- or
- $\boxed{\text{TAN}^{-1}}$

or $\boxed{\text{2nd}}$ $\boxed{\text{TAN}}$ 1.9633124 $\boxed{\text{ENTER}}$

Display: 63.008286

θ = 63.008286°

43. sin θ = .73254290

Since sin θ is positive, there will be one angle in quadrant I and one angle in quadrant II. If θ' is the reference angle, then the two angles are θ' and 180° − θ'.

Enter: .73254290 $\boxed{\text{INV}}$ $\boxed{\text{SIN}}$

or $\boxed{\text{2nd}}$ $\boxed{\text{SIN}}$ .73254290 $\boxed{\text{ENTER}}$

Display: 47.100000

θ' = 47.1°

180° − θ' = 180° − 47.1° = 132.9°

45.   sin 50° + sin 40° = sin 90°

sin 50° = .76604444

sin 40° = .64278761, so

sin 50° + sin 40°

= .76604444 + .64278761

= 1.4088321

sin 90° = 1, so

1.4088321 ≠ 1

False

47.   sin 240° = 2 sin 120° · cos 120°

$\sin 240° = -\sin 60° = -\dfrac{\sqrt{3}}{2}$

2 sin 120° · cos 120°

= 2(sin 60°)(−cos 60°)

$= 2\left(\dfrac{\sqrt{3}}{2}\right)\left(-\dfrac{1}{2}\right)$

$= -\dfrac{\sqrt{3}}{2}$

True

49. No, cot 25° = 1/tan 25°.

Enter: 25 $\boxed{\text{TAN}}$ $\boxed{1/x}$

or $\boxed{(}$ $\boxed{\text{TAN}}$ 25 $\boxed{)}$ $\boxed{x^{-1}}$ $\boxed{\text{ENTER}}$

Tan⁻¹ 25 is calculating the angle whose tangent is 25.

51. θ = 1997°

cos 1997° = −.956304756

sin 1997° = −.292371705

Since sine and cosine are both negative, the angle θ is in quadrant III.

53. A = 58° 30′, c = 748

$$A + B = 90°$$
$$B = 90° - A$$
$$B = 90° - 58° 30′$$
$$B = 31° 30′$$

$$\sin A = \frac{a}{c}$$
$$a = c \sin A$$
$$a = 748 \sin 58° 30′$$
$$a = 638$$

$$\cos A = \frac{b}{c}$$
$$b = c \cos A$$
$$b = 748 \cos 58° 30′$$
$$b = 391$$

55. A = 39.72°, b = 38.97 m

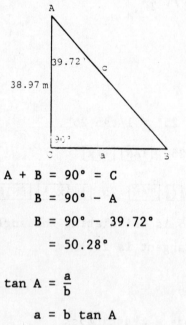

$$A + B = 90° = C$$
$$B = 90° - A$$
$$B = 90° - 39.72°$$
$$= 50.28°$$

$$\tan A = \frac{a}{b}$$
$$a = b \tan A$$
$$a = 38.97 \tan 39.72°$$
$$a = 32.38 \text{ m}$$

$$\cos A = \frac{b}{c}$$
$$c = \frac{b}{\cos A}$$
$$c = \frac{38.97}{\cos 39.72°}$$
$$c = 50.66 \text{ m}$$

57. Let x = height of the tower.

$$\tan 38° 20′ = \frac{x}{93.2}$$
$$x = 93.2 \tan 38° 20′$$
$$x \approx 73.693005$$

The height of the tower is 73.7 ft.

59. The diagonal d of the rectangle is the hypotenuse of a right triangle whose legs are the adjacent sides of the rectangle.

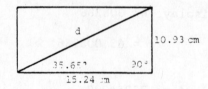

By the Pythagorean theorem, the length d of the diagonal is

$$d = \sqrt{(15.24)^2 + (10.93)^2} \approx 18.75 \text{ cm.}$$

An alternative method gives the same answer.

$$\sin 35.65° = \frac{10.93}{d}$$
$$d = \frac{10.93}{\sin 35.65°}$$
$$\approx 18.75 \text{ cm}$$

61. Draw triangle ABC and extend the
north-south line to a point X south
of A. Extend the north-south line
to a point Y, north of C.

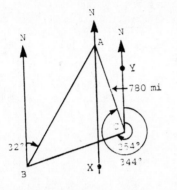

Angle ACB = 344° – 254° = 90°, so
ABC is a right triangle.
Angle BAX = 32° since it is an
alternate interior angle to 32°.

Angle YCA = 360° – 344° = 16°

Angle XAC = 16° since it is an
alternate interior angle to
angle YCA.

Angle BAC = 32° + 16° = 48°

In triangle ABC,

$$\cos A = \frac{AC}{AB}$$

$$AB = \frac{AC}{\cos A}$$

$$AB = \frac{780}{\cos 48°}$$

$$\approx 1200.$$

The distance from A to B is 1200 m.

63. Suppose A is the car heading south
at 55 mph, B is the car heading west
at 40 mph, and point C is the intersection from which they start.
After two hours, by d = rt,

$$AC = 55(2) = 110$$
$$BC = 40(2) = 80.$$

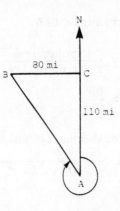

Since angle ACB is a right angle,
use the Pythagorean theorem to find
the distance from A to B.

$$(AB)^2 = 80^2 + 110^2$$
$$(AB)^2 = 6400 + 12,100$$
$$AB = \sqrt{18,500} \approx 140 \text{ mi}$$

An alternative method follows.
Since the bearing of A from B is
324°,

angle CAB = 360° – 324° = 36°.

$$\sin CAB = \frac{BC}{AB}$$

$$AB = \frac{BC}{\sin CAB} = \frac{80}{\sin 36°}$$

$$\approx 140 \text{ mi}$$

67. In right triangle OAB,

$$\cot \theta = \frac{\text{side adjacent}}{\text{side opposite}}$$

$$= \frac{AB}{OA}.$$

Using the distance formula,

$$OA = \sqrt{(0 - 0)^2 + (1 - 0)^2}$$
$$OA = 1.$$

Therefore, cot θ = AB.

69. In right triangle OAB,

$$\csc \theta = \frac{\text{hypotenuse}}{\text{side opposite}}$$

$$= \frac{OB}{OA}.$$

Using the distance formula,

$$OA = \sqrt{(0 - 0)^2 + (1 - 0)^2}$$

$$OA = 1.$$

Therefore, $\csc \theta = OB$.

71. **(a)** From the figure in the text,

$$\cos \theta = \frac{x_Q - x_P}{d}$$

$$X_Q = x_P + d \cos \theta.$$

Similarly,

$$\sin \theta = \frac{y_Q - y_P}{d}$$

$$y_Q = y_P + d \sin \theta.$$

(b) Let $(x_P, y_P) = (123.62, 337.95)$,

$\theta = 17° \, 19' \, 22"$, and $d = 193.86$.

$$x_Q = x_P + d \cos \theta$$

$$= 337.95 + 193.86 \cos 17° \, 19' \, 22"$$

$$= 123.62 + 193.86 \cos 17.3228°$$

$$\approx 308.69.$$

$$y_Q = y_P + d \sin \theta$$

$$= 337.95 + 193.86 \sin 17° \, 19' \, 22"$$

$$= 337.95 + 193.86 \sin 17.3228°$$

$$\approx 395.67$$

The coordinates of Q are
$(308.69, 395.67)$.

CHAPTER 2   TEST

[2.1]   Solve each equation. Assume that all angles are acute angles.

1. $\sin 6\theta = \cos 9\theta$

2. $\tan (11\alpha + 8°) = \cot (12\alpha - 10°)$

[2.2]   Find the reference angle for each of the following.

3. 201°

4. 517°

[2.2]   5. Can 90° be the measure of a reference angle? Explain.

[2.2]   Find the exact values of the sine, cosine, and tangent functions for each of the following angles. Do not use a calculator. Rationalize denominators when applicable.

6. 240°

7. 675°

[2.2]   Evaluate. Do not use a calculator.

8. $\tan^2 45° + \sin^2 150°$

9. $2 \cos^2 210° + \csc^2 300°$

10. $\sec^2 180° + 3 \cot^2 120°$

[2.3]   Use a calculator to find decimal approximations for each of the following.

11. $\sin 108° \ 20'$

12. $\sec 41° \ 30'$

13. $\cot 128.8°$

[2.3]   Find a value of θ in the interval [0°, 90°) that satisfies each statement. Give answers to the nearest tenth of a degree.

14. $\sin \theta = .91382416$

15. $\tan \theta = 1.3865127$

16. $\cos \theta = .10875886$

[2.4]   Solve each right triangle. The right angle is at C.

17. a = 28.4 cm, b = 34.7 cm

18. B = 38°, a = 17 ft

[2.5] Solve each problem.

19. The angle of elevation from a park bench to the top of a tower is 9.8°. The tower is 31.2 ft high. How far is the bench from the top of the tower?

20. The bearing from A to B is 35°. The bearing from B to C is 152°. The bearing from A to C is 62°. If the distance from B to C is 18 km, find the distance from A to C.

CHAPTER 2   TEST ANSWERS

1. 6°   2. 4°   3. 21°   4. 23°   5. No, a reference angle is

an acute angle.   6. $-\sqrt{3}/2$; $-1/2$; $\sqrt{3}$ 7. $-\sqrt{2}/2$; $\sqrt{2}/2$; -1

8. 5/4   9. 17/6   10. 2   11. .94924264   12. 1.3351924

13. -.80402064   14. 66.0°   15. 54.2°   16. 83.8°

17. c = 44.8 cm; A = 39.3°; B = 50.7°   18. A = 52°; b = 13 ft; c = 22 ft

19. 183.3 ft   20. 35 km

**CHAPTER 3 RADIAN MEASURE AND THE
                CIRCULAR FUNCTIONS**

Section 3.1

1. Since θ is in quadrant I,

    $$0 < \theta < \frac{\pi}{2}$$

    $$0 < \theta < 1.57.$$

    Therefore, θ to the nearest whole
    number is 1 radian.

3. Since θ is in quadrant II,

    $$\frac{\pi}{2} < \theta < \pi$$

    $$1.57 < \theta < 3.14.$$

    Since θ is closer to π, the radian
    measure to the nearest whole number
    is 3 radians.

5. $60° = 60°\left(\frac{\pi}{180°}\right) = \frac{\pi}{3}$

    This method is the formula method.
    The proportion method may also be
    used as follows.

    $$\frac{\text{Radian measure}}{\pi} = \frac{\text{Degree measure}}{180}$$

    $$\frac{\text{Radian measure}}{\pi} = \frac{60}{180}$$

    $$\text{Radian measure} = \frac{60\pi}{180} = \frac{\pi}{3}$$

7. $90° = 90°\left(\frac{\pi}{180°}\right) = \frac{\pi}{2}$

9. $150° = 150°\left(\frac{\pi}{180°}\right) = \frac{5\pi}{6}$

11. $300° = 300°\left(\frac{\pi}{180°}\right) = \frac{5\pi}{3}$

13. $450° = 450°\left(\frac{\pi}{180°}\right) = \frac{5\pi}{2}$

21. $\frac{\pi}{3} = \frac{\pi}{3}\left(\frac{180°}{\pi}\right) = 60°$

    This method is the formula method.
    The proportion method may also be
    used.

    $$\frac{\text{Radian measure}}{\pi} = \frac{\text{Degree measure}}{180}$$

    $$\frac{\frac{\pi}{3}}{\pi} = \frac{\text{Degree measure}}{180}$$

    $$\frac{1}{3} = \frac{\text{Degree measure}}{180}$$

    $$\text{Degree measure} = \frac{180}{3} = 60$$

    In $\frac{\pi}{3}$ radians, there are 60°.

23. $\frac{7\pi}{4} = \frac{7\pi}{4}\left(\frac{180°}{\pi}\right) = 315°$

25. $\frac{11\pi}{6} = \frac{11\pi}{6}\left(\frac{180°}{\pi}\right) = 330°$

27. $-\frac{\pi}{6} = -\frac{\pi}{6}\left(\frac{180°}{\pi}\right) = -30°$

29. $\frac{7\pi}{10} = \frac{7\pi}{10}\left(\frac{180°}{\pi}\right) = 126°$

31. $\frac{4\pi}{15} = \frac{4\pi}{15}\left(\frac{180°}{\pi}\right) = 48°$

33. $\frac{17\pi}{20} = \frac{17\pi}{20}\left(\frac{180°}{\pi}\right) = 153°$

35. $39° = 39°\left(\frac{\pi}{180°}\right) = .68$

37. $42° \ 30' = 42.5° \left(\frac{\pi}{180°}\right) = .742$

39. $139° \ 10' = 139\frac{1}{6}° \left(\frac{\pi}{180°}\right) = 2.43$

41. $64.29° = 64.29° \left(\frac{\pi}{180°}\right)$

$= 1.122$

43. $56° \ 25' = 56\frac{25}{60}° \left(\frac{\pi}{180°}\right)$

$= .9847$

45. $47.6925° = 47.6925° \left(\frac{\pi}{180°}\right)$

$= .832391$

47. $2 \text{ radians} = 2\left(\frac{180°}{\pi}\right) = 114.5916°$

$= 114° + (.5916)(60')$

$= 114° \ 35'$

49. $1.74 \text{ radians} = 1.74\left(\frac{180°}{\pi}\right)$

$= 99.6947°$

$= 99° + (.6947)(60')$

$= 99° \ 42'$

51. $.3417 \text{ radians} = .3417\left(\frac{180°}{\pi}\right)$

$= 19.5780°$

$= 19° \ 35'$

53. $5.01095 \text{ radians} = 5.01095\left(\frac{180°}{\pi}\right)$

$= 287.1063°$

$= 287° \ 6'$

55. The value of sin 30 is not $\frac{1}{2}$, because 30, without a degree symbol, is seen as a radian measure. If you want to have an answer of $\frac{1}{2}$, you must say sin 30°.

57. $\sin \frac{\pi}{3} = \sin \left(\frac{1}{3} \cdot 180°\right)$ *Substitute 180° for π.*

$= \sin 60°$

$= \frac{\sqrt{3}}{2}$

59. $\tan \frac{\pi}{4} = \tan \left(\frac{1}{4} \cdot 180°\right)$

$= \tan 45°$

$= 1$

61. $\sec \frac{\pi}{6} = \sec \left(\frac{1}{6} \cdot 180°\right)$

$= \sec 30°$

$= \frac{2\sqrt{3}}{3}$

63. $\sin \frac{\pi}{2} = \sin \left(\frac{1}{2} \cdot 180°\right)$

$= \sin 90°$

$= 1$

65. $\tan \frac{2\pi}{3} = \tan \left(\frac{2}{3} \cdot 180°\right)$

$= \tan 120°$

$= -\sqrt{3}$

67. $\sin \frac{5\pi}{6} = \sin \left(\frac{5}{6} \cdot 180°\right)$

$= \sin 150°$

$= \frac{1}{2}$

69. $\cos 3\pi = \cos (3 \cdot 180°)$

$= \cos 540°$

$= -1$

71. $\sin \dfrac{4\pi}{3} = \sin \left(\dfrac{4}{3} \cdot 180°\right)$

$= \sin 240°$

$= -\dfrac{\sqrt{3}}{2}$

73. $\sin \left(-\dfrac{7\pi}{6}\right) = \sin \left(-\dfrac{7}{6} \cdot 180°\right)$

$= \sin (-210°)$

$= \dfrac{1}{2}$

75.   Answers start at 30° and proceed counterclockwise around circle.

$30° = 30°\left(\dfrac{\pi}{180°}\right) = \dfrac{\pi}{6}$

$\dfrac{\pi}{4}$ radian $= \dfrac{\pi}{4}\left(\dfrac{180°}{\pi}\right) = 45°$

$60° = 60°\left(\dfrac{\pi}{180°}\right) = \dfrac{\pi}{3}$

$\dfrac{2\pi}{3}$ radians $= \dfrac{2\pi}{3}\left(\dfrac{180°}{\pi}\right) = 120°$

$\dfrac{3\pi}{4}$ radians $= \dfrac{3\pi}{4}\left(\dfrac{180°}{\pi}\right) = 135°$

$150° = 150°\left(\dfrac{\pi}{180°}\right) = \dfrac{5\pi}{6}$

$180° = 180°\left(\dfrac{\pi}{180°}\right) = \pi$

$210° = 210°\left(\dfrac{\pi}{180°}\right) = \dfrac{7\pi}{6}$

$225° = 225°\left(\dfrac{\pi}{180°}\right) = \dfrac{5\pi}{4}$

$\dfrac{4\pi}{3}$ radians $= \dfrac{4\pi}{3}\left(\dfrac{180°}{\pi}\right) = 240°$

$\dfrac{5\pi}{3}$ radians $= \dfrac{5\pi}{3}\left(\dfrac{180°}{\pi}\right) = 300°$

$315° = 315°\left(\dfrac{\pi}{180°}\right) = \dfrac{7\pi}{4}$

$330° = 330°\left(\dfrac{\pi}{180°}\right) = \dfrac{11\pi}{6}$

77.   Since π is associated with radians, the calculator was in radian mode.

79.   (a) The hour hand completes 2 revolutions in 24 hr. One revolution is 2π.

$\theta = 2(2\pi) = 4\pi$

(b) In 4 hr it rotates through

$\theta = \left(\dfrac{4}{12}\right)(2\pi) = \dfrac{2\pi}{3}.$

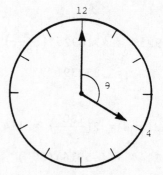

81.   (a) S = 1367, N = 80

$\Delta S = .034S \sin \left[\dfrac{2\pi(82.5 - N)}{365.25}\right]$

$= .034(1367) \sin \left[\dfrac{2\pi(82.5 - 80)}{365.25}\right]$

$\approx 1.998 \text{ w/m}^2$

(b) S = 1367, N = 1268

$\Delta S = .034S \sin \left[\dfrac{2\pi(82.5 - N)}{365.25}\right]$

$= .034(1367) \sin \left[\dfrac{2\pi(82.5 - 1268)}{365.25}\right]$

$\approx -46.461 \text{ w/m}^2$

(c) The maximum value of ΔS would occur when

$\sin \left[\dfrac{2\pi(82.5 - N)}{365.25}\right] = 1.$

In this case ΔS would be equal to

$.034S = .034(1367) = 46.478$ w/m^2.

(d) $\Delta S = 0$, so

$$\sin\left[\frac{2\pi(82.5 - N)}{365.25}\right] = 0.$$

This occurs when N = 82.5 or 165. Since N represents a day number, which should be a natural number, we might interpret day 82.5 as noon on the 82nd day.

Section 3.2

1. $r = 4$, $\theta = \frac{\pi}{2}$

    $s = r\theta = 4\left(\frac{\pi}{2}\right) = 2\pi$

3. $s = 6\pi$, $\theta = \frac{3\pi}{4}$

    $s = r\theta$

    $r = \frac{s}{\theta} = \frac{6\pi}{\frac{3\pi}{4}}$

    $\qquad = 6\pi \cdot \frac{4}{3\pi}$

    $\qquad = 8$

5. $r = 3$, $s = 3$

    $s = r\theta$

    $\theta = \frac{s}{r} = \frac{3}{3} = 1$

7. $r = 12.3$ cm, $\theta = \frac{2\pi}{3}$ radians

    $s = r\theta = 12.3\left(\frac{2\pi}{3}\right)$

    $\qquad = 8.2\pi$ cm

    $\qquad \approx 25.8$ cm

9. $r = 253$ m, $\theta = \frac{2\pi}{5}$ radians

    $s = r\theta = 253\left(\frac{2\pi}{5}\right) = 101.2\pi$ m

    $\qquad \approx 318$ m

11. $r = 4.82$ m, $\theta = 60°$

    Convert θ to radians.

    $\theta = 60° = \frac{\pi}{3}$

    $s = r\theta = 4.82\left(\frac{\pi}{3}\right)$

    $\qquad = 1.61\pi$

    $\qquad \approx 5.05$ m

13. $r = 2$ in, $s = 5$ in

    $s = r\theta$

    $\theta = \frac{s}{r} = \frac{5}{2}$ or 2.5 radians

15. $\theta = \frac{\pi}{5}$ radian, $s = 4$ in

    $s = r\theta$

    $r = \frac{s}{\theta} = \frac{4}{\frac{\pi}{5}}$

    $\qquad = 4 \cdot \frac{5}{\pi}$

    $\qquad = \frac{20}{\pi}$ in

17. The formula for arc length is $s = r\theta$. Substituting 2r for r we obtain $s = 2r\theta$. The length of the arc is doubled.

For Exercises 19 and 21, note that since 6400 has two significant digits and the angles are given to the nearest degree, we can have only two significant digits in the answers.

19. 9° N, 40° N

$$\theta = 40° - 9° = 31°$$

$$= 31°\left(\frac{\pi}{180°}\right)$$

$$= \frac{31\pi}{180}$$

$$s = r\theta$$

$$= 6400\left(\frac{31\pi}{180}\right)$$

$$\approx 3500 \text{ km}$$

21. 41° N, 12° S

12° S = -12° N

$$\theta = 41° - (-12°) = 53°$$

$$= 53°\left(\frac{\pi}{180°}\right)$$

$$= \frac{53\pi}{180}$$

$$s = r\theta = 6400\left(\frac{53\pi}{180}\right)$$

$$\approx 5900 \text{ km}$$

23. r = 6400 km, s = 1200 km

$$s = r\theta$$

$$1200 = 6400\theta$$

$$\theta = \frac{3}{16}$$

Convert 3/16 radian to degrees.

$$\theta = \frac{3}{16}\left(\frac{180°}{\pi}\right) \approx 11°$$

The north—south distance between the two cities is 11°.

Let x = the latitude of Madison.

$$x - 33° = 11°$$

$$x = 44° \text{ N}$$

25. (a) The number of inches lifted is the arc length in a circle with r = 9.27 in and θ = 71° 50′.

$$71° 50′ = \left(71° + \frac{50°}{60}\right)\left(\frac{\pi}{180°}\right)$$

$$s = r\theta$$

$$= 9.27\left(71° + \frac{50°}{60}\right)\left(\frac{\pi}{180°}\right)$$

$$\approx 11.6$$

The weight will rise 11.6 in.

(b) When the weight is raised 6 in,

$$s = r\theta$$

$$\theta = \frac{s}{r}$$

$$= \frac{6}{9.27}\left(\frac{180°}{\pi}\right)$$

$$= 37.085° = 37° + (.085)(60′)$$

$$= 37° 5′.$$

The pulley must be rotated through 37° 5′.

27. A rotation of

$$\theta = 60.0° = \frac{\pi}{3}$$

on the smaller wheel moves through an arc length of

$$s = r\theta = 5.23\left(\frac{\pi}{3}\right) \approx 5.48 \text{ cm.}$$

Since both wheels move together, the larger wheel moves 5.48 cm, which rotates it through an angle θ, where

$$5.48 = 8.16\theta$$

$$\theta = \frac{5.48}{8.16} = .671 \text{ radians}$$

$$= .671\left(\frac{180°}{\pi}\right) \approx 38.5°.$$

The larger wheel rotates through 38.5°.

29. The chain moves a distance equal to the arc length on the larger gear. So, for the large gear and pedal,

$$s = r\theta$$

$$= 4.72(180°)\frac{\pi}{180°}$$

$$= 4.72\pi.$$

Thus, the chain moves 4.72π in. The small gear rotates through an angle

$$\theta = \frac{s}{r}$$

$$= \frac{4.72\pi}{1.38}$$

$$= 3.42\pi.$$

θ for the wheel and θ for the small gear are the same, or 3.42π. So, for the wheel,

$$s = r\theta$$

$$= 13.6(3.42\pi)$$

$$\approx 146 \text{ in.}$$

The bicycle will move 146 in.

31. Let t = the length of the train. t is approximately the arc length subtended by 3° 20′.

$$\theta = 3° \ 20' = 3\frac{1}{3}°$$

$$\theta = \left(3\frac{1}{3}°\right)\left(\frac{\pi}{180°}\right)$$

$$= \frac{\pi}{54}$$

$$t \approx r\theta = 3.5\left(\frac{\pi}{54}\right)$$

$$\approx .20 \text{ km}$$

The train is about .20 km long.

33. Let d = the diameter of the moon. d is approximately the arc length subtended by 1/2°.

$$\theta = \left(\frac{1}{2}°\right)\left(\frac{\pi}{180°}\right)$$

$$= \frac{\pi}{360}$$

$$d \approx r\theta = 240,000\left(\frac{\pi}{360}\right)$$

$$\approx 2100 \text{ mi}$$

The diameter of the moon is about 2100 mi.

35. $r = 6$, $\theta = \frac{\pi}{3}$

$$A = \frac{1}{2}r^2\theta$$

$$= \frac{1}{2}(6)^2\left(\frac{\pi}{3}\right)$$

$$= 6\pi$$

37. $A = 3$ units^2, $r = 2$

$$A = \frac{1}{2}r^2\theta$$

$$3 = \frac{1}{2}(2)^2\theta$$

$$3 = 2\theta$$

$$\theta = \frac{3}{2} \text{ or } 1.5 \text{ radians}$$

39. $r = 29.2$ m, $\theta = \dfrac{5\pi}{6}$

$A = \dfrac{1}{2}r^2\theta = \dfrac{1}{2}(29.2)^2\left(\dfrac{5\pi}{6}\right)$

$= 355.27\pi$

≈ 1120 m²

41. $r = 52$ cm, $\theta = \dfrac{3\pi}{10}$

$A = \dfrac{1}{2}(52)^2\left(\dfrac{3\pi}{10}\right)$

$= 405.6\pi$

≈ 1300 cm²

43. $r = 12.7$ cm, $\theta = 81°$

Convert θ to radians.

$\theta = 81°\left(\dfrac{\pi}{180°}\right) = \dfrac{9\pi}{20}$

$A = \dfrac{1}{2}(12.7)^2\left(\dfrac{9\pi}{20}\right)$

$= 36.29\pi$

≈ 114 cm²

45. $A = 16$ in², $r = 3.0$ in

$A = \dfrac{1}{2}r^2\theta$

$16 = \dfrac{1}{2}(3)^2\theta$

$16 = \dfrac{9}{2}\theta$

$\theta \approx 3.6$ radians

47. $A = 64$ m², $\theta = \dfrac{\pi}{6}$ radian

$A = \dfrac{1}{2}r^2\theta$

$64 = \dfrac{1}{2}r^2\left(\dfrac{\pi}{6}\right)$

$64 = \dfrac{\pi}{12}r^2$

$\dfrac{768}{\pi} = r^2$

$r \approx 16$ m

49. $A = \dfrac{1}{2}r^2\theta$

Substituting $\theta = 2\pi$,

$A = \dfrac{1}{2}r^2(2\pi) = \pi r^2.$

The area of a circle of radius r is πr^2.

51. $x^2 + y^2 = 4$ is the equation of a circle of radius 2. $y = (\sqrt{3}/3)x$ is the equation of a line with slope $\sqrt{3}/3$. Therefore,

$\tan\theta = \dfrac{\sqrt{3}}{3}$

$\theta = \dfrac{\pi}{6}.$

Substitute $r = 2$ and $\theta = \pi/6$ in the formula for the area of a sector.

$A = \dfrac{1}{2}r^2\theta$

$= \dfrac{1}{2}(2)^2\left(\dfrac{\pi}{6}\right)$

$= \dfrac{\pi}{3}$

53.   **(a)** Let θ = the central angle for each region.

In degrees,

$\theta = \dfrac{360°}{26} \approx 13.85°.$

(b) For each sector, $r = 25$ m (since the diameter is 50 m). In radians,

$$\theta = \frac{2\pi}{26} \approx .24166.$$

$$A = \frac{1}{2}r^2\theta$$
$$= \frac{1}{2}(25)^2(.24166)$$
$$\approx 76 \text{ m}^2.$$

55. Use Volume V = Area of base × height.

$$\text{Area of base} = \frac{1}{2}r^2\theta$$
$$\text{height} = h$$
$$V = \left(\frac{1}{2}r^2\theta\right)(h)$$
$$= \frac{r^2\theta h}{2}$$

(θ is in radians.)

57. **(a)** Substitute 950,000 for A in the formula for area of a circle.

$$A = \pi r^2$$
$$950,000 = \pi r^2$$
$$r = \sqrt{\frac{950,000}{\pi}}$$
$$\approx 550 \text{ m}$$

(b) $35° = 35\left(\frac{\pi}{180}\right) = \frac{7\pi}{36}$ radian

Substitute 950,000 for A and $7\pi/36$ for θ in the formula for area of a sector.

$$A = \frac{1}{2}r^2\theta$$
$$950,000 = \frac{1}{2}r^2\left(\frac{7\pi}{36}\right)$$
$$r = \sqrt{\frac{68,400,000}{7\pi}}$$
$$\approx 1800 \text{ m}$$

Section 3.3

1. Since there are three reference angles of $\pi/3$, $\pi/4$, and $\pi/6$ and four quadrants, there are 12 exact values. Since there are 5 quadrantal angles of 0, $\pi/2$, π, $3\pi/2$, and 2π, 12 + 5 or 17 exact values can be determined.

3. Find $\sin\frac{7\pi}{6}$.

In radians, the reference angle for $\frac{7\pi}{6}$ is

$$\frac{7\pi}{6} - \pi = \frac{\pi}{6}.$$

$\frac{7\pi}{6}$ is in quadrant III, where the sine is negative.

$$\sin\frac{7\pi}{6} = -\sin\frac{\pi}{6} = -\frac{1}{2}.$$

Or, convert to degrees.

$$\frac{7\pi}{6} = \frac{7}{6}(180°) = 210°$$
$$\sin\frac{7\pi}{6} = \sin 210° = -\frac{1}{2}$$

5. Find $\tan\frac{3\pi}{4}$.

In radians, the reference angle for $\frac{3\pi}{4}$ is

$$\pi - \frac{3\pi}{4} = \frac{\pi}{4}.$$

$\frac{3\pi}{4}$ is in quadrant II, where the tangent is negative.

$$\tan\frac{3\pi}{4} = -\tan\frac{\pi}{4} = -1$$

Or, convert to degrees.

$$\frac{3\pi}{4} = \frac{3}{4}(180°) = 135°$$

$$\tan \frac{3\pi}{4} = \tan 135° = -1$$

7. Find $\cos \frac{7\pi}{6}$.

The reference angle for $\frac{7\pi}{6}$ is

$$\frac{7\pi}{6} - \pi = \frac{\pi}{6}.$$

$\frac{7\pi}{6}$ is in quadrant III, where the cosine is negative.

$$\cos \frac{7\pi}{6} = -\cos \frac{\pi}{6} = -\frac{\sqrt{3}}{2}$$

Or, convert to degrees.

$$\frac{7\pi}{6} = \frac{7}{6}(180°) = 210°$$

$$\cos \frac{7\pi}{6} = \cos 210° = -\frac{\sqrt{3}}{2}$$

9. Find $\sec \frac{2\pi}{3}$.

The reference angle for $\frac{2\pi}{3}$ is

$$\pi - \frac{2\pi}{3} = \frac{\pi}{3}.$$

$\frac{2\pi}{3}$ is in quadrant II, where the secant is negative.

$$\sec \frac{2\pi}{3} = -\sec \frac{\pi}{3} = -2$$

Or, convert to degrees.

$$\frac{2\pi}{3} = \frac{2}{3}(180°) = 120°$$

$$\sec \frac{2\pi}{3} = \sec 120° = -2$$

11. Find $\cot \frac{5\pi}{6}$.

The reference angle for $\frac{5\pi}{6}$ is

$$\pi - \frac{5\pi}{6} = \frac{\pi}{6}.$$

$\frac{5\pi}{6}$ is in quadrant II, where the cotangent is negative.

$$\cot \frac{5\pi}{6} = -\cot \frac{\pi}{6} = -\sqrt{3}$$

Or, convert to degrees.

$$\frac{5\pi}{6} = \frac{5}{6}(180°) = 150°$$

$$\cot \frac{5\pi}{6} = \cot 150° = -\sqrt{3}$$

13. Find $\sin \left(-\frac{5\pi}{6}\right)$.

$-\frac{5\pi}{6}$ is coterminal with $\frac{7\pi}{6}$.

The reference angle for $\frac{7\pi}{6}$ is

$$\frac{7\pi}{6} - \pi = \frac{\pi}{6}.$$

$-\frac{5\pi}{6}$ is in quadrant III, where the sine is negative.

$$\sin \left(-\frac{5\pi}{6}\right) = \sin \frac{7\pi}{6} = -\sin \frac{\pi}{6} = -\frac{1}{2}$$

Or, convert to degrees.

$$-\frac{5\pi}{6} = -\frac{5}{6}(180°) = -150°$$

$$\sin \left(-\frac{5\pi}{6}\right) = \sin (-150°) = -\frac{1}{2}$$

15. Find $\sec \dfrac{23\pi}{6}$.

$\dfrac{23\pi}{6}$ is not between 0 and 2π. Subtract 2π.

$$\dfrac{23\pi}{6} - 2\pi = \dfrac{23\pi}{6} - \dfrac{12\pi}{6} = \dfrac{11\pi}{6}$$

$\dfrac{23\pi}{6}$ is coterminal with $\dfrac{11\pi}{6}$. The reference angle for $\dfrac{11\pi}{6}$ is

$$2\pi - \dfrac{11\pi}{6} = \dfrac{\pi}{6}.$$

$\dfrac{23\pi}{6}$ is in quadrant IV, where the secant is positive.

$$\sec \dfrac{23\pi}{6} = \sec \dfrac{11\pi}{6} = \sec \dfrac{\pi}{6} = \dfrac{2\sqrt{3}}{3}$$

Or, converting to degrees,

$$\dfrac{11\pi}{6} = \dfrac{11}{6}(180°) = 330°$$

$$\sec \dfrac{23\pi}{6} = \sec \dfrac{11\pi}{6} = \sec 330° = \dfrac{2\sqrt{3}}{3}.$$

17. Find $\cos \dfrac{13\pi}{4}$.

$$\dfrac{13\pi}{4} - 2\pi = \dfrac{13\pi}{4} - \dfrac{8\pi}{4} = \dfrac{5\pi}{4}$$

$\dfrac{13\pi}{4}$ is coterminal with $\dfrac{5\pi}{4}$.

The reference angle for $\dfrac{5\pi}{4}$ is

$$\dfrac{5\pi}{4} - \pi = \dfrac{\pi}{4}.$$

$\dfrac{13\pi}{4}$ is in quadrant III, where the cosine is negative.

$$\cos \dfrac{13\pi}{4} = \cos \dfrac{5\pi}{4} = -\cos \dfrac{\pi}{4} = -\dfrac{\sqrt{2}}{2}$$

Converting to degrees,

$$\dfrac{5\pi}{4} = \dfrac{5}{4}(180°) = 225°$$

$$\cos \dfrac{13\pi}{4} = \cos \dfrac{5\pi}{4} = \cos 225° = -\dfrac{\sqrt{2}}{2}.$$

Your calculator must be set in radian mode for calculator exercises in this section. Keystroke sequences may vary based on the type and/or model of calculator being used.

19. sin .8203

   Enter: .8203 $\boxed{\text{SIN}}$

   or $\quad$ $\boxed{\text{SIN}}$ .8203 $\boxed{\text{ENTER}}$

   sin .8203 = .73135046

21. cos .6429

   Enter: .6429 $\boxed{\text{COS}}$

   or $\quad$ $\boxed{\text{COS}}$ .6429 $\boxed{\text{ENTER}}$

   cos .6429 = .80036052

23. sin 1.5097

   Enter: 1.5097 $\boxed{\text{SIN}}$

   or $\quad$ $\boxed{\text{SIN}}$ 1.5097 $\boxed{\text{ENTER}}$

   sin 1.5097 = .99813420

25. csc 1.3875

   Enter: 1.3875 $\boxed{\text{SIN}}$ $\boxed{\text{1/x}}$

   or $\quad$ $\boxed{(}$ $\boxed{\text{SIN}}$ 1.3875 $\boxed{)}$ $\boxed{x^{-1}}$

   $\boxed{\text{ENTER}}$

   csc 1.3875 = 1.0170372

27. sin 7.5835

   Enter: 7.5835 $\boxed{\text{SIN}}$

   or $\quad$ $\boxed{\text{SIN}}$ 7.5835 $\boxed{\text{ENTER}}$

   sin 7.5835 = .96364232

29. cot 7.4526

Enter: 7.4526 [TAN] [1/x]

or   [(] [TAN] 7.4526 [)] [x⁻¹]

[ENTER]

cot 7.4526 = .42442278

31. tan 4.0230

Enter: 4.0230 [TAN]

or   [TAN] 4.0230 [ENTER]

tan 4.0230 = 1.2131367

33. cos 4.2528

Enter: 4.2528 [COS]

or   [COS] 4.2528 [ENTER]

cos 4.2528 = −.44357977

35. sin (−2.2864)

Enter: 2.2864 [+/−] [SIN]

or   [SIN] [(−)] 2.2864 [ENTER]

sin (−2.2864) = −.75469733

37. cos (−3.0602)

Enter: 3.0602 [+/−] [COS]

or   [COS] [(−)] 3.0602 [ENTER]

cos (−3.0602) = −.99668945

39. tan s = .21264138

Enter: .21264138 [INV] [TAN]

or   [2nd] [TAN] .21264138 [ENTER]

Display: .20952066

s = .20952066

41. sin s = .99184065

Enter: .99184065 [INV] [SIN]

or   [2nd] [SIN] .99184065 [ENTER]

Display: 1.4429646

s = 1.4429646

43. cot s = .62084613

Enter: .62084613 [1/x] [INV] [TAN]

or   [2nd] [TAN] .62084613 [x⁻¹]

[ENTER]

Display: 1.0151896

s = 1.0151896

45. cos s = .57834328

Enter: .57834328 [INV] [COS]

or   [2nd] [COS] .57834328 [ENTER]

Display: .95409991

s = .95409991

47. cot s = .09637041

Enter: .09637041 [1/x] [INV] [TAN]

or   [2nd] [TAN] .09637041 [x⁻¹]

[ENTER]

Display: 1.4747226

s = 1.4747226

49. tan s = 1.6213129

Enter: 1.6213129 [INV] [TAN]

or   [2nd] [TAN] 1.6213129 [ENTER]

Display: 1.0181269

s = 1.0181269

51. From the screen, $\sin s = \sqrt{3}/2$. The reference angle for s must be $\pi/3$ since $\sin \pi/3 = \sqrt{3}/2$. Since $\pi/3$ is in the interval $[0, \pi/2]$, $s = \pi/3$.

53. $\left[\dfrac{\pi}{2}, \pi\right]$; $\cos s = -\dfrac{1}{2}$

Because $\cos s = -1/2$, the reference angle for s must be $\pi/3$ since $\cos \pi/3 = 1/2$. For s to be in the interval $[\pi/2, \pi]$, we must subtract the reference angle from π. Therefore,

$$s = \pi - \frac{\pi}{3} = \frac{2\pi}{3}.$$

55. $\left[\pi, \dfrac{3\pi}{2}\right]$; $\sin s = -\dfrac{1}{2}$

Because $\sin s = -1/2$, the reference angle for s must be $\pi/6$ since $\sin \pi/6 = 1/2$. For s to be in the interval $[\pi, 3\pi/2]$, we must add the reference angle to π. Therefore,

$$s = \pi + \frac{\pi}{6} = \frac{7\pi}{6}.$$

57. $\left[\dfrac{3\pi}{2}, 2\pi\right]$; $\cos s = \dfrac{\sqrt{3}}{2}$

Because $\cos s = \sqrt{3}/2$, the reference angle for s must be $\pi/6$ since $\cos \dfrac{\pi}{6} = \sqrt{3}/2$. For s to be in the interval $[3\pi/2, 2\pi]$, we must subtract the reference angle from 2π. Therefore,

$$s = 2\pi - \frac{\pi}{6} = \frac{11\pi}{6}.$$

59. Since $X = \cos s$ and $Y = \sin s$,

$$\tan s = \frac{\sin s}{\cos s} = \frac{Y}{X}$$

$$\tan s = \frac{.35154706}{.9361702}$$

$$= .375516183.$$

Use radian mode.

Enter: .375516183 $\boxed{\text{INV}}$ $\boxed{\text{TAN}}$

or $\boxed{\text{2nd}}$ $\boxed{\text{TAN}}$ .375516183 $\boxed{\text{ENTER}}$

Display: .359223138

$s = .359223138$

Since s is in quadrant II,

$$\pi - s \approx 2.782.$$

For Exercises 61–67, have calculator in radian mode.

61. s = the length of an arc on the unit circle = 3.4

$x = \cos s$ $\qquad\qquad$ $y = \sin s$

$x = \cos 3.4$ $\qquad\quad$ $y = \sin 3.4$

$x = -.96679819$ $\quad$ $y = -.25554110$

$\qquad$ $(-.96679819, -.25554110)$

63. s = -3.9

$x = \cos s$ $\qquad\qquad$ $y = \sin s$

$x = \cos (-3.9)$ $\qquad$ $y = \sin (-3.9)$

$x = -.72593230$ $\quad$ $y = .68776616$

$\qquad$ $(-.72593230, .68776616)$

65. s = 49

$\cos 49 = .30059254$

$\sin 49 = -.95375265$

Since cosine is positive and sine is negative, an angle of 49 radians lies in quadrant IV.

67. $s = 79$

$\cos 79 = -.89597095$

$\sin 79 = -.44411267$

Since cosine and sine are both negative, an angle of 79 radians lies in quadrant III.

69. $T(x) = 37 \sin \left[\frac{2\pi}{365}(x - 101)\right] + 25$

(a) March 1 (day 60)

$T(60) = 37 \sin \left[\frac{2\pi}{365}(60 - 101)\right] + 25$

$\approx 1°F$

(b) April 1 (day 91)

$T(91) = 37 \sin\left[\frac{2\pi}{365}(91 - 101)\right] + 25$

$\approx 19°F$

(c) Day 150

$T(150) \approx 53°F$

(d) June 15 is day 166.

$T(166) \approx 58°F$

(e) September 1 is day 244.

$T(244) \approx 48°F$

(f) October 31 is day 304.

$T(304) \approx 12°F$

71. $\cos x = \cos (x + 1)$

If x is in quadrant I, then x + 1 must be in quadrant IV. Therefore,

$$x + 1 = 2\pi - x$$
$$2x = 2\pi - 1$$
$$x = \pi - .5.$$

On the interval $0 \le x \le 2\pi$, this would also be true if we use an angle of $4\pi - x$ which is coterminal to $2\pi - x$. Therefore,

$$x + 1 = 4\pi - x$$
$$2x = 4\pi - 1$$
$$x = 2\pi - .5.$$

For the statement $\cos x = \cos (x + 1)$ to be true

$$x = \pi - .5 \text{ or } x = 2\pi - .5.$$

73. To solve this problem make sure your calculator is in radian mode. $23.44° \approx .4901$ radian and $44.88° \approx .7833$ radian.

Shortest Day:

$$\cos (.1309H) = -\tan D \tan L$$
$$= -\tan (-.4091) \tan (.7833)$$
$$\approx .4317$$
$$.1309H \approx 1.124$$
$$H \approx 8.6 \text{ hr}$$

Longest Day:

$$\cos (.1309H) = -\tan D \tan L$$
$$\cos (.1309H) = -\tan (.4091) \tan (.7833)$$
$$\approx -.4317$$
$$.1309H \approx 2.017$$
$$H \approx 15.4 \text{ hr}$$

Section 3.4

1. The circumference of the unit circle is 2π.

   $\omega = 1$ radian per sec, $\theta = 2\pi$ radians

   $\omega = \dfrac{\theta}{t}$

   $t = \dfrac{\theta}{\omega}$

   $t = \dfrac{2\pi}{1} = 2\pi$ sec

5. $\omega = \dfrac{2\pi}{3}$ radians per sec, $t = 3$ sec

   $\omega = \dfrac{\theta}{t}$

   $\dfrac{2\pi}{3} = \dfrac{\theta}{3}$

   $\theta = 2\pi$ radians

7. $\theta = \dfrac{3\pi}{4}$ radians, $t = 8$ sec

   $\omega = \dfrac{\theta}{t} = \dfrac{\frac{3\pi}{4}}{8}$

   $= \dfrac{3\pi}{32}$ radian per sec

9. $\theta = \dfrac{2\pi}{9}$ radian, $\omega = \dfrac{5\pi}{27}$ radian per min

   $\omega = \dfrac{\theta}{t}$

   $\dfrac{5\pi}{27} = \dfrac{\frac{2\pi}{9}}{t}$

   $t = \dfrac{2\pi}{9} \cdot \dfrac{27}{5\pi} = \dfrac{6}{5}$ min

11. $\theta = 3.871142$ radians, $t = 21.4693$ sec

    $\omega = \dfrac{\theta}{t}$

    $\omega = \dfrac{3.871142}{21.4693}$

    $\approx .180311$ radian per sec

13. $r = 12$ m, $\omega = \dfrac{2\pi}{3}$ radians per sec

    $v = r\omega$

    $v = 12\left(\dfrac{2\pi}{3}\right) = 8\pi$ m per sec

15. $v = 9$ m per sec, $r = 5$ m

    $v = r\omega$

    $9 = 5\omega$

    $\omega = \dfrac{9}{5}$ radians per sec

17. $r = 107.692$ m per sec, $r = 58.7413$ m

    $v = r\omega$

    $107.692 = 58.7413\omega$

    $\omega \approx 1.83333$ radians per sec

19. $r = 6$ cm, $\omega = \dfrac{\pi}{3}$ radians per sec,

    $t = 9$ sec

    $s = r\omega t$

    $s = 6\left(\dfrac{\pi}{3}\right)9$

    $= 18\pi$ cm

21. $s = 6\pi$ cm, $r = 2$ cm, $\omega = \dfrac{\pi}{4}$

radian per sec

$s = r\omega t$

$6\pi = 2\left(\dfrac{\pi}{4}\right)t$

$t = \dfrac{6\pi}{\dfrac{\pi}{2}} = 12$ sec

23. $s = \dfrac{3\pi}{4}$ km, $r = 2$ km, $t = 4$ sec

$s = r\omega t$

$\dfrac{3\pi}{4} = 2(\omega)4$

$\omega = \dfrac{\dfrac{3\pi}{4}}{8}$

$= \dfrac{3\pi}{32}$ radian per sec

25.   Both d and s are distances. d is
a linear distance, whereas s is a
circular distance, that is, the
distance along an arc. In both
cases, t is time where r and v are
both rates, r being a rate along a
straight line and v being a rate
along an arc.

27.   The hour hand of a clock moves
through an angle of 2π radians
in 12 hours, so

$\omega = \dfrac{\theta}{t} = \dfrac{2\pi}{12}$

$= \dfrac{\pi}{6}$ radian per hr.

29.   The second hand of a clock moves
through an angle of 2π radians in
60 sec, so

$\omega = \dfrac{\theta}{t} = \dfrac{2\pi}{60}$

$= \dfrac{\pi}{30}$ radian per sec.

31.   The minute hand of a clock moves
through an angle of 2π radians in
60 min, and, at the tip of the
minute hand, $r = 7$ cm, so

$v = \dfrac{r\theta}{t} = \dfrac{7(2\pi)}{60}$

$= \dfrac{7\pi}{30}$ cm per min.

33.   The flywheel making 42 rotations
per min turns through an angle
$42(2\pi)$ or 84π radians in 1 min
with $r = 2$ m.

$v = \dfrac{r\theta}{t} = \dfrac{2(84\pi)}{1} = 168\pi$ m per min

35.   At 500 rotations per min, the
propeller turns through an
angle of

$\theta = 500(2\pi) = 1000\pi$ radians

in 1 min with $r = 3/2 = 1.5$ m.

$v = \dfrac{r\theta}{t} = \dfrac{1.5(1000\pi)}{1}$

$= 1500\pi$ m per min

37.   The larger pulley rotates 80 times
per minute. Thus, for the larger
pulley,

$\omega = 2\pi(80) = 160\pi$ radians per min

or $\dfrac{160}{60} = \dfrac{8\pi}{3}$ radians per sec.

The diameter of the larger pulley is 4 m, so its radius is 2 m. To find the speed of the belt in m per sec, use the formula

$v = r\omega$.

$v = 2\left(\frac{8\pi}{3}\right) = \frac{16\pi}{3}$ m per sec

To find the angular velocity of the smaller pulley, use $v = \frac{16\pi}{3}$ m per sec and r = 1 m.

$$v = r\omega$$

$$\frac{16\pi}{3} = 1 \cdot \omega$$

$$\omega = \frac{16\pi}{3} \text{ radians per sec}$$

39. Since s = 56 cm of belt go around in t = 18 sec, the linear velocity is

$$v = \frac{s}{t}$$

$$= \frac{56}{18}$$

$$= \frac{28}{9} \text{ cm per sec.}$$

$v = r\omega$, r = 12.96 cm

$$\frac{28}{9} = (12.96)\omega$$

$$\omega = \frac{\frac{28}{9}}{12.96}$$

$$\approx .24 \text{ radian per sec}$$

41. v = 1.46 m per sec and 46 revolutions per min is

$$\omega = \frac{46(2\pi)}{60} = \frac{23\pi}{15} \text{ radians per sec.}$$

So,

$$v = r\omega$$

$$1.46 = r\left(\frac{23\pi}{15}\right)$$

$$r = \frac{1.46}{\frac{23\pi}{15}} \approx .303 \text{ m.}$$

43. **(a)** The earth completes one revolution per day, so it turns through $\theta = 2\pi$ radians in time t = 1 day = 24 hr.

$$\omega = \frac{\theta}{t} = \frac{2\pi}{1}$$

$$= 2\pi \text{ radians per day}$$

$$\omega = \frac{2\pi}{24}$$

$$= \frac{\pi}{12} \text{ radian per hr}$$

(b) At the poles, r = 0, so

$$v = r\omega = 0.$$

(c) At the equator, r = 6400 km.

$$v = r\omega$$

$$= 6400(2\pi)$$

$$= 12,800\pi \text{ km per day}$$

$$= \frac{12,800\pi}{24} \text{ km per hr}$$

$$\approx 533\pi \text{ km per hr}$$

(d) Salem rotates about the axis in a circle of radius r at angular velocity $\omega = 2\pi$ radians per day.

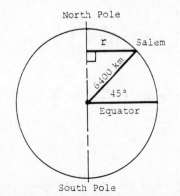

$$\sin 45° = \frac{r}{6400}$$

$$r = 6400 \sin 45°$$

$$= 6400\left(\frac{\sqrt{2}}{2}\right)$$

$$r = 3200\sqrt{2} \text{ km}$$

$$v = r\omega$$

$$= 3200\sqrt{2}(2\pi)$$

$$\approx 9050\pi \text{ km per day}$$

$$= \frac{9050\pi}{24} \text{ km per hr}$$

$$\approx 377\pi \text{ km per hr}$$

45. In one minute, the propeller makes 5000 revolutions. Each revolution is 2π radians, so

$$5000(2\pi) = 10,000\pi \text{ radians per min.}$$

Since there are 60 sec in a minute,

$$\omega = \frac{10,000\pi}{60}$$

$$= \frac{500\pi}{3}$$

$$\approx 523.6 \text{ radians per sec.}$$

Chapter 3 Review Exercises

3. To find a coterminal angle, add or subtract multiples of 2π. Three of the many possible answers are $2 - 2\pi$, $2 - 4\pi$, and $2 - 6\pi$.

5. $45° = 45°\left(\frac{\pi}{180°}\right) = \frac{\pi}{4}$

7. $80° = 80°\left(\frac{\pi}{180°}\right) = \frac{4\pi}{9}$

9. $330° = 330°\left(\frac{\pi}{180°}\right) = \frac{11\pi}{6}$

11. $1020° = 1020°\left(\frac{\pi}{180°}\right) = \frac{17\pi}{3}$

13. $\frac{5\pi}{4} = \frac{5\pi}{4}\left(\frac{180°}{\pi}\right) = 225°$

15. $\frac{8\pi}{3} = \frac{8\pi}{3}\left(\frac{180°}{\pi}\right) = 480°$

17. $-\frac{11\pi}{18} = -\frac{11\pi}{18}\left(\frac{180°}{\pi}\right) = -110°$

19. $\frac{14\pi}{15} = \frac{14\pi}{15}\left(\frac{180°}{\pi}\right) = 168°$

21. $\tan\frac{\pi}{3} = \tan 60° = \sqrt{3}$

23. $\sin\left(-\frac{5\pi}{6}\right) = \sin\left(-\frac{5}{6}\cdot 180°\right)$

$$= \sin(-150°)$$

$$= \sin(-150° + 360°)$$

$$= \sin 210°$$

$$= -\frac{1}{2}$$

25. $\tan\left(-\frac{7\pi}{3}\right) = \tan\left(-\frac{7}{3}\cdot 180°\right)$

$$= \tan(-420°)$$

$$= \tan[-420° + 2(360°)]$$

$$= \tan 300°$$

$$= -\sqrt{3}$$

27. $\csc\left(-\frac{11\pi}{6}\right) = \csc\left(-\frac{11}{6}\cdot 180°\right)$

$$= \csc(-330°)$$

$$= \csc(-330° + 360°)$$

$$= \csc 30°$$

$$= 2$$

29. $r = 15.2$ cm, $\theta = \dfrac{3\pi}{4}$

$s = r\theta = 15.2\left(\dfrac{3\pi}{4}\right)$

$\qquad = 11.4\pi$

$\qquad \approx 35.8$ cm

31. $r = 8.973$ cm,

$\theta = 49.06°$

$\quad = 49.06°\left(\dfrac{\pi}{180°}\right)$

$s = r\theta$

$\quad = 8.973(49.06°)\left(\dfrac{\pi}{180°}\right)$

$\quad \approx 7.683$ cm

33. $r = 38.0$ m, $\theta = 21°\ 40' = 21\dfrac{2}{3}°$

$\qquad\qquad = \dfrac{65°}{3}\left(\dfrac{\pi}{180°}\right)$

$A = \dfrac{1}{2}r^2\theta$

$\quad = \dfrac{1}{2}(38.0)^2\left(\dfrac{65°}{3}\right)\left(\dfrac{\pi}{180°}\right)$

$\quad \approx 273$ m²

35. $28°$ N, $12°$ S

$12°$ S $= -12°$ N

$\theta = 28° - (-12°) = 40°$

$\quad = 40°\left(\dfrac{\pi}{180°}\right) = \dfrac{2\pi}{9}$

$s = r\theta$

$\quad = 6400\left(\dfrac{2\pi}{9}\right)$

$\quad \approx 4500$ km

37. $s = 1.5$, $r = 2$

$s = r\theta$

$\theta = \dfrac{s}{r} = \dfrac{1.5}{2} = \dfrac{3}{4}$

$A = \dfrac{1}{2}r^2\theta$

$\quad = \dfrac{1}{2}(2)^2\left(\dfrac{3}{4}\right)$

$\quad = \dfrac{3}{2}$ or 1.5 units²

39. (a) The hour hand of a clock moves through an angle of 2π radians in 12 hr, so

$$\omega = \frac{\theta}{t} = \frac{2\pi}{12}$$

$$= \frac{\pi}{6} \text{ radian per hr.}$$

In 2 hr the angle would be

$$2\left(\frac{\pi}{6}\right) \text{ or } \frac{\pi}{3} \text{ radians.}$$

(b) The distance s the tip of the hour hand travels during the time period from 1 o'clock to 3 o'clock, is the arc length when $\theta = \pi/3$ and $r = 6$ in.

$$s = r\theta$$

$$= 6\left(\frac{\pi}{3}\right) = 2\pi \text{ in}$$

In Exercises 41–53, be sure the calculator is set in radians. Keystroke sequences may vary based on the type and/or model of calculator being used.

41. sin 1.0472

Enter: 1.0472 [SIN]

or [SIN] 1.0472 [ENTER]

sin 1.0472 = .86602663

43. cos (−.2443)

Enter: .2443 [+/−] [COS]

or [COS] [(−)] .2443 [ENTER]

cos (−.2443) = .97030688

45. tan 7.3159 = 1.6755332

47. sec .4864

Enter: .4864 $\boxed{\text{COS}}$ $\boxed{1/x}$

or $\boxed{(}$ $\boxed{\text{COS}}$ .4864 $\boxed{)}$ $\boxed{x^{-1}}$

         $\boxed{\text{ENTER}}$

sec .4864 = 1.1311944

49. cos s = .92500448

Enter: .92500448 $\boxed{\text{INV}}$ $\boxed{\text{COS}}$

or $\boxed{\text{2nd}}$ $\boxed{\text{COS}}$ .92500448 $\boxed{\text{ENTER}}$

Display: .38974894

s = .38974894

51. sin s = .49244294

Enter: .49244294 $\boxed{\text{INV}}$ $\boxed{\text{SIN}}$

or $\boxed{\text{2nd}}$ $\boxed{\text{SIN}}$ .49244294 $\boxed{\text{ENTER}}$

Display: .51489440

s = .51489440

53. cot s = .50221761

Enter: .50221761 $\boxed{1/x}$ $\boxed{\text{INV}}$ $\boxed{\text{COS}}$

or $\boxed{\text{2nd}}$ $\boxed{\text{TAN}}$ .50221761 $\boxed{x^{-1}}$

         $\boxed{\text{ENTER}}$

Display: 1.1053762

s = 1.1053762

55. $\left[0, \frac{\pi}{2}\right]$, cos s = $\frac{\sqrt{2}}{2}$

Because cos s = $\sqrt{2}/2$, the reference angle for s must be $\pi/4$ since cos $\pi/4 = \sqrt{2}/2$. For s to be in the interval $[0, \pi/2]$, s must be the reference angle. Therefore, s = $\pi/4$.

57. $\left[\pi, \frac{3\pi}{2}\right]$, sec s = $-\frac{2\sqrt{3}}{3}$

Because sec s = $-2\sqrt{3}/3$, the reference angle for s must be $\pi/6$ since sec $\pi/6 = 2\sqrt{3}/3$. For s to be in the interval $[\pi, 3\pi/2]$, we must add the reference angle to π. Therefore,

$$s = \pi + \frac{\pi}{6} = \frac{7\pi}{6}.$$

59. 2 radians is in quadrant II so cosine must be negative. In quadrant II cosine decreases from 0 to -1. Therefore, cos 2 is closest to -.4.

61. X = cos s, y = sin s

Therefore,

$$\tan s = \frac{Y}{X}$$

$$= \frac{6.3018678}{9}$$

$$\approx .7002.$$

Use degree mode.

Enter: .7002 $\boxed{\text{INV}}$ $\boxed{\text{TAN}}$

or $\boxed{\text{2nd}}$ $\boxed{\text{TAN}}$ .7002 $\boxed{\text{ENTER}}$

Display: 34.9997102

s = 35°

s = $35°\left(\frac{\pi}{180°}\right) = \frac{7\pi}{36}$ radian

63. $\theta = \dfrac{5\pi}{12}$, $\omega = \dfrac{8\pi}{9}$ radians per sec

$\omega = \dfrac{\theta}{t}$

$\dfrac{8\pi}{9} = \dfrac{\dfrac{5\pi}{12}}{t}$

$t = \dfrac{5\pi}{12} \cdot \dfrac{9}{8\pi}$

$\quad = \dfrac{45}{96} = \dfrac{15}{32}$ sec

65. $t = 8$ sec, $\theta = \dfrac{2\pi}{5}$ radians

$\omega = \dfrac{\theta}{t}$

$\quad = \dfrac{\dfrac{2\pi}{5}}{8}$

$\quad = \dfrac{2\pi}{40} = \dfrac{\pi}{20}$ radians per sec

67. $r = 11.46$ cm, $\omega = 4.283$ radians
per sec, $t = 5.813$ sec

$v = r\omega$

$\quad = 11.46(4.283)$

$\quad = 49.08$ cm per sec

$v = \dfrac{s}{t}$

$49.08 = \dfrac{s}{5.813}$

$\quad\ s = 49.08(5.813)$

$\quad\quad = 285.3$ cm

69.   **(a)** $\sin \theta = \dfrac{h}{d}$

$\quad\quad d = \dfrac{h}{\sin \theta}$

$\quad\ d = h\csc \theta$

d is larger than h by a factor of
$\csc \theta$.

(b) $\csc \theta = 2$

$\quad \sin \theta = \dfrac{1}{2}$

$\quad\quad \theta = \dfrac{\pi}{6}$

d is double h when the sun is 30°
above the horizon.

(c) $\csc \dfrac{\pi}{2} = 1$ and $\csc \dfrac{\pi}{3} \approx 1.15$

When the sun is lower in the sky
($\theta = \pi/3$), sunlight is filtered by
more atmosphere. There is less
ultraviolet light reaching the
earth's surface, and therefore,
there is less likelihood of becoming
sunburn. In this case, sunlight
passes through 15% more atmosphere.

CHAPTER 3   TEST

Do not use a calculator for Problems 1–14.

[3.1]   Convert each degree measure to radians. Leave answers as multiples of π.

   1. 45°   2. 360°   3. 240°   4. 495°

[3.1]   Convert each radian measure to degrees.

   5. $\frac{3\pi}{2}$ 6. $-\frac{\pi}{3}$ 7. $\frac{7\pi}{3}$ 8. $\frac{7\pi}{18}$

[3.1]   Find the exact value.

   9. $\sin \frac{2\pi}{3}$ 10. $\cos \left(-\frac{3\pi}{4}\right)$ 11. $\cot \frac{\pi}{3}$

   12. $\csc \frac{3\pi}{2}$ 13. $\tan \left(-\frac{5\pi}{6}\right)$ 14. $\sec \frac{5\pi}{3}$

[3.2]   Solve the problem.

   15. The radius of a circle is 1.35 cm. Find the length of the arc of the circle intercepted by a central angle of $3\pi/4$ radians.

   16. Find the area of a sector of a circle intercepted by a central angle of 210° in a circle of radius 12.5 ft.

   17. Find the distance between Kampala, Uganda, and Belem, Brazil, if the two cities are on the equator and Kampala is at 30° E longitude and Belem is at 50° W longitude. Assume that the radius of the earth is 6400 km.

[3.3]   Use a calculator to find an approximation for each circular function value. Be sure that your calculator is set in radian mode.

   18. sin .9104   19. cos 1.415   20. tan 3.005

[3.3]   Find the value of s in the interval $[0, \pi/2]$ that makes each statement true.

     21.   cot s = 1.3821596   **22.**   cos s = .92846507

     23.   sec s = 3.8421657

[3.4]   **24.**   Find the linear velocity of a point on the tip of the minute hand of a clock, if the hand is 5 cm long.

[3.4]   **25.**   Explain how you change radians per second to radians per hour.

CHAPTER 3   TEST ANSWERS

1. $\pi/4$ **2.** 2π **3.** $4\pi/3$ **4.** $11\pi/4$ **5.** $270°$ **6.** $-60°$

7. $420°$ **8.** $70°$ **9.** $\sqrt{3}/2$ **10.** $-\sqrt{2}/2$ **11.** $\sqrt{3}/3$ **12.** -1

13. $\sqrt{3}/3$ **14.** 2 **15.** 3.18 cm   **16.** 286 ft^2 **17.** 8900 km

18. $.78974917$ **19.** $.15516683$ **20.** $-.13744854$ **21.** $.62632786$

22. $.38053767$ **23.** 1.3074946 **24.** $\pi/6$ cm per min

25. Multiply radians per second by 3600 to get radians per hour.

CUMULATIVE REVIEW EXERCISES (Chapters 1-3)

1. Find the distance between the points $(-2, 5\sqrt{3})$ and $(8, -\sqrt{3})$.

2. Find the midpoint of the line segment connecting the two points
   $(\sqrt{2}, 5)$ and $(4 - \sqrt{2}, 7)$.

3. Write $\{x \mid 2 \le x < 7\}$ in interval notation.

4. Find the domain and range of the function $y = \sqrt{x - 2}$.

5. Calculate $12° \ 43' + 21° \ 34'$.

6. Convert $82° \ 12' \ 45''$ to decimal degrees.

7. Convert $28.123°$ to degrees, minutes, and seconds.

8. Find the angle of smallest positive measure coterminal with the angle $723°$.

9. Find the measures of the numbered angles in the figure, given that the lines
   m and n are parallel.

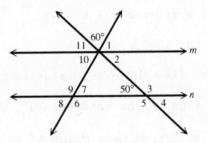

10. Find the value of x in the pair of similar triangles.

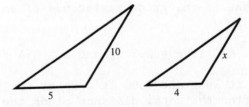

11. What is a scalene triangle?

12. The terminal side of an angle θ in standard position goes through the
    point $(-12, -5)$. Find the values of the six trigonometric functions
    of θ.

13. Find the six trigonometric function values of the angle θ in standard
    position, if the terminal side of θ is defined by $3x + 4y = 0$, $x \le 0$.

14. Find the signs of the six trigonometric functions for the angle 288.54°.

15. Find $\sin^2 270° + 2 \cot 90°$.

16. Identify the possible quadrant(s) for angle β if $\sec \beta < 0$ and $\csc \beta > 0$.

17. Suppose that angle θ is in quadrant IV and $\sin \theta = -3/4$. Use the definitions of the trigonometric functions to find the exact values of the other five trigonometric functions.

18. Suppose that angle α is in quadrant III and $\cos \alpha = -.2$. Use the Pythagorean and quotient identities to find $\sin \alpha$ and $\tan \alpha$.

19. Find the values of the six trigonometric functions for angle A in the right triangle.

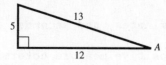

20. Give the exact value of $\sin 60°$.

21. Find an acute angle θ for which $\sin \theta = \cos 4\theta$.

22. Find the reference angle for 170°.

23. Use the reference angle to find the exact value of $\cos (-150°)$.

24. Find all values of θ, if θ is in the interval $[0°, 360°)$ and $\cos \theta = 1/2$.

25. Use a calculator to find an approximate value of $\sin 12° \, 12'$.

26. Use a calculator to find a value of θ in the interval $[0°, 90°]$ for which $\tan \theta = 1.1917536$.

27. To the nearest pound, what is the grade resistance of an 1800–pound car traveling on a 5° uphill grade?

28. Solve the right triangle ABC if $a = 2.3$ ft and $b = 3.4$ ft.

29. A swimming pool is 40.0 ft long and 4.00 ft deep at one end. If it is 8.00 ft deep at the other end, find the total distance along the bottom.

30. Very accurate measurements have shown that the distance between California's Owens Valley Radio Observatory and the Haystack Observatory in Massachusetts is 2441.2938 mi. Suppose that the two observatories focus on a distant star and find that angles E and E' in the figure are both 89.99999°. Find the distance to the star from Haystack. Assume that the earth is flat.

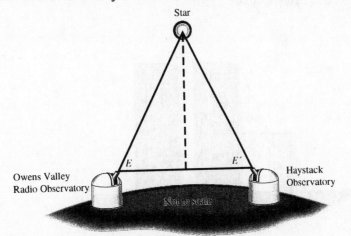

31. A sailboat is sailing due east at 10 mph. A powerboat is several miles south of the sailboat and capable of traveling at 20 mph. (See the figure.) What bearing will allow the powerboat to reach the sailboat in the shortest time? (Hint: Let the distances traveled by the two boats be 10x and 20x, where x is the number of hours until they meet.)

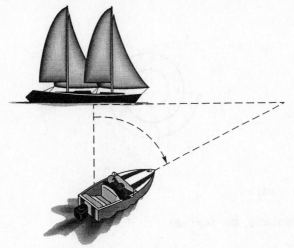

32. Find the value of h in the triangle.

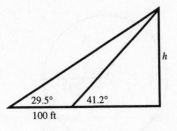

33. The angle of elevation from the top of an office building in New York City to the top of the World Trade Center is 68°, while the angle of depression from the top of the office building to the bottom of the World Trade Center is 63°. (See the figure.) The office building is 290 ft from the World Trade Center. Find the height of the World Trade Center.

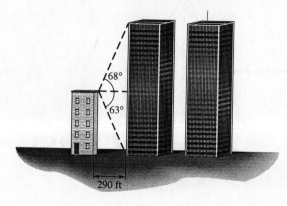

34. Give the degree and radian measures for the angle shown in the figure.

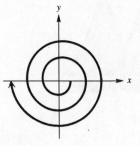

35. Convert 135° to radians.

36. Convert $13\pi/6$ radians to degrees.

37. Find the exact value of $\cos 5\pi/6$ without using a calculator.

38. Find the length of the arc intercepted in the figure.

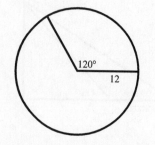

39. Find the radius of a circle in which a central angle of 1.5 radians intercepts an arc of length 9 ft.

40. Grand Portage, Minnesota has a latitude of 44° N and lies on the same north-south line as New Orleans, Louisiana which has a latitude of 30° N. Find the distance between the two cities assuming that the radius of the earth is 4.0×10^3 mi.

41. A gear of radius 3 in interlocks and turns a larger gear of radius 5 in. If the smaller gear makes 10 complete turns, how many turns will the larger gear make?

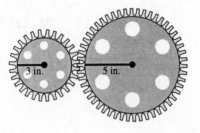

42. What is the area of a sector of a circle of radius 1.8 ft and central angle $\pi/6$ radian?

43. Use a calculator to find an approximation of cot 1.0821.

44. Use a calculator to find the value of s in the interval $[0, \pi/2]$ for which cot s = 1.2345678.

45. Suppose an arc of length 1.7 lies on the unit circle $x^2 + y^2 = 1$, starting at the point (1, 0) and terminating at the point (x, y). Use a calculator to find the approximate coordinates for (x, y).

46. Refer to the figure for Exercise 41. If the smaller gear turns at a rate of 5 revolutions per second, what is the linear velocity of the larger gear? What is its angular velocity?

47. A model airplane flies in a horizontal circle of radius 30 ft attached to a horizontal string. The airplane makes a complete revolution every 6 sec.

     (a) Find the angular velocity of the airplane.

     (b) How far does the airplane travel in 3 min?

     (c) Find the linear velocity of the airplane.

48. The tires of a bicycle have a radius of 13 in and are turning at the rate of 200 revolutions per minute. How fast is the bicycle traveling in miles per hour?

**SOLUTIONS TO CUMULATIVE REVIEW EXERCISES
(Chapters 1–3)**

1. Let $(x_1, y_1) = (-2, 5\sqrt{3})$ and
    $(x_2, y_2) = (8, -\sqrt{3})$.

$$d = \sqrt{(x_2 - x_1)^2 + (y_2 - y_1)^2}$$
$$= \sqrt{[8 - (-2)]^2 + (-\sqrt{3} - 5\sqrt{3})^2}$$
$$= \sqrt{10^2 + (-6\sqrt{3})^2}$$
$$= \sqrt{100 + 108}$$
$$= \sqrt{208}$$
$$= \sqrt{16 \cdot 13}$$
$$= 4\sqrt{13}$$

2. Let $(x_1, y_1) = (\sqrt{2}, 5)$ and
    $(x_2, y_2) = (4 - \sqrt{2}, 7)$.

$$m = \left(\frac{x_1 + x_2}{2}, \frac{y_1 + y_2}{2}\right)$$
$$= \left(\frac{\sqrt{2} + (4 - \sqrt{2})}{2}, \frac{5 + 7}{2}\right)$$
$$= \left(\frac{4}{2}, \frac{12}{2}\right)$$
$$= (2, 6)$$

3. $\{x \mid 2 \le x < 7\}$ This is a half–open interval between 2 and 7 which includes 2, but not 7. In interval notation this is $[2, 7)$.

4. $y = \sqrt{x - 2}$

    Domain: $x - 2 \ge 0$
    $\qquad\qquad x \ge 2$

    The domain is $[2, \infty)$.
    Range: $y \ge 0$, since a square root is always nonnegative.
    The range is $[0, \infty)$.

5. $12°\ 43' + 21°\ 34' = 33°\ 77'$
    $\qquad\qquad\qquad\qquad = 33°\ 60' + 17'$
    $\qquad\qquad\qquad\qquad = 34°\ 17'$

6. $82°\ 12'\ 45'' = 82° + \dfrac{12}{60}° + \dfrac{45}{3600}°$
    $\qquad\qquad\qquad = 82.2125°$

7. $28.123° = 28° + .123(60')$
    $\qquad\qquad = 28° + 7.38'$
    $\qquad\qquad = 28° + 7' + .38(60'')$
    $\qquad\qquad = 28°\ 7'\ 23''$

8. $723° - 2(360°) = 3°$

9. angle 2 = 50° (alternate interior
    $\qquad\qquad$ angle with 50° angle)
    angle 1 = 180° - 50° - 60° = 70°
    $\qquad\qquad$ (supplemental angle with
    $\qquad\qquad$ 60° and angle 2)
    angle 3 = 180° - 50° = 130°
    $\qquad\qquad$ (supplemental angle with
    $\qquad\qquad$ 50°)
    angle 4 = 50° (vertical angle with
    $\qquad\qquad$ 50°)
    angle 5 = 130° (vertical angle with
    $\qquad\qquad$ angle 3)
    angle 10 = 70° (vertical angle with
    $\qquad\qquad$ angle 1)
    angle 11 = 50° (vertical angle with
    $\qquad\qquad$ angle 2)
    angle 7 = 70° (alternate interior
    $\qquad\qquad$ angle with angle 10)
    angle 6 = 180° - 70° = 110°
    $\qquad\qquad$ (supplemental angle with
    $\qquad\qquad$ angle 7)
    angle 8 = 70° (vertical angle with
    $\qquad\qquad$ angle 7)
    angle 9 = 110° (vertical angle with
    $\qquad\qquad$ angle 6)

10. Corresponding sides of similar triangles are proportional.

$$\frac{5}{4} = \frac{10}{x}$$

$$5x = 40$$

$$x = 8$$

11. A scalene triangle is a triangle with no sides equal.

12. $(-12, -5)$

$x = -12, \; y = -5$

$r = \sqrt{(-12)^2 + (-5)^2}$

$\quad = \sqrt{144 + 25}$

$\quad = \sqrt{169}$

$\quad = 13$

$\sin \theta = \dfrac{y}{r} = -\dfrac{5}{13}$

$\cos \theta = \dfrac{x}{r} = -\dfrac{12}{13}$

$\tan \theta = \dfrac{y}{x} = \dfrac{5}{12}$

$\cot \theta = \dfrac{x}{y} = \dfrac{12}{5}$

$\sec \theta = \dfrac{r}{x} = -\dfrac{13}{12}$

$\csc \theta = \dfrac{r}{y} = -\dfrac{13}{5}$

13. Since $x \le 0$, the graph of the line $3x + 4y = 0$ is shown to the left of the y-axis. A point on this line is $(-4, 3)$ since $3(-4) + 4(3) = 0$.

$r = \sqrt{(-4)^2 + 3^2} = \sqrt{16 + 9} = \sqrt{25} = 5$

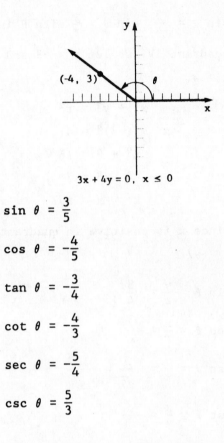

$3x + 4y = 0, \; x \le 0$

$\sin \theta = \dfrac{3}{5}$

$\cos \theta = -\dfrac{4}{5}$

$\tan \theta = -\dfrac{3}{4}$

$\cot \theta = -\dfrac{4}{3}$

$\sec \theta = -\dfrac{5}{4}$

$\csc \theta = \dfrac{5}{3}$

14. Since $288.54°$ is in quadrant IV, cosine and secant are positive; the other trigonometric functions are negative.

15. $\sin^2 270° + 2 \cot 90°$

$\quad = (-1)^2 + 2(0)$

$\quad = 1$

16. Secant is negative in quadrants II and III. Cosecant is positive in quadrants I and II. Since $\sec \beta < 0$ and $\csc \beta > 0$, β lies in quadrant II.

17. $\sin \theta = -\frac{3}{4}$, $\sin \theta = \frac{y}{r}$ with θ in quadrant IV, so let $y = -3$ and $r = 4$.

$$x^2 + y^2 = r^2$$
$$x^2 + (-3)^2 = 4^2$$
$$x^2 + 9 = 16$$
$$x^2 = 7$$
$$x = \pm\sqrt{7}$$

Since x is positive in quadrant IV, $x = \sqrt{7}$.

$$\sin \theta = \frac{y}{r} = -\frac{3}{4}$$

$$\cos \theta = \frac{x}{r} = \frac{\sqrt{7}}{4}$$

$$\tan \theta = \frac{y}{x} = -\frac{3}{\sqrt{7}} = -\frac{3\sqrt{7}}{7}$$

$$\cot \theta = \frac{x}{y} = -\frac{\sqrt{7}}{3}$$

$$\sec \theta = \frac{r}{x} = \frac{4}{\sqrt{7}} = \frac{4\sqrt{7}}{7}$$

$$\csc \theta = \frac{r}{y} = -\frac{4}{3}$$

18. $\sin^2 \alpha + \cos^2 \alpha = 1$

$\sin^2 \alpha + (-.2)^2 = 1$

$$\sin^2 \alpha + \frac{4}{100} = 1$$

$$\sin^2 \alpha = \frac{96}{100}$$

$$\sin \alpha = \pm\frac{4\sqrt{6}}{10} = \pm\frac{2\sqrt{6}}{5}$$

Since α is in quadrant III, $\sin \alpha < 0$, and

$$\sin \alpha = -\frac{2\sqrt{6}}{5}.$$

$$\tan \alpha = \frac{\sin \alpha}{\cos \alpha} = \frac{\frac{-2\sqrt{6}}{5}}{-.2}$$

$$= -\frac{2\sqrt{6}}{5} \cdot -\frac{10}{2}$$

$$= 2\sqrt{6}$$

19. $\sin A = \dfrac{\text{side opposite}}{\text{hypotenuse}} = \dfrac{5}{13}$

$\cos A = \dfrac{\text{side adjacent}}{\text{hypotenuse}} = \dfrac{12}{13}$

$\tan A = \dfrac{\text{side opposite}}{\text{side adjacent}} = \dfrac{5}{12}$

$\cot A = \dfrac{\text{side adjacent}}{\text{side opposite}} = \dfrac{12}{5}$

$\sec A = \dfrac{\text{hypotenuse}}{\text{side adjacent}} = \dfrac{13}{12}$

$\csc A = \dfrac{\text{hypotenuse}}{\text{side opposite}} = \dfrac{13}{5}$

20.

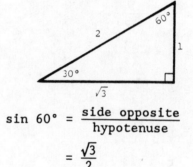

$\sin 60° = \dfrac{\text{side opposite}}{\text{hypotenuse}}$

$$= \frac{\sqrt{3}}{2}$$

21. $\sin \theta = \cos 4\theta$

Since sine and cosine are cofunctions,

$$\theta + 4\theta = 90°$$
$$5\theta = 90°$$
$$\theta = 18°.$$

22. The reference angle for 170° is $180° - 170° = 10°$.

23. cos (-150°)

The reference angle is 30°. Since the cosine is negative in quadrant III,

$$\cos(-150°) = -\cos 30° = -\frac{\sqrt{3}}{2}.$$

24. cos θ = 1/2

cos θ > 0 for θ in quadrant I or IV. Since cos θ = 1/2, the reference angle θ' must be 60°.
In quadrant I, θ = 60°.
In quadrant IV, θ = 360° - 60° = 300°.

In Exercises 25 and 26, be sure your calculator is in degree mode. Keystroke sequences may vary based on the type and/or model of calculator being used.

25. sin 12° 12′

$$12° \ 12′ = \left(12 + \frac{12}{60}\right)°$$
$$= 12.2°$$

Enter: 12.2 SIN

or SIN 12.2 ENTER

Display: .2113247965

26. tan θ = 1.1917536

Enter: 1.1917536 $\left\{\begin{array}{c} \boxed{\text{INV}} \ \boxed{\text{TAN}} \\ \text{or} \\ \boxed{\text{ARCTAN}} \\ \text{or} \\ \boxed{\text{TAN}^{-1}} \end{array}\right.$

or 2nd TAN 1.1917536 ENTER

Display: = 50

θ = 50°

27. W = F sin θ
   = 1800(sin 5°)
   ≈ 157 lb

28. a = 2.3 ft, b = 3.4 ft
$c^2 = a^2 + b^2$
$c^2 = (2.3)^2 + (3.4)^2$
$c^2 = 16.85$
$c ≈ 4.1$ ft

$\sin A = \dfrac{a}{c}$

$\sin A = \dfrac{2.3}{4.1}$

$A = 34.1°$

$A + B = 90°$
$34.1° + B = 90°$
$B = 55.9°$

29. Name points in the figure as follows.

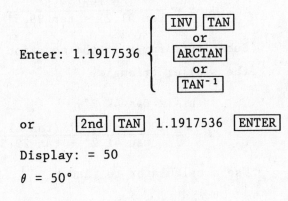

BC = DC - BC
BC = 8.00 - 4.00
BC = 4.00
AB = 40.0

Using the Pythagorean theorem gives

$$x^2 = (40.0)^2 + (4.00)^2$$
$$x^2 = 1600.0 + 16.0$$
$$x^2 = 1616.0$$
$$x ≈ 40.1995.$$

The distance along the bottom is 40.2 ft.

30. Let point S represent the star.

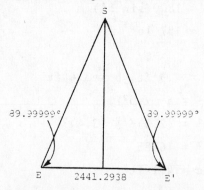

Triangle SEE′ has equal base angles each 89.99999° so the triangle is isosceles. The altitude of the triangle bisects the base of the triangle and forms two right triangles. In one of the right triangles, the side adjacent to the 89.99999° angle is 2441.2938/2, or 1220.6469 mi.

$$\cos E' = \frac{1220.6469}{SE'}$$

$$SE' = \frac{1220.6469}{\cos 89.99999°}$$

$$= 6.993792 \times 10^9 \text{ mi}$$

The distance from the Haystack Observatory to the star is

$$6.993792 \times 10^9 \text{ mi.}$$

31. Let x = the number of hours until the boats meet.

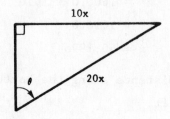

$$\sin \theta = \frac{10x}{20x}$$

$$\sin \theta = \frac{1}{2}$$

$$\theta = 30°$$

A bearing of 30° will allow the powerboat to reach the sailboat in the shortest time.

32. Let x be the side adjacent to 41.2° in the smaller right triangle. In the larger right triangle,

$$\tan 29.5° = \frac{h}{100 + x}$$

$$h = (100 + x) \tan 29.5°.$$

In the smaller right triangle,

$$\tan 41.2° = \frac{h}{x}$$

$$h = x \tan 41.2°.$$

Substitute for h in this equation and solve for x.

$$(100 + x) \tan 29.5°$$
$$= x \tan 41.2°$$
$$100 \tan 29.5° + x \tan 29.5°$$
$$= x \tan 41.2°$$
$$100 \tan 29.5° = x \tan 41.2° - x \tan 29.5°$$
$$100 \tan 29.5° = x(\tan 41.2° - \tan 29.5°)$$
$$x = \frac{100 \tan 29.5°}{\tan 41.2° - \tan 29.5°}$$

Substitute for x in the equation for the smaller triangle.

$$h = x \tan 41.2°.$$

$$= \frac{100 \tan 29.5° (\tan 41.2°)}{\tan 41.2° - \tan 29.5°}$$

Use a calculator to find h.

$$h \approx 160$$

The height is 160 ft.

33. Let x + y = the height of the World Trade Center.

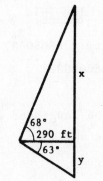

$$\tan 68° = \frac{x}{290}$$

$$x = 290 \tan 68°$$

$$\tan 63° = \frac{y}{290}$$

$$y = 290 \tan 63°$$

$$x + y = 290 \tan 68° + 290 \tan 63°$$

$$\approx 1300$$

The height of the World Trade Center is 1300 ft.

34. The angle is drawn clockwise; there fore, it is negative.

$$-[2(360°) + 180°] = -900°$$

$$-900 = -900°\left(\frac{\pi}{180°}\right)$$

$$= -5\pi \text{ radians}$$

35. $135° = 135°\left(\frac{\pi}{180°}\right) = \frac{3\pi}{4}$ radians

36. $\frac{13\pi}{6} = \frac{13\pi}{6}\left(\frac{180°}{\pi}\right) = 390°$

37. $\cos \frac{5\pi}{6} = \cos \left(\frac{5}{6} \cdot 180°\right)$

$$= \cos 150°$$

$$= -\frac{\sqrt{3}}{2}$$

38. $r = 12$, $\theta = 120°$

Convert θ to radians.

$$\theta = 120° = \frac{2\pi}{3}$$

$$s = r\theta = 12\left(\frac{2\pi}{3}\right)$$

$$= 8\pi$$

39. $s = 9$ ft, $\theta = 1.5$ radians

$$s = r\theta$$

$$r = \frac{s}{\theta}$$

$$= \frac{9}{1.5}$$

$$= 6 \text{ ft}$$

40. 44° N, 30° N

$$r = 4 \times 10^3 \text{ mi} = 4000 \text{ mi}$$

$$\theta = 44° - 30° = 14°$$

$$= 14°\left(\frac{\pi}{180°}\right)$$

$$= \frac{7\pi}{90}$$

The distance between the two cities is

$$s = r\theta$$

$$= 4000\left(\frac{7\pi}{90}\right)$$

$$\approx 980 \text{ mi.}$$

41. A rotation of 10 complete turns

$$\theta = 10(2\pi) = 20\pi$$

on the smaller gear moves through an arc length of

$$s = r\theta = 3(20\pi) = 188.50 \text{ in.}$$

Since both gears move together, the larger gear moves 188.50 in, which rotates it through an angle θ, where

$$188.50 = 5\theta$$

$$\theta = \frac{188.50}{5} = 37.7 \text{ radians,}$$

and

$$\frac{37.7}{2\pi} = 6.$$

The larger gear will make 6 turns.

42. $r = 1.8 \text{ ft}, \theta = \frac{\pi}{6}$ radian

$$A = \frac{1}{2}r^2\theta$$

$$= \frac{1}{2}(1.8)^2\left(\frac{\pi}{6}\right)$$

$$= .27\pi$$

$$\approx .85 \text{ ft}^2$$

In Exercises 43-45, be sure your calculator is set in radian mode. Keystroke sequences may vary based on the type and/or model of calculator being used.

43. cot 1.0821

Enter: 1.0821 [TAN] [1/x]

or   [( ] [TAN] 1.0821 [ )] [x⁻¹]

        [ENTER]

Display: .53171474

cot 1.0821 = .53171474

44. cot s = 1.2345678

Enter: 1.2345678 [1/x] [INV] [TAN]

or   [2nd] [TAN] 1.2345678 [x⁻¹]

        [ENTER]

Display: .68080887

s = .68080887

45. s = the length of an arc on the unit circle = 1.7.

$x = \cos s$ $y = \sin s$

$x = \cos 1.7$ $y = \sin 1.7$

$x = -.12884449$ $y = .99166481$

        $(-.12884449, .99166481)$

46. $r = 3 \text{ in}, \omega = 5$ revolutions per sec

$$\omega = 5(2\pi) = 10\pi \text{ radians per sec}$$

$$v = r\omega$$

$$v = 3(10\pi)$$

$$= 30\pi \text{ in per sec}$$

$$v = 30\pi, r = 5 \text{ in}$$

$$v = r\omega$$

$$\omega = \frac{v}{r}$$

$$\omega = \frac{30\pi}{5}$$

$$\omega = 6\pi \text{ radians per sec}$$

47. (a) $r = 30 \text{ ft}, \theta = \frac{2\pi}{6} = \frac{\pi}{3}$

$$\omega = \frac{\theta}{t} = \frac{\frac{\pi}{3}}{1}$$

$$= \frac{\pi}{3} \text{ radians per sec}$$

(b) $s = r\theta$

$\theta = \dfrac{\pi}{3}$ radians per sec, $r = 30$ ft

$\theta = \dfrac{\pi}{3} \cdot \dfrac{60 \text{ sec}}{1 \text{ min}} \cdot 3 \text{ min}$

$\quad = 60\pi$

$s = 30(60\pi)$

$\quad = 1800\pi$

$\quad \approx 5650$ ft

(c) $r = 30$ ft, $\omega = \dfrac{\pi}{3}$ radians per sec

$v = r\omega$

$\quad = 30\left(\dfrac{\pi}{3}\right)$

$\quad = 10\pi$

$\quad \approx 31.4$ ft per sec

48. $r = 13$ in, $\omega = 200(2\pi) = 400\pi$
radians per min

$v = r\omega$

$\quad = 13(400\pi)$

$\quad = 5200\pi$ in per min

$\quad = 312{,}000\pi$ in per hr

$\quad = 26{,}000\pi$ ft per hr

$\quad \approx 4.924\pi$ mph

$\quad \approx 15.5$ mph

CHAPTER 4   GRAPHS OF THE CIRCULAR
               FUNCTIONS

Section 4.1

5. $\cos \pi/8 = .92387953$

$\sin \pi/8 = .38268343$

7.   The highest point on the graph of
$y = \sin x$ is the point that is the
greatest distance above the horizontal or x-axis. This occurs when
$y = 1$. Since $x^2 + y^2 = 1$, the point
is $(0, 1)$.

For Exercises 9-27, see the answer graphs
in the back of the textbook. Some of the
data for plotting the graphs are given
here.

9. $y = 2 \cos x$

Amplitude: $|2| = 2$

x	0	$\frac{\pi}{2}$	π	$\frac{3\pi}{2}$	2π
2 cos x	2	0	-2	0	2

This table gives five values for
graphing one period of the function.
Repeat this cycle for the interval
$[-2\pi, 0]$.

11. $y = \frac{2}{3} \sin x$

Amplitude: $\left|\frac{2}{3}\right| = \frac{2}{3}$

x	0	$\frac{\pi}{2}$	π	$\frac{3\pi}{2}$	2π
$\frac{2}{3}$ sin x	0	.7	0	-.7	0

13. $y = -\cos x$

Amplitude: $|-1| = 1$

x	0	$\frac{\pi}{2}$	π	$\frac{3\pi}{2}$	2π
-cos x	-1	0	1	0	-1

15. $y = -2 \sin x$

Amplitude: $|-2| = 2$

x	0	$\frac{\pi}{2}$	π	$\frac{3\pi}{2}$	2π
-2 sin x	0	-2	0	2	0

17. $y = \sin \frac{1}{2}x$

Period: $\frac{2\pi}{\frac{1}{2}} = 4\pi$

Amplitude: $|1| = 1$

Divide the interval $[0, 4\pi]$ into
four equal parts to get x-values
that will yield minimum and maximum points and x-intercepts. Then
make a table.

x	0	π	2π	3π	4π
$\frac{1}{2}$x	0	$\frac{\pi}{2}$	π	$\frac{3\pi}{2}$	2π
sin $\frac{1}{2}$x	0	1	0	-1	0

19. $y = \cos \frac{1}{3}x$

Period: $\frac{2\pi}{\left(\frac{1}{3}\right)} = 6\pi$

Amplitude: 1

x	0	$\frac{3\pi}{4}$	3π	$\frac{9\pi}{2}$	6π
$\frac{1}{3}x$	0	$\frac{\pi}{2}$	π	$\frac{3\pi}{2}$	2π
$\cos \frac{1}{3}x$	1	0	-1	0	1

21. $y = \sin 3x$

Period: $\frac{2\pi}{3}$

Amplitude: $|1| = 1$

x	0	$\frac{\pi}{6}$	$\frac{\pi}{3}$	$\frac{\pi}{2}$	$\frac{2\pi}{3}$
$3x$	0	$\frac{\pi}{2}$	π	$\frac{3\pi}{2}$	2π
$\sin 3x$	0	1	0	-1	0

23. $y = 2 \sin \frac{1}{4}x$

Period: $\dfrac{2\pi}{\frac{1}{4}} = 8\pi$

Amplitude: 2

x	0	2π	4π	6π	8π
$\frac{1}{4}x$	0	$\frac{\pi}{2}$	π	$\frac{3\pi}{2}$	2π
$2 \sin \frac{1}{4}x$	0	2	0	-2	0

25. $y = -2 \cos 3x$

Period: $\frac{2\pi}{3}$

Amplitude: 2

x	0	$\frac{\pi}{6}$	$\frac{\pi}{3}$	$\frac{\pi}{2}$	$\frac{2\pi}{3}$
$3x$	0	$\frac{\pi}{2}$	π	$\frac{3\pi}{2}$	2π
$-2 \cos 3x$	-2	0	2	0	-2

27. $y = \cos \pi x$

Period: $\frac{2\pi}{\pi} = 2$

Amplitude: 1

x	0	$\frac{1}{2}$	1	$\frac{3}{2}$	2
$\cos \pi x$	1	0	-1	0	1

29. $y = \sin x$

The graph is a sinusoidal curve with an amplitude of 1 and a period of 2π. Since $\sin 0 = 0$, the point $(0, 0)$ is on the graph. This matches with graph G.

31. $y = -\sin x$

The graph is a sinusoidal curve with an amplitude of 1 and a period of 2π. Because $a = -1$, the graph is a reflection of $y = \sin x$ in the x-axis. This matches with graph E.

33. $y = \sin 2x$

The graph is a sinusoidal curve with a period of π and an amplitude of 1. Since $\sin 2(0) = 0$, the point $(0, 0)$ is on the graph. This matches with graph B.

35. $y = 2 \sin x$

The graph is a sinusoidal curve with a period of 2π and an amplitude of 2. Since $2 \sin 0 = 0$, the point $(0, 0)$ is on the graph. This matches with graph F.

37. The graph has an amplitude of 4 and a period of 4π. Then

$$y = a \sin bx$$

where $a = 4$ and $b = 1/2$, so

$$y = 4 \sin \frac{1}{2}x.$$

39. Y_1 is a graph of the form $y = a \cos bx$. The table shows an amplitude of 3 and a period of π. Therefore, $a = 3$ and $b = 2$, so

$$Y_1 = 3 \cos 2x.$$

41. The graph of $y = \sin 2x$ has an amplitude of 1 and a period of π. Over the interval $[0, 2\pi]$ there will be three complete cycles.
The graph of $y = 2 \sin x$ has an amplitude of 2 and a period of 2π. Over the interval $[0, 2\pi]$ there will be one complete cycle. Thus, $\sin 2x \neq 2 \sin x$, and, in general, $\sin bx \neq b \sin x$.

45. (a) The highest temperature is 80°; the lowest is 50°.

(b) The amplitude is

$$\frac{1}{2}(80° - 50°) = 15°.$$

(c) The period is about 35,000 yr.

(d) The trend of the temperature now is downward.

47. (a) The amplitude of the graph is $1/3$; the period is $3/2$. Since

$$\frac{2\pi}{k} = \frac{3}{2},$$

$$k = \frac{4\pi}{3}.$$

The equation is

$$y = \frac{1}{3} \sin \frac{4\pi t}{3}.$$

(b) It takes $3/2$ sec for a complete movement of the arm.

49. $-1 \leq y \leq 1$

Amplitude: 1
Period: 4 squares = 4(30°)
$$= 120° \text{ or } \frac{2\pi}{3}$$

51. $E = 3.8 \cos 40\pi t$

(a) Amplitude: 3.8
Period: $\frac{2\pi}{40\pi} = \frac{1}{20}$

(b) Frequency = number of cycles

per second $= \dfrac{1}{\text{period}} = 20$

(c) $t = .02$
$E = 3.8 \cos 40\pi(.02) \approx -3.074$

$t = .04$
$E = 3.8 \cos 40\pi(.04) \approx 1.174$

$t = .08$
$E = 3.8 \cos 40\pi(.08) \approx -3.074$
$t = .12$
$E = 3.8 \cos 40\pi(.12) \approx -3.074$
$t = .14$
$E = 3.8 \cos 40\pi(.14) \approx 1.174$

(d) See the answer graph in the back of the textbook.

53. **(a)** The graph of C has a general upward trend similar to L (in Exercise 52) except that both the carbon dioxide levels and the seasonal oscillations are larger for C than L. See the answer graph in the back of the textbook.

(b) The oscillations are caused by the seasonal changes in vegetation as they alternate between photosynthesis and respiration. Since the seasons are more dramatic in Alaska than in Hawaii, the oscillations are larger farther north. In the winter there is very little sunlight in Alaska, whereas there is almost continuous sunlight in the summer.

(c) To solve this problem, horizontally translate the graph of C a distance of 1970 units to the right. The new C function can be written

$$C(x) = .04(x - 1970)^2 + .6(x - 1970) + 330 + 7.5 \sin [2\pi(x - 1970)],$$

where x is the actual year. This function would now be valid for $1970 \le x \le 1995$.

Section 4.2

1. The minimum value of $y = \cos 4x$ is -1. The graph of $y = 2 + \cos 4x$ is the same as the graph of $y = \cos 4x$ with a vertical translation of 2 units up. Therefore, the minimum value is 1.

3. A phase shift of $3\pi/2$, or $3\pi/2$ units to the right, converts the graph of $y = \sin x$ to the graph of $y = \cos x$.

5. $y = \sin \left(x - \frac{\pi}{4}\right)$

The graph is a sinusoidal curve $y = \sin x$ shifted $\pi/4$ units to the right. This matches graph D.

7. $y = \cos \left(x - \frac{\pi}{4}\right)$

The graph is a sinusoidal curve $y = \cos x$ shifted $\pi/4$ units to the right. This matches graph H.

9. $y = 1 + \sin x$

The graph is a sinusoidal curve $y = \sin x$ translated vertically 1 unit up. This matches graph B.

11. $y = 1 + \cos x$

The graph is a sinusoidal curve $y = \cos x$ translated vertically 1 unit up. This matches graph F.

13. The function $y = -4 + \sin 4\,(x + \pi/2)$ has amplitude *1*, period $\pi/2$, phase shift $\pi/2$ units to the *left*, and has vertical translation 4 units *down*.

15. $y = 2 \sin (x - \pi)$

The amplitude is $|2|$, which is 2.
The period is $2\pi/1$, which is 2π.
There is no vertical translation.
The phase shift is π units to the right.

17. $y = 4 \cos \left(\dfrac{x}{2} + \dfrac{\pi}{2}\right)$

$y = 4 \cos \dfrac{1}{2}(x + \pi)$

The amplitude is $|4|$, which is 4.
The period is $\dfrac{2\pi}{\frac{1}{2}}$, which is 4π.

There is no vertical translation.
The phase shift is π units to the left.

19. $y = 3 \cos 2\left(x - \dfrac{\pi}{4}\right)$

The amplitude is $|3|$, which is 3.

The period is $2\pi/2$, which is π.
There is no vertical translation.
The phase shift is $\pi/4$ units to the right.

21. $y = 2 - \sin \left(3x - \dfrac{\pi}{5}\right)$

$y = -1 \sin 3\left(x - \dfrac{\pi}{15}\right) + 2$

The amplitude is $|-1|$, which is 1.
The period is $2\pi/3$.

The vertical translation is 2 units up. The phase shift is $\pi/15$ units to the right.

For Exercises 23–45, see the answer graphs in the back of the textbook.

23. $y = \cos \left(x - \dfrac{\pi}{2}\right)$

The amplitude is 1.
The period is 2π.
There is no vertical translation.
The phase shift is $\pi/2$ units to the right.

25. $y = \sin \left(x + \dfrac{\pi}{4}\right)$

The amplitude is 1.
The period is 2π.
There is no vertical translation.
The phase shift is $\pi/4$ units to the left.

27. $y = 2 \cos \left(x - \dfrac{\pi}{3}\right)$

The amplitude is 2.
The period is 2π.
There is no vertical translation.
The phase shift is $\pi/3$ units to the right.

29. $y = \dfrac{3}{2} \sin 2\left(x + \dfrac{\pi}{4}\right)$

The amplitude is 3/2.
The period is $2\pi/2$, which is π.
There is no vertical translation.
The phase shift is $\pi/4$ units to the left.

31. $y = -4 \sin (2x - \pi)$

$y = -4 \sin 2\left(x - \dfrac{\pi}{2}\right)$

The amplitude is $|-4|$, which is 4.
The period is $2\pi/2$, which is π.
There is no vertical translation.
The phase shift is $\pi/2$ units to the right.

33. $y = \dfrac{1}{2} \cos \left(\dfrac{1}{2}x - \dfrac{\pi}{4}\right)$

$y = \dfrac{1}{2} \cos \dfrac{1}{2}\left(x - \dfrac{\pi}{2}\right)$

The amplitude is $1/2$.

The period is $\dfrac{2\pi}{\dfrac{1}{2}}$, which is 4π.

This is no vertical translation.
The phase shift is $\pi/2$ units to the right.

35. $y = -3 + 2 \sin x$

The amplitude is 2.
The period is 2π.
The vertical translation is 3 units down.
There is no phase shift.

37. $y = 1 - \dfrac{2}{3} \sin \dfrac{3}{4}x$

The amplitude is $|-2/3|$, which is $2/3$.

The period is $\dfrac{2\pi}{\dfrac{3}{4}}$, which is $8\pi/3$.

The vertical translation is 1 unit up.
There is no phase shift.

39. $y = 1 - 2 \cos \dfrac{1}{2}x$

The amplitude is $|-2|$, which is 2.

The period is $\dfrac{2\pi}{\dfrac{1}{2}}$, which is 4π.

The vertical translation is 1 unit up.
There is no phase shift.

41. $y = -2 + \dfrac{1}{2} \sin 3x$

The amplitude is $1/2$.

The period is $2\pi/3$.
The vertical translation is 2 units down.
There is no phase shift.

43. $y = -3 + 2 \sin \left(x + \dfrac{\pi}{2}\right)$

The amplitude is 2.
The period is 2π.
The vertical translation is 3 units down.
The phase shift is $\pi/2$ units to the left.

45. $y = \dfrac{1}{2} + \sin 2\left(x + \dfrac{\pi}{4}\right)$

The amplitude is 1.
The period is $2\pi/2$, which is π.
The vertical translation is $1/2$ unit up.
The phase shift is $\pi/4$ units to the left.

49. The graph has an amplitude of 3, a period of π, and a phase shift of $\pi/4$ units to the right. In the equation

$$y = a \sin b(x - d),$$

$a = 3$, $b = 2$, and $d = \pi/4$; therefore,

$$y = 3 \sin 2\left(x - \frac{\pi}{4}\right).$$

(*Note*: There are other correct answers.)

51. The graph has an amplitude of 3, a period of 2, and a phase shift of $\pi/2$ units to the left. In the equation

$$y = a \cos b(x - d),$$

$a = 3$, $b = 2$, and $d = -\pi/2$; therefore,

$$y = 3 \cos 2\left(x + \frac{\pi}{2}\right).$$

(*Note*: There are other correct answers.)

53. From the table, the function Y_1 has an amplitude of 1, a period of 3π, a phase shift of $\pi/2$ units to the right, and a vertical translation of 2 up. The graph has been reflected about the x-axis. Therefore,

$$Y_1 = 2 - \sin\left[\frac{2}{3}\left(x - \frac{\pi}{2}\right)\right].$$

55. **(a)** See the answer graph in the back of the textbook. The amplitude is 17.5, the period is $\dfrac{2\pi}{\pi/6} = 12$ mo or 1 yr, the phase shift is 4, and the vertical translation is 67.5.

(b) The average temperature during the month of December is given by

$$f(12) = 17.5 \sin\left[\frac{\pi}{6}(8)\right] + 67.5$$
$$\approx 52°F,$$

which is equal to the actual value.

(c) The graph of f is periodic and parallels the seasons. The minimum temperature of

$$-17.5 + 67.5 = 50°F$$

occurs during January (at $x = 1$). The maximum temperature of

$$17.5 + 67.5 = 85°F$$

occurs during July (at $x = 7$). Both values are equal to the actual values.

(d) The graph is centered vertically about the line $y = 67.5$. The sine function in the formula for f is shifted up, that is, vertically translated, by an amount approximately equal to the average *yearly* temperature. From the graph it appears that the average yearly temperature is about 67.5°F. (The actual average yearly temperature in Austin is 68°F.)

57. (a) The maximum average monthly temperature is 90°F, and the minimum
average monthly temperature is 51°F.
The average of these two values is

$$\frac{90 + 51}{2} = 70.5°F,$$

which is very close to the actual
value of 70°F.

(b) See the answer graph in the back
of the textbook.

(c) Let the amplitude a be
$(90 - 51)/2 = 19.5$. Since the period is 12, let $b = \pi/6$. Let $c =$
$(90 + 51)/2 = 70.5$. The minimum
temperature occurs in January.
Thus, when $x = 1$, $b(x - d)$ must
equal an odd multiple of π since the
cosine function is minimum at these
values. Solving for d,

$$\frac{\pi}{6}(1 - d) = -\pi$$

$$d = 7.$$

d can be adjusted slightly to give
a better visual fit. Try $d = 7.2$.
Thus,

$f(x) = a \cos b(x - d) + c$

$$= 19.5 \cos \left[\frac{\pi}{6}(x - 7.2)\right] + 70.5.$$

(d) Plotting the data with

$$f(x) = 19.5 \cos \left[\frac{\pi}{6}(x - 7.2)\right] + 70.5$$

on the same coordinate axes gives a
good fit. See the answer graph in
the back of the textbook.

Section 4.3

3. $\csc x = \dfrac{1}{\sin x}$

Therefore,

$$3 \csc x = 3\left(\frac{1}{\sin x}\right) = \frac{3}{\sin x}.$$

The statement is false.

5. From the graph, $y = \tan x$ is symmetric about the origin. Therefore,
$\tan (-x) = -\tan x$ for all x in the
domain of tan x.
The statement is true.

7. $y = -\csc x$

The graph is the reflection of the
graph of $y = \csc x$ about the x-axis.
This matches with graph B.

9. $y = -\tan x$

The graph is the reflection of the
graph of $y = \tan x$ about the x-axis.
This matches with graph E.

11. $y = \tan \left(x - \frac{\pi}{4}\right)$

The graph is the graph of $y = \tan x$
shifted $\pi/4$ units to the right.
This matches with graph D.

For Exercises 13-41, see the answer graphs in the back of the textbook.

13. $y = \csc \left(x - \frac{\pi}{4} \right)$

We graph this function by first graphing the corresponding reciprocal function

$$y = \sin \left(x - \frac{\pi}{4} \right).$$

The period is 2π.

15. $y = \sec \left(x + \frac{\pi}{4} \right)$

We graph this function by first graphing the corresponding reciprocal function

$$y = \cos \left(x + \frac{\pi}{4} \right).$$

The period is 2π.

17. $y = \sec \left(\frac{1}{2}x + \frac{\pi}{3} \right)$

$y = \sec \frac{1}{2}\left(x + \frac{2\pi}{3} \right)$

We graph this function by first graphing the corresponding reciprocal function

$$y = \cos \frac{1}{2}\left(x + \frac{2\pi}{3} \right).$$

The period is 4π.

19. $y = 2 + 3 \sec \left(2x - \pi \right)$

$y = 2 + 3 \sec 2\left(x - \frac{\pi}{2} \right)$

We graph this function by first graphing the corresponding reciprocal function

$$y = 2 + 3 \cos 2\left(x - \frac{\pi}{2} \right).$$

The period is π.

21. $y = 1 - \frac{1}{2} \csc \left(x - \frac{3\pi}{4} \right)$

We graph this function by first graphing the corresponding reciprocal function

$$y = 1 - \frac{1}{2} \sin \left(x - \frac{3\pi}{4} \right).$$

The period is 2π.

23. $y = 2 \tan x$

Two adjacent vertical asymptotes are $x = -\pi/2$ and $x = \pi/2$. The graph is "stretched" because $a = 2$ and $|2| > 1$.

25. $y = \frac{1}{2} \cot x$

Two adjacent vertical asymptotes are $x = 0$ and $x = \pi$. The graph is "compressed" because $a = 1/2$ and $|1/2| < 1$.

27. $y = \cot 3x$

Two adjacent vertical asymptotes are

$$3x = 0 \quad \text{and} \quad 3x = \pi$$

or $\quad x = 0 \quad \text{and} \quad x = \frac{\pi}{3}.$

29. $y = \tan \left(2x - \pi \right)$

$y = \tan 2\left(x - \frac{\pi}{2} \right)$

The period is $\pi/2$.

Two adjacent vertical asymptotes are

$$2\left(x - \frac{\pi}{2} \right) = -\frac{\pi}{2} \quad \text{and} \quad 2\left(x - \frac{\pi}{2} \right) = \frac{\pi}{2}$$

or $\quad x - \frac{\pi}{2} = -\frac{\pi}{4} \quad \text{and} \quad x - \frac{\pi}{2} = \frac{\pi}{4}$

or $\quad x = \frac{\pi}{4} \quad \text{and} \quad x = \frac{3\pi}{4}.$

31. $y = \cot\left(3x + \frac{\pi}{4}\right)$

$y = \cot 3\left(x + \frac{\pi}{12}\right)$

The period is $\pi/3$.
Two adjacent vertical asymptotes are

$3\left(x + \frac{\pi}{12}\right) = 0$ and $3\left(x + \frac{\pi}{12}\right) = \pi$

or $x + \frac{\pi}{12} = 0$ and $x + \frac{\pi}{12} = \frac{\pi}{3}$

or $x = -\frac{\pi}{12}$ and $x = \frac{\pi}{4}$.

33. $y = 1 + \tan x$

This is the graph of $y = \tan x$ translated vertically 1 unit up.

35. $y = 1 - \cot x$

This is the graph of $y = \cot x$ reflected about the x–axis and then translated vertically 1 unit up.

37. $y = -1 + 2 \tan x$
This is the graph of $y = 2 \tan x$ translated vertically 1 unit down.

39. $y = -1 + \frac{1}{2} \cot (2x - 3\pi)$

$y = -1 + \frac{1}{2} \cot 2\left(x - \frac{3\pi}{2}\right)$

The period is $\pi/2$.
Two adjacent vertical asymptotes are

$2\left(x - \frac{3\pi}{2}\right) = 0$ and $2\left(x - \frac{3\pi}{2}\right) = \pi$

or $x - \frac{3\pi}{2} = 0$ and $x - \frac{3\pi}{2} = \frac{\pi}{2}$

or $x = \frac{3\pi}{2}$ and $x = 2\pi$.

This is the graph of $y = \frac{1}{2} \cot 2x$ translated vertically 1 unit down.

41. $y = \frac{2}{3} \tan \left(\frac{3}{4}x - \pi\right) - 2$

$y = -2 + \frac{2}{3} \tan \frac{3}{4}\left(x - \frac{4\pi}{3}\right)$

The period is $\dfrac{\pi}{\frac{3}{4}} = \dfrac{4\pi}{3}$.

Two adjacent vertical asymptotes are

$\frac{3}{4}\left(x - \frac{4\pi}{3}\right) = -\frac{\pi}{2}$ and $\frac{3}{4}\left(x - \frac{4\pi}{3}\right) = \frac{\pi}{2}$

or $x - \frac{4\pi}{3} = -\frac{2\pi}{3}$ and $x - \frac{4\pi}{3} = \frac{2\pi}{3}$

or $x = \frac{2\pi}{3}$ and $x = 2\pi$.

This is the graph of $y = \frac{2}{3} \tan \frac{3\pi}{4}$ translated vertically 2 units down.

43. The domain of the tangent function is

$\left\{x \mid x \neq \frac{\pi}{2} + n\pi, \text{ where } n \text{ is any integer}\right\},$

and the range is $(-\infty, \infty)$. For the function

$f(x) = -4 \tan (2x + \pi)$

$= -4 \tan 2\left(x + \frac{\pi}{2}\right),$

the period is $\pi/2$. Therefore, the domain is

$\left\{x \mid x \neq \frac{\pi}{4} + \frac{\pi}{2}n, \text{ where } n \text{ is any integer}\right\}.$

This can also be written as

$\left\{x \mid x \neq (2n + 1)\frac{\pi}{4}, \text{ where } n \text{ is any integer}\right\}.$

The range remains $(-\infty, \infty)$.

45. $d = \tan 4 \tan 2\pi t$

    (a) $d = \tan 2\pi(0) = 4 \tan 0 = 0$ m

    (b) $d = 4 \tan 2\pi(.4)$

        $= 4 \tan (.8\pi)$

        $= 4 \tan (2.5133)$

        $\approx 4(-.7265) \approx -2.9$ m

    (c) $d = 4 \tan 2\pi(.8) = 4 \tan 1.6\pi$

        $= 4 \tan 5.0265$

        $\approx 4(-3.0782) \approx -12.3$ m

    (d) $d = 4 \tan 2\pi(1.2)$

        $= 4 \tan (2.4\pi)$

        $= 4 \tan (7.5398)$

        $\approx 4(3.0774) \approx 12.3$ m

    (e) $t = .25$ leads to $\tan \pi/2$, which is undefined.

Chapter 4 Review Exercises

1. If x is an angle in quadrant I, then $f(x + \pi)$ is an angle in quadrant III. $-f(x)$ makes the function value negative. The trigonometric functions sin x, csc x, cos x, and sec x are negative in quadrant III.
If x is an angle in quadrant II, then $f(x + \pi)$ is an angle in quadrant IV. $-f(x)$ makes the function negative for sin x and csc x and positive for cos x and sec x, which is the same as in quadrant IV. Therefore, sin x, csc x, cos x, and sec x satisfy the condition.

3. The range for sin x and cos x is $[-1, 1]$. The range for tan x and cot x is $(-\infty, \infty)$. Since 1/2 falls in these intervals, those trigonometric functions can attain the value 1/2.

5. $y = 2 \sin x$

Amplitude: 2

Period: 2π

Vertical translation: none

Phase shift: none

7. $y = -\dfrac{1}{2} \cos 3x$

Amplitude: $\dfrac{1}{2}$

Period: $\dfrac{2\pi}{3}$

Vertical translation: none

Phase shift: none

9. $y = 1 + 2 \sin \dfrac{1}{4}x$

Amplitude: 2

Period: $\dfrac{2\pi}{\frac{1}{4}} = 8\pi$

Vertical translation: up 1 unit

Phase shift: none

11. $y = 3 \cos \left(x + \dfrac{\pi}{2}\right)$

Amplitude: 3

Period: 2π

Vertical translation: none

Phase shift: $\dfrac{\pi}{2}$ units to the left

13. $y = \frac{1}{2} \csc \left(2x - \frac{\pi}{4}\right)$

    Amplitude: not applicable

    Period: $\frac{2\pi}{2} = \pi$

    Vertical translation: none

    Phase shift: $\frac{\pi}{8}$ units to the right

15. $y = \frac{1}{3} \tan \left(3x - \frac{\pi}{3}\right)$

    $y = \frac{1}{3} \tan 3\left(x - \frac{\pi}{9}\right)$

    Amplitude: not applicable

    Period: $\frac{\pi}{3}$

    Vertical translation: none

    Phase shift: $\frac{\pi}{9}$ units to the right

17. The tangent function has a period of π and x–intercepts at integral multiples of π.

19. The cosine function has a period of 2π and has the value 0 when $x = \pi/2$.

21. The cotangent function has a period of π and decreases on the interval $0 < x < \pi$.

For Exercises 25–43, see the answer graphs in the back of the textbook.

25. $y = 3 \sin x$

    Period: 2π

    Amplitude: 3

27. $y = -\tan x$

    Period: π

    Asymptotes: $x = -\frac{\pi}{2}$, $x = \frac{\pi}{2}$

29. $y = 2 + \cot x$

    Period: π

    Vertical translation: up 2 units

    Asymptotes: $x = 0$, $x = \pi$

31. $y = \sin 2x$

    Period: $\frac{2\pi}{2} = \pi$

    Amplitude: 1

33. $y = 3 \cos 2x$

    Amplitude: 3

    Period: $\frac{2\pi}{2} = \pi$

35. $y = \cos \left(x - \frac{\pi}{4}\right)$

    Amplitude: 1

    Period: 2π

    Phase shift: $\frac{\pi}{4}$ units to the right

37. $y = \sec \left(2x + \frac{\pi}{3}\right)$

    $= \sec \left[2\left(x + \frac{\pi}{6}\right)\right]$

    Period: $\frac{2\pi}{2} = \pi$

    Phase shift: $\frac{\pi}{6}$ units to the left

39. $y = 1 + 2 \cos 3x$

Amplitude: 2

Period: $\dfrac{2\pi}{3}$

Vertical translation: 1 unit up

41. $y = 2 \sin \pi x$

Amplitude: 2

Period: 2

43. $y = 1 - 2 \sec \left(x - \dfrac{\pi}{4}\right)$

No amplitude

Period: $\dfrac{2\pi}{1} = 2$

Vertical translation: 1 unit up

Phase shift: $\dfrac{\pi}{4}$ units to the right

45. The shorter leg of the right triangle has length $h_2 - h_1$.

(a) $\cot \theta = \dfrac{d}{h_2 - h_1}$. So,

$$d = (h_2 - h_1) \cot \theta.$$

(b) When $h_2 = 55$ and $h_1 = 5$,

$$d = (55 - 5) \cot \theta = 50 \cot \theta.$$

The period is π, but the graph
wanted is d for $0 < \theta < \pi/2$.
The asymptote is the line $\theta = 0$.

When $\theta = \dfrac{\pi}{4}$,

$$d = 50(1) = 50.$$

See the answer graph in the back of
the textbook.

47. (a) From the graph one period is
about 20 yr.

(b) The population of hares fluc-
tuates between a minimum of about
5000 and a maximum of about 150,000.

49. (a) Let January correspond to x = 1,
February to x = 2, ... , and
December of the second year to
x = 24. The data appear to follow
the pattern of a translated sine
graph. See the answer graph in the
back of the textbook.

(b) The maximum average monthly temperature is 75°F, and the minimum is
25°F. Let the amplitude a be
$(75 - 25)/2 = 25$. Since the period
is 12, let $b = \pi/6$. The data are
centered vertically around the line
$y = (75 + 25)/2 = 50$, so let c = 50.
The minimum temperature occurs in
January. Thus, when x = 1,
$b(x - d) = -\pi/2$ since the sine function is minimum at $-\pi/2$. Solving
for d,

$$\frac{\pi}{6}(1 - d) = -\frac{\pi}{2}$$

$$d = 4.$$

d can be adjusted slightly to give a
better visual fit. Try d = 4.2.
Thus,

$$f(x) = a \sin b(x - d) + c$$

$$= 25 \sin \left[\frac{\pi}{6}(x - 4.2)\right] + 50.$$

(c) a is equal to the amplitude, which is equal to half the temperature variation between the maximum and minimum average monthly temperatures. b determines the oscillations in the graph of f and the period of the sine function. c, which represents the vertical translation of the sine graph, also represents the average *yearly* temperature. d is the phase shift or horizontal translation of the sine graph. One could think of d ≈ 4.2 as the first time when the average *monthly* temperature equals the average *yearly* temperature. In this example, the average yearly temperature is 50°F which occurs just after the April (x = 4) average of 48°F. Graphically the phase shift is equal to the x-coordinate where the graph of f crosses the line y = 50.

(d) Plotting the data with

$$f(x) = 25 \sin \left[\frac{\pi}{6}(x - 4.2) \right] + 50$$

on the same coordinate axes give a good fit. See the answer graph in the back of the textbook.

CHAPTER 4   TEST

[4.1]   1. Give the amplitude for $y = -\frac{2}{3} \cos x$.

Give the period for each function.

[4.2]   2. $y = 3 \sin 4x$ [4.3]   3. $y = \cot (2x - \pi)$

Give the vertical translation (if any) for each function.

[4.2]   4. $y = 4 + 2 \cos \left(\frac{1}{2}x - \pi\right)$ [4.3]   5. $y = \sec (3x - \pi)$

Give the phase shift (if any) for each function.

[4.2]   6. $y = -1 - \sin \left(x + \frac{\pi}{2}\right)$ [4.3]   7. $y = 3 + \csc (4x - \pi)$

[4.3]   8. Which circular function has a period of π and is increasing
            on the interval $-\pi/2 < x < \pi/2$?

Graph each function over a one-period interval.

[4.1]   9. $y = \frac{7}{4} \sin x$ [4.2]   10. $y = -2 - 2 \cos x$

[4.3] 11. $y = -\sec x$ [4.3]   12. $y = 2 \cot x$

[4.2] 13. $y = -\sin (x + \pi)$ [4.3]   14. $y = 1 + \tan \left(\frac{1}{2}x - \frac{\pi}{2}\right)$

[4.3] 15. $y = \csc (x - \pi)$ [4.2]   16. $y = 3 \cos \left(x - \frac{\pi}{2}\right) - 1$

CHAPTER 4 TEST ANSWERS

1. 2/3   **2.** $\pi/2$ **3.** $\pi/2$ **4.** 4 units up

5. No vertical translation   **6.** $\pi/2$ units to the left

7. $\pi/4$ units to the right   **8.** tangent

9.

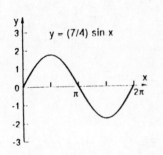

10.

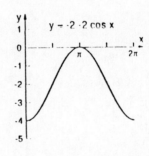

11.

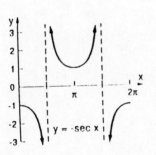

12.

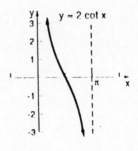

13.

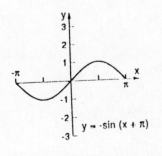

14.

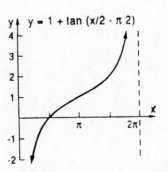

15.

16.

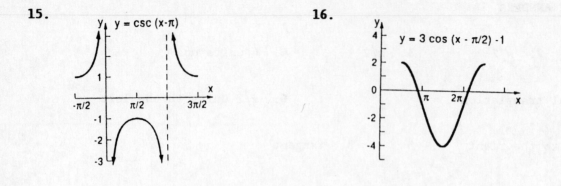

CHAPTER 5 TRIGONOMETRIC IDENTITIES

Section 5.1

1. By a negative-angle identity,

$$\tan(-X) = -\tan X.$$

From the graphing calculator screen, $\tan X = 2.6$, so

$$\tan(-X) = -\tan X = -2.6.$$

3. By a Pythagorean identity,

$$\tan^2 X + 1 = \sec^2 X.$$

From the graphing calculator screen, $\tan^2 X$ or $(\tan X)^2 = 1.5$, so

$$\tan^2 X + 1 = \sec^2 X$$
$$1.5 + 1 = \sec^2 X$$
$$2.5 = \sec^2 X.$$

By a reciprocal identity

$$\sec X = \frac{1}{\cos X},$$

so

$$2.5 = (\sec X)^2$$
$$2.5 = \left(\frac{1}{\cos X}\right)^2.$$

5. $\cos s = \frac{3}{4}$, s in quadrant I

$$\sin^2 s + \cos^2 s = 1$$
$$\sin^2 s + \left(\frac{3}{4}\right)^2 = 1$$
$$\sin^2 s = 1 - \frac{9}{16} = \frac{7}{16}$$
$$\sin s = \pm\frac{\sqrt{7}}{4}$$

Since s is in quadrant I,

$$\sin s = \frac{\sqrt{7}}{4}.$$

7. $\cos s = \frac{\sqrt{5}}{5}$, $\tan s < 0$ implies s is in quadrant IV.

$$\sin^2 s + \cos^2 s = 1$$
$$\sin^2 s + \left(\frac{\sqrt{5}}{5}\right)^2 = 1$$
$$\sin^2 s = 1 - \frac{5}{25} = \frac{20}{25}$$
$$= \frac{4}{5}$$
$$\sin s = \pm\frac{2}{\sqrt{5}}$$

Since s is in quadrant IV,

$$\sin s = -\frac{2}{\sqrt{5}}$$
$$= -\frac{2\sqrt{5}}{5}.$$

9. $\sec s = \frac{11}{4}$, $\tan s < 0$ implies s is in quadrant IV.

Since $\sec s = \frac{1}{\cos s}$,

$$\frac{11}{4} = \frac{1}{\cos s}$$
$$\frac{4}{11} = \cos s.$$

$$\sin^2 s + \cos^2 s = 1$$
$$\sin^2 s + \left(\frac{4}{11}\right)^2 = 1$$
$$\sin^2 s + \frac{16}{121} = 1$$
$$\sin s = \frac{105}{121}$$
$$\sin s = \pm\frac{\sqrt{105}}{11}$$

Since s is in quadrant IV,

$$\sin s = -\frac{\sqrt{105}}{11}.$$

11. Since $\csc s = \dfrac{1}{\sin s}$, we know that if csc s is positive, sin s is also positive. Also, if csc s is nega- tive, sin s is also negative. There- fore, a quadrant does not have to be stipulated, since the sine and cosecant have the same sign no matter which quadrant s is in.

13. Find $\tan \theta$ if $\cos \theta = -\dfrac{2}{5}$, and $\sin \theta < 0$.

$$\sin^2 \theta + \cos^2 \theta = 1$$

$$\sin^2 \theta + \left(-\frac{2}{5}\right)^2 = 1$$

$$\sin^2 \theta = 1 - \frac{4}{25} = \frac{21}{25}$$

$$\sin \theta = \pm\sqrt{\frac{21}{25}} = \pm\frac{\sqrt{21}}{5}$$

Since $\sin \theta < 0$,

$$\sin \theta = -\frac{\sqrt{21}}{5}.$$

$$\tan \theta = \frac{\sin \theta}{\cos \theta}$$

$$= \frac{-\dfrac{\sqrt{21}}{5}}{-\dfrac{2}{5}}$$

$$= \frac{\sqrt{21}}{2}$$

15. $\sin \theta = \dfrac{2}{3}$, θ in quadrant II

$$\sin^2 \theta + \cos^2 \theta = 1$$

$$\cos^2 \theta = 1 - \sin^2 \theta$$

$$= 1 - \left(\frac{2}{3}\right)^2 = \frac{5}{9}$$

$$\cos \theta = \pm\frac{\sqrt{5}}{3}$$

Since θ is in quadrant II, $\cos \theta = -\sqrt{5}/3$.

$$\tan \theta = \frac{\sin \theta}{\cos \theta} = \frac{\dfrac{2}{3}}{-\dfrac{\sqrt{5}}{3}}$$

$$= -\frac{2}{\sqrt{5}} = -\frac{2\sqrt{5}}{5}$$

$$\cot \theta = \frac{1}{\tan \theta}$$

$$= \frac{1}{-\dfrac{2}{\sqrt{5}}} = -\frac{\sqrt{5}}{2}$$

$$\sec \theta = \frac{1}{\cos \theta} = \frac{1}{-\dfrac{\sqrt{5}}{3}}$$

$$= -\frac{3}{\sqrt{5}} = -\frac{3\sqrt{5}}{5}$$

$$\csc \theta = \frac{1}{\sin \theta} = \frac{1}{\dfrac{2}{3}} = \frac{3}{2}$$

17. $\tan \theta = -\dfrac{1}{4}$, θ in quadrant IV

$$\sec^2 \theta = 1 + \tan^2 \theta$$

$$= 1 + \left(-\frac{1}{4}\right)^2$$

$$= \frac{17}{16}$$

$$\sec \theta = \pm\frac{\sqrt{17}}{4}$$

Since θ is in quadrant IV, $\sec \theta = \sqrt{17}/4$.

$$\cos \theta = \frac{1}{\sec \theta}$$

$$= \frac{1}{\dfrac{\sqrt{17}}{4}} = \frac{4}{\sqrt{17}} = \frac{4\sqrt{17}}{17}$$

$$\sin \theta = \tan \theta \cdot \cos \theta$$

$$= -\frac{1}{4} \cdot \frac{4}{\sqrt{17}} = -\frac{1}{\sqrt{17}} = -\frac{\sqrt{17}}{17}$$

$$\csc \theta = \frac{1}{\sin \theta} = \frac{1}{-\frac{1}{\sqrt{17}}} = -\sqrt{17}$$

$$\cot \theta = \frac{1}{\tan \theta} = \frac{1}{-\frac{1}{4}} = -4$$

19. $\cot \theta = \frac{4}{3}$, $\sin \theta > 0$

Since $\cot \theta > 0$ and $\sin \theta > 0$, θ is in quadrant I and all the functions are positive.

$$\tan \theta = \frac{1}{\cot \theta} = \frac{1}{\frac{4}{3}} = \frac{3}{4}$$

$$\sec^2 \theta = 1 + \tan^2 \theta$$
$$= 1 + \left(\frac{3}{4}\right)^2 = \frac{25}{16}$$

$$\sec \theta = \frac{5}{4}$$

$$\cos \theta = \frac{1}{\sec \theta} = \frac{1}{\frac{5}{4}} = \frac{4}{5}$$

$$\sin^2 \theta + \cos^2 \theta = 1$$
$$\sin^2 \theta = 1 - \cos^2 \theta$$
$$= 1 - \left(\frac{4}{5}\right)^2 = \frac{9}{25}$$

$$\sin \theta = \frac{3}{5}$$

$$\csc \theta = \frac{1}{\sin \theta} = \frac{5}{3}$$

21. $\sec \theta = \frac{4}{3}$, $\sin \theta < 0$

Since $\sec \theta > 0$ and $\sin \theta < 0$, θ is in quadrant IV.

$$\cos \theta = \frac{1}{\sec \theta}$$
$$= \frac{1}{\frac{4}{3}} = \frac{3}{4}$$

$$\sin^2 \theta + \cos^2 \theta = 1$$
$$\sin^2 \theta = 1 - \cos^2 \theta$$
$$= 1 - \left(\frac{3}{4}\right)^2 = \frac{7}{16}$$

Since $\sin \theta < 0$,

$$\sin \theta = -\frac{\sqrt{7}}{4}.$$

$$\tan \theta = \frac{\sin \theta}{\cos \theta} = \frac{-\frac{\sqrt{7}}{4}}{\frac{3}{4}} = -\frac{\sqrt{7}}{3}$$

$$\cot \theta = \frac{1}{\tan \theta} = -\frac{1}{\frac{\sqrt{7}}{3}} = -\frac{3\sqrt{7}}{7}$$

$$\csc \theta = \frac{1}{\sin \theta} = \frac{1}{-\frac{\sqrt{7}}{4}} = -\frac{4\sqrt{7}}{7}$$

23. $\frac{\cos x}{\sin x} = \cot x$ **(b)**

25. $\cos (-x) = \cos x$ **(e)**

27. $1 = \sin^2 x + \cos^2 x$ **(a)**

29. $\sec^2 x - 1 = \tan^2 x = \frac{\sin^2 x}{\cos^2 x}$ **(a)**

31. $1 + \sin^2 x$
$$= \csc^2 x - \cot^2 x + \sin^2 x$$ **(d)**

33. It is incorrect to write $1 + \cot^2 = \csc^2$ because there is no angle given as argument to the functions cotangent and cosecant. You need to take the cotangent of something for computational purposes. A correct way to write this statement is $1 + \cot^2 \theta = \csc^2 \theta$.

35. Find $\sin \theta$ if $\cos \theta = x/(x + 1)$.

$$\cos \theta = \frac{x}{x + 1}$$

Since $\sin^2 \theta + \cos^2 \theta = 1$,

$$\sin^2 \theta = 1 - \cos^2 \theta$$

$$\sin^2 \theta = 1 - \left(\frac{x}{x + 1}\right)^2$$

$$= \frac{(x + 1)^2 - x^2}{(x + 1)^2}$$

$$= \frac{x^2 + 2x + 1 - x^2}{(x + 1)^2}$$

$$= \frac{2x + 1}{(x + 1)^2}$$

$$\sin \theta = \frac{\pm\sqrt{2x + 1}}{x + 1}.$$

37. $\cot \theta \sin \theta = \dfrac{\cos \theta}{\sin \theta} \cdot \sin \theta$

$$= \cos \theta$$

39. $\cos \theta \csc \theta = \cos \theta \cdot \dfrac{1}{\sin \theta}$

$$= \frac{\cos \theta}{\sin \theta}$$

$$= \cot \theta$$

41. $\sin^2 \theta (\csc^2 \theta - 1)$

$$= \sin^2 \theta \left(\frac{1}{\sin^2 \theta} - 1\right)$$

$$= \frac{\sin^2 \theta}{\sin^2 \theta} - \sin^2 \theta$$

$$= 1 - \sin^2 \theta$$

$$= \cos^2 \theta$$

43. $(1 - \cos \theta)(1 + \sec \theta)$

$$= 1 + \sec \theta - \cos \theta - \cos \theta \sec \theta$$

$$= 1 + \sec \theta - \cos \theta - \cos \theta \left(\frac{1}{\cos \theta}\right)$$

$$= 1 + \sec \theta - \cos \theta - 1$$

$$= \sec \theta - \cos \theta$$

45. $\dfrac{\cos^2 \theta - \sin^2 \theta}{\sin \theta \cos \theta}$

$$= \frac{\cos^2 \theta}{\sin \theta \cos \theta} - \frac{\sin^2 \theta}{\sin \theta \cos \theta}$$

$$= \frac{\cos \theta}{\sin \theta} - \frac{\sin \theta}{\cos \theta}$$

$$= \cot \theta - \tan \theta$$

47. $\tan \theta + \cot \theta$

$$= \frac{\sin \theta}{\cos \theta} + \frac{\cos \theta}{\sin \theta}$$

$$= \frac{\sin^2 \theta + \cos^2 \theta}{\cos \theta \sin \theta}$$

$$= \frac{1}{\cos \theta \sin \theta}$$

$$= \frac{1}{\cos \theta} \cdot \frac{1}{\sin \theta}$$

$$= \sec \theta \csc \theta$$

49. $\sin \theta (\csc \theta - \sin \theta)$

$$= \sin \theta \csc \theta - \sin^2 \theta$$

$$= 1 - \sin^2 \theta$$

$$= \cos^2 \theta$$

51. $\sin^2 \theta + \tan^2 \theta + \cos^2 \theta$

$$= (\sin^2 \theta + \cos^2 \theta) + \tan^2 \theta$$

$$= 1 + \tan^2 \theta$$

$$= \sec^2 \theta$$

53. Express $\sin \theta$ in terms of $\cot \theta$ and in terms of $\sec \theta$.

$$\sin \theta = \frac{1}{\csc \theta} = \frac{1}{\pm\sqrt{1 + \cot^2 \theta}}$$

$$= \frac{\pm\sqrt{1 + \cot^2 \theta}}{1 + \cot^2 \theta}$$

$$\sin \theta = \cos \theta \tan \theta$$

$$= \frac{1}{\sec \theta} \cdot \pm\sqrt{\sec^2 \theta - 1}$$

$$= \frac{\pm\sqrt{\sec^2 \theta - 1}}{\sec \theta}$$

55. Express $\tan \theta$ in terms of $\sin \theta$, $\cos \theta$, $\sec \theta$, and $\csc \theta$.

$$\tan \theta = \frac{\sin \theta}{\cos \theta}$$

$$= \frac{\sin \theta}{\pm\sqrt{1 - \sin^2 \theta}}$$

$$= \frac{\pm\sin \theta}{\sqrt{1 - \sin^2 \theta}}$$

$$= \frac{\pm\sin \theta \sqrt{1 - \sin^2 \theta}}{1 - \sin^2 \theta}$$

$$\tan \theta = \frac{\sin \theta}{\cos \theta}$$

$$= \frac{\pm\sqrt{1 - \cos^2 \theta}}{\cos \theta}$$

$$\tan \theta = \pm\sqrt{\sec^2 \theta - 1}$$

$$\tan \theta = \frac{1}{\cot \theta}$$

$$= \frac{1}{\pm\sqrt{\csc^2 \theta - 1}}$$

$$= \frac{\pm\sqrt{\csc^2 \theta - 1}}{\csc^2 \theta - 1}$$

57. Express $\sec \theta$ in terms of $\sin \theta$, $\tan \theta$, $\cot \theta$, and $\csc \theta$.

$$\sec \theta = \frac{1}{\cos \theta}$$

$$= \frac{1}{\pm\sqrt{1 - \sin^2 \theta}}$$

$$= \frac{\pm\sqrt{1 - \sin^2 \theta}}{1 - \sin^2 \theta}$$

$$\sec \theta = \pm\sqrt{\tan^2 \theta + 1}$$

Since $\tan \theta = \dfrac{1}{\cot \theta}$,

$$\sec \theta = \pm\sqrt{\tan^2 \theta + 1}$$

$$= \pm\sqrt{\left(\frac{1}{\cot^2 \theta}\right) + 1}$$

$$= \frac{\pm\sqrt{1 + \cot^2 \theta}}{\cot \theta}.$$

$$\sec \theta = \frac{1}{\cos \theta}$$

$$= \frac{1}{\pm\sqrt{1 - \sin^2 \theta}}$$

$$= \frac{1}{\pm\sqrt{1 - \dfrac{1}{\csc^2 \theta}}}$$

$$= \frac{\sqrt{\csc^2 \theta}}{\pm\sqrt{\csc^2 \theta - 1}}$$

$$= \frac{\pm\csc \theta \sqrt{\csc^2 \theta - 1}}{\csc^2 \theta - 1}$$

59. Let $\cos x = \dfrac{1}{5}$. Then x is in quadrant I or quadrant IV. Then,

$$\sin x = \pm\sqrt{1 - \cos^2 x}$$

$$= \pm\sqrt{1 - \left(\frac{1}{5}\right)^2}$$

$$= \pm\sqrt{\frac{24}{25}} = \pm\frac{2\sqrt{6}}{5}.$$

$$\tan x = \frac{\sin x}{\cos x} = \frac{\pm\dfrac{2\sqrt{6}}{5}}{\dfrac{1}{5}} = \pm 2\sqrt{6}$$

$$\sec x = \frac{1}{\cos x} = 5$$

If x is in quadrant I,

$$\frac{\sec x - \tan x}{\sin x} = \frac{5 - 2\sqrt{6}}{\dfrac{2\sqrt{6}}{5}}$$

$$= \frac{25 - 10\sqrt{6}}{2\sqrt{6}}$$

$$= \frac{25\sqrt{6} - 60}{12}.$$

If x is in quadrant IV,

$$\frac{\sec x - \tan x}{\sin x} = \frac{5 - (-2\sqrt{6})}{-\frac{2\sqrt{6}}{5}}$$

$$= \frac{25 + 10\sqrt{6}}{-2\sqrt{6}}$$

$$= \frac{-25\sqrt{6} - 60}{12}.$$

61. Odd functions have graphs that are symmetric with respect to the origin, while even functions have graphs that are symmetric with respect to the y-axis.

The graphs of cot x and csc x are symmetric with respect to the origin. Thus, cot x and csc x are odd functions.

The graph of sec x is symmetric with respect to the y-axis. Thus, sec x is an even function.

63. $y = \sin(-2x)$

$y = -\sin(2x)$

64. It is the negative of $\sin(2x)$.

65. $y = \cos(-4x)$

$y = \cos(4x)$

66. It is the same function.

67. (a) $y = \sin(-4x)$

$y = -\sin 4x$

(b) $y = \cos(-2x)$

$y = \cos(2x)$

(c) $y = -5 \sin(-3x)$

$y = -5[-\sin(3x)]$

$y = 5 \sin(3x)$

69. $y = \csc(-x)$

$= \frac{1}{\sin(-x)} = \frac{1}{-\sin x} = -\csc x$

Since $\csc(-x) = -\csc x$, the graph of $y = \csc(-x)$ is a reflection across the x-axis as compared to the graph of $y = \csc x$.

71. $y = \cot(-x)$

$= \frac{\cos(-x)}{\sin(-x)} = \frac{\cos x}{-\sin x} = -\cot x$

Since $\cot(-x) = -\cot x$, the graph of $y = \cot(-x)$ is a reflection across the x-axis as compared to the graph of $y = \cot x$.

73. Is $\sin x = \sqrt{1 - \cos^2 x}$ an identity?

Let $x = \frac{7\pi}{6}$.

$$\sin \frac{7\pi}{6} = -\frac{1}{2}$$

$$\sqrt{1 - \cos^2 \frac{7\pi}{6}} = \sqrt{1 - \left(-\frac{\sqrt{3}}{2}\right)^2}$$

$$= \sqrt{1 - \frac{3}{4}}$$

$$= \sqrt{\frac{1}{4}}$$

$$= \frac{1}{2}$$

Since $-\frac{1}{2} \neq \frac{1}{2}$, this is not an identity.

Section 5.2

1. $\tan \theta + \dfrac{1}{\tan \theta}$

    $= \dfrac{\sin \theta}{\cos \theta} + \dfrac{\cos \theta}{\sin \theta} = \dfrac{\sin^2 \theta + \cos^2 \theta}{\sin \theta \cos \theta}$

    $= \dfrac{1}{\sin \theta \cos \theta}$

    or $\csc \theta \sec \theta$

3. $\cot s (\tan s + \sin s)$

    $= \dfrac{\cos s}{\sin s} \left(\dfrac{\sin s}{\cos s} + \sin s\right)$

    $= 1 + \cos s$

5. $\dfrac{1}{\csc^2 \theta} + \dfrac{1}{\sec^2 \theta}$

    $= \sin^2 \theta + \cos^2 \theta$

    $= 1$

7. $\dfrac{\cos x}{\sec x} + \dfrac{\sin x}{\csc x}$

    $= \cos x (\cos x) + \sin x (\sin x)$

    $= \cos^2 x + \sin^2 x$

    $= 1$

9. $(1 + \sin t)^2 + \cos^2 t$

    $= 1 + 2 \sin t + \sin^2 t + \cos^2 t$

    $= 2 + 2 \sin t$

11. $\dfrac{1}{1 + \cos x} - \dfrac{1}{1 - \cos x}$

    $= \dfrac{1 - \cos x - 1 - \cos x}{1 - \cos^2 x}$

    $= \dfrac{-2 \cos x}{\sin^2 x}$

    or $-2 \left(\dfrac{\cos x}{\sin x}\right) \left(\dfrac{1}{\sin x}\right)$

    $= -2 \cot x \csc x$

13. $\sin^2 \gamma - 1$

    $= (\sin \gamma - 1)(\sin \gamma + 1)$

15. $(\sin x + 1)^2 - (\sin x - 1)^2$

    Let $a = \sin x + 1$ and $b = \sin x - 1$.

    $(\sin x + 1)^2 - (\sin x - 1)^2$

    $= a^2 - b^2 = (a - b)(a + b)$

    $= [(\sin x + 1) - (\sin x - 1)] \cdot$

    $\quad [(\sin x + 1) + (\sin x - 1)]$

    $= 2(2 \sin x)$

    $= 4 \sin x$

17. $2 \sin^2 x + 3 \sin x + 1$

    Let $a = \sin x$.

    $2 \sin^2 x + 3 \sin x + 1$

    $= 2a^2 + 3a + 1$

    $= (2a + 1)(a + 1)$

    $= (2 \sin x + 1)(\sin x + 1)$

19. $\cos^4 x + 2 \cos^2 x + 1$

    Let $\cos^2 x = a$.

    $\cos^4 x + 2 \cos^2 x + 1$

    $= a^2 + 2a + 1 = (a + 1)^2$

    $= (\cos^2 x + 1)^2$

21. $\sin^3 x - \cos^3 x$

    Let $a = \sin x, b = \cos x$.

    $\sin^3 x - \cos^3 x$

    $= a^3 - b^3$

    $= (a - b)(a^2 + ab + b^2)$

    $= (\sin x - \cos x) \cdot$

    $\quad (\sin^2 x + \sin x \cos x + \cos^2 x)$

    $= (\sin x - \cos x)(1 + \sin x \cos x)$

23. $\tan \theta \cos \theta = \dfrac{\sin \theta}{\cos \theta} \cos \theta = \sin \theta$

25. $\sec r \cos r = \dfrac{1}{\cos r} \cos r = 1$

27. $\dfrac{\sin \beta \tan \beta}{\cos \beta} = \tan \beta \tan \beta = \tan^2 \beta$

29. $\sec^2 x - 1 = \dfrac{1}{\cos^2 x} - 1$

$$= \dfrac{1 - \cos^2 x}{\cos^2 x}$$

$$= \dfrac{\sin^2 x}{\cos^2 x} = \tan^2 x$$

31. $\dfrac{\sin^2 x}{\cos^2 x} + \sin x \csc x$

$$= \tan^2 x + \sin x \dfrac{1}{\sin x}$$

$$= \tan^2 x + 1$$

$$= \sec^2 x$$

33. Verify $\dfrac{\cot \theta}{\csc \theta} = \cos \theta$.

$$\dfrac{\cot \theta}{\csc \theta} = \dfrac{\dfrac{\cos \theta}{\sin \theta}}{\dfrac{1}{\sin \theta}} = \cos \theta$$

35. Verify $\dfrac{1 - \sin^2 \beta}{\cos \beta} = \cos \beta$.

$$\dfrac{1 - \sin^2 \beta}{\cos \beta} = \dfrac{\cos^2 \beta}{\cos \beta} = \cos \beta$$

37. Verify $\cos^2 \theta \, (\tan^2 \theta + 1) = 1$.

$$\cos^2 \theta \, (\tan^2 \theta + 1)$$

$$= \cos^2 \theta \left(\dfrac{\sin^2 \theta}{\cos^2 \theta} + 1 \right)$$

$$= \cos^2 \theta \left(\dfrac{\sin^2 \theta + \cos^2 \theta}{\cos^2 \theta} \right)$$

$$= 1$$

39. Verify $\cot s + \tan s = \sec s \csc s$.

$$\cot s + \tan s = \dfrac{\cos s}{\sin s} + \dfrac{\sin s}{\cos s}$$

$$= \dfrac{\cos^2 s + \sin^2 s}{\sin s \cos s}$$

$$= \dfrac{1}{\sin s \cos s}$$

$$= \sec s \csc s$$

41. Verify $\dfrac{\cos \alpha}{\sec \alpha} + \dfrac{\sin \alpha}{\csc \alpha} = \sec^2 \alpha - \tan^2 \alpha$.

Work with the left side.

$$\dfrac{\cos \alpha}{\sec \alpha} + \dfrac{\sin \alpha}{\csc \alpha} = \dfrac{\cos \alpha}{\dfrac{1}{\cos \alpha}} + \dfrac{\sin \alpha}{\dfrac{1}{\sin \alpha}}$$

$$= \cos^2 \alpha + \sin^2 \alpha$$

$$= 1$$

Now work with the right side.

$$\sec^2 \alpha - \tan^2 \alpha = \sec^2 \alpha - (\sec^2 \alpha - 1)$$

$$= 1$$

43. Verify $\sin^4 \theta - \cos^4 \theta = 2 \sin^2 \theta - 1$.

$$\sin^4 \theta - \cos^4 \theta$$

$$= (\sin^2 \theta - \cos^2 \theta)(\sin^2 \theta + \cos^2 \theta)$$

$$= \sin^2 \theta - \cos^2 \theta$$

$$= \sin^2 \theta - (1 - \sin^2 \theta)$$

$$= 2 \sin^2 \theta - 1$$

45. Verify $(1 - \cos^2 \alpha)(1 + \cos^2 \alpha) = 2 \sin^2 \alpha - \sin^4 \alpha$.

$$(1 - \cos^2 \alpha)(1 + \cos^2 \alpha)$$

$$= \sin^2 \alpha \, [1 + (1 - \sin^2 \alpha)]$$

$$= \sin^2 \alpha \, (2 - \sin^2 \alpha)$$

$$= 2 \sin^2 \alpha - \sin^4 \alpha$$

47. Verify $\dfrac{\cos \theta + 1}{\tan^2 \theta} = \dfrac{\cos \theta}{\sec \theta - 1}$.

Work with the left side.

$$\dfrac{\cos \theta + 1}{\tan^2 \theta} = \dfrac{\cos \theta + 1}{\sec^2 \theta - 1}$$

$$= \dfrac{\cos \theta + 1}{\dfrac{1}{\cos^2 \theta} - 1}$$

$$= \dfrac{\cos \theta + 1}{\dfrac{1 - \cos^2 \theta}{\cos^2 \theta}}$$

$$= \dfrac{\cos^2 \theta \, (\cos \theta + 1)}{(1 - \cos \theta)(1 + \cos \theta)}$$

$$= \dfrac{\cos^2 \theta}{1 - \cos \theta}$$

Now work with the right side.

$$\frac{\cos \theta}{\sec \theta - 1} = \frac{\cos \theta}{\dfrac{1}{\cos \theta} - 1} = \frac{\cos \theta}{\dfrac{1 - \cos \theta}{\cos \theta}} = \frac{\cos^2 \theta}{1 - \cos \theta}$$

49. Verify $\dfrac{1}{1 - \sin \theta} + \dfrac{1}{1 + \sin \theta} = 2 \sec^2 \theta$.

$$\frac{1}{1 - \sin \theta} + \frac{1}{1 + \sin \theta} = \frac{(1 + \sin \theta) + (1 - \sin \theta)}{(1 - \sin \theta)(1 + \sin \theta)} = \frac{2}{1 - \sin^2 \theta}$$

$$= \frac{2}{\cos^2 \theta} = 2 \sec^2 \theta$$

51. Verify $\dfrac{\tan s}{1 + \cos s} + \dfrac{\sin s}{1 - \cos s} = \cot s + \sec s \csc s$.

$$\frac{\tan s}{1 + \cos s} + \frac{\sin s}{1 - \cos s} = \frac{\tan s (1 - \cos s) + \sin s (1 + \cos s)}{1 - \cos^2 s}$$

$$= \frac{\tan s - \sin s + \sin s + \sin s \cos s}{\sin^2 s} = \frac{\tan s}{\sin^2 s} + \frac{\cos s}{\sin s}$$

$$= \frac{\sin s}{\cos s} \cdot \frac{1}{\sin^2 s} + \cot s = \frac{1}{\cos s} \cdot \frac{1}{\sin s} + \cot s$$

$$= \sec s \csc s + \cot s$$

53. Verify $\dfrac{\cot \alpha + 1}{\cot \alpha - 1} = \dfrac{1 + \tan \alpha}{1 - \tan \alpha}$.

$$\frac{\cot \alpha + 1}{\cot \alpha - 1} = \frac{\dfrac{1}{\tan \alpha} + 1}{\dfrac{1}{\tan \alpha} - 1} = \frac{\dfrac{1 + \tan \alpha}{\tan \alpha}}{\dfrac{1 - \tan \alpha}{\tan \alpha}} = \frac{1 + \tan \alpha}{1 - \tan \alpha}$$

55. Verify $\sin^2 \alpha \sec^2 \alpha + \sin^2 \alpha \csc^2 \alpha = \sec^2 \alpha$.

$$\sin^2 \alpha \sec^2 \alpha + \sin^2 \alpha \csc^2 \alpha = \frac{\sin^2 \alpha}{\cos^2 \alpha} + 1 = \tan^2 \alpha + 1 = \sec^2 \alpha$$

57. Verify $\sec^4 x - \sec^2 x = \tan^4 x + \tan^2 x$.

Simplify left side.

$$\sec^4 x - \sec^2 x = \sec^2 x(\sec^2 x - 1) = \sec^2 x \tan^2 x$$

Simplify right side.

$$\tan^4 x + \tan^2 x = \tan^2 x(\tan^2 x + 1) = \tan^2 x \sec^2 x$$

59. Verify $\sin \theta + \cos \theta = \dfrac{\sin \theta}{1 - \dfrac{\cos \theta}{\sin \theta}} + \dfrac{\cos \theta}{1 - \dfrac{\sin \theta}{\cos \theta}}$.

Work with the right side.

$$\frac{\sin \theta}{1 - \dfrac{\cos \theta}{\sin \theta}} + \frac{\cos \theta}{1 - \dfrac{\sin \theta}{\cos \theta}} = \frac{\sin^2 \theta}{\sin \theta - \cos \theta} + \frac{\cos^2 \theta}{\cos \theta - \sin \theta}$$

$$= \frac{\sin^2 \theta}{\sin \theta - \cos \theta} + \frac{\cos^2 \theta}{-(\sin \theta - \cos \theta)}$$

$$= \frac{\sin^2 \theta - \cos^2 \theta}{\sin \theta - \cos \theta}$$

$$= \frac{(\sin \theta + \cos \theta)(\sin \theta - \cos \theta)}{(\sin \theta - \cos \theta)}$$

$$= \sin \theta + \cos \theta$$

61. Verify $\dfrac{\sec^4 s - \tan^4 s}{\sec^2 s + \tan^2 s} = \sec^2 s - \tan^2 s$.

$$\frac{\sec^4 s - \tan^4 s}{\sec^2 s + \tan^2 s} = \frac{(\sec^2 s - \tan^2 s)(\sec^2 s + \tan^2 s)}{\sec^2 s + \tan^2 s} = \sec^2 s - \tan^2 s$$

63. Verify $\dfrac{\tan^2 t - 1}{\sec^2 t} = \dfrac{\tan t - \cot t}{\tan t + \cot t}$.

Work with the right side.

$$\frac{\tan t - \cot t}{\tan t + \cot t} = \frac{\tan t - \dfrac{1}{\tan t}}{\tan t + \dfrac{1}{\tan t}} = \frac{\dfrac{\tan^2 t - 1}{\tan t}}{\dfrac{\tan^2 t + 1}{\tan t}} = \frac{\tan^2 t - 1}{\tan^2 t + 1} = \frac{\tan^2 t - 1}{\sec^2 t}$$

65. Verify $\dfrac{1 + \cos x}{1 - \cos x} - \dfrac{1 - \cos x}{1 + \cos x} = 4 \cot x \csc x$.

$$\frac{1 + \cos x}{1 - \cos x} - \frac{1 - \cos x}{1 + \cos x} = \frac{1 + \cos^2 x + 2 \cos x - 1 - \cos^2 x + 2 \cos x}{1 - \cos^2 x}$$

$$= \frac{4 \cos x}{\sin^2 x} = 4 \cot x \csc x$$

67. Verify $(\sec \alpha + \csc \alpha)(\cos \alpha - \sin \alpha) = \cot \alpha - \tan \alpha$.

$$(\sec \alpha + \csc \alpha)(\cos \alpha - \sin \alpha) = \left(\frac{1}{\cos \alpha} + \frac{1}{\sin \alpha}\right)(\cos \alpha - \sin \alpha)$$

$$= 1 + \cot \alpha - \tan \alpha - 1 = \cot \alpha - \tan \alpha$$

69. $\cos \theta + \sin \theta = 1$ is true when $\theta = 90°$. However, an identity is a statement that is true not just for a particular value of θ, but for all values of θ. Notice if $\theta = 180°$, $\cos 180° + \sin 180° = -1 + 0 = -1 \neq 1$. Since the statement is not true for all values of θ, it is not an identity.

71.

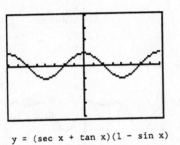

$$y = (\sec x + \tan x)(1 - \sin x)$$

The calculator graph of

$$y = (\sec x + \tan x)(1 - \sin x)$$

appears to be the same as the graph of $y = \cos x$.

$(\sec \theta + \tan \theta)(1 - \sin \theta)$

$= \left(\dfrac{1}{\cos \theta} + \dfrac{\sin \theta}{\cos \theta}\right)(1 - \sin \theta)$

$= \dfrac{(1 + \sin \theta)(1 - \sin \theta)}{\cos \theta}$

$= \dfrac{1 - \sin^2 \theta}{\cos \theta}$

$= \dfrac{\cos^2 \theta}{\cos \theta}$

$= \cos \theta$

73.

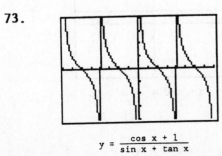

$$y = \dfrac{\cos x + 1}{\sin x + \tan x}$$

The calculator graph of

$$y = \frac{\cos x + 1}{\sin x + \tan x}$$

appears to be the same as the graph of $y = \cot x$.

$\dfrac{\cos \theta + 1}{\sin \theta + \tan \theta}$

$= \dfrac{1 + \cos \theta}{\sin \theta + \dfrac{\sin \theta}{\cos \theta}}$

$= \dfrac{1 + \cos \theta}{\sin \theta \left(1 + \dfrac{1}{\cos \theta}\right)}$

$= \dfrac{(1 + \cos \theta)\cos \theta}{\sin \theta \left(1 + \dfrac{1}{\cos \theta}\right)\cos \theta}$

$= \dfrac{(1 + \cos \theta)\cos \theta}{\sin \theta (\cos \theta + 1)}$

$= \dfrac{\cos \theta}{\sin \theta}$

$= \cot \theta$

75. Is $\dfrac{2 + 5\cos s}{\sin s} = 2\csc s + 5\cot s$ an identity?

Graph each side of the equation separately on a calculator.

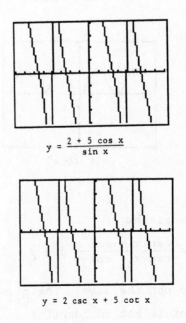

$$y = \dfrac{2 + 5\cos x}{\sin x}$$

$$y = 2\csc x + 5\cot x$$

The graphs of

$$y = \frac{2 + 5 \cos x}{\sin x}$$

and $y = 2 \csc x + 5 \cot x$

appear to be the same. The given
equation may be an identity. Verify
it.

$$\frac{2 + 5 \cos s}{\sin s} = \frac{2}{\sin s} + \frac{5 \cos s}{\sin s}$$

$$= 2 \csc s + 5 \cot s$$

The given statement is an identity.

77. Is $\frac{\tan s - \cot s}{\tan s + \cot s} = 2 \sin^2 s$ an

identity?
Graph each side of the equation
separately on a calculator.

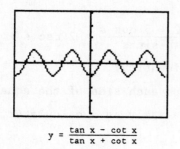

$$y = \frac{\tan x - \cot x}{\tan x + \cot x}$$

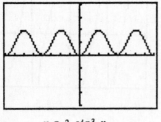

$$y = 2 \sin^2 x$$

The graphs of

$y = \frac{\tan x - \cot x}{\tan x + \cot x}$ and $y = 2 \sin^2 x$

are not the same. The given statement is not an identity.

79. Is $\frac{1 - \tan^2 s}{1 + \tan^2 s} = \cos^2 s - \sin s$ an

identity?
Graph each side of the equation
separately.

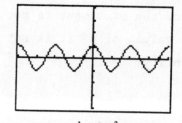

$$y = \frac{1 - \tan^2 x}{1 + \tan^2 x}$$

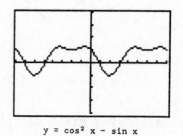

$$y = \cos^2 x - \sin x$$

The calculator graphs of

$y = \frac{1 - \tan^2 x}{1 + \tan^2 x}$ and $y = \cos^2 x - \sin x$

are not the same. The given statement is not an identity.

81. Is $\sin^2 s + \cos^2 s = \frac{1}{2}(1 - \cos 4s)$

an identity?

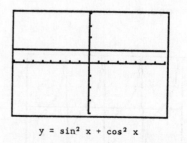

$$y = \sin^2 x + \cos^2 x$$

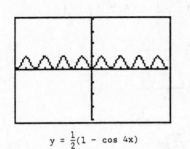

$y = \frac{1}{2}(1 - \cos 4x)$

The calculator graphs of

$$y = \sin^2 x + \cos^2 x$$

and $y = \frac{1}{2}(1 - \cos 4x)$

are not the same. The given statement is not an identity.

83. Is $\tan^2 x - \sin^2 x = (\tan x \sin x)^2$ an identity?

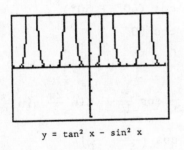

$y = \tan^2 x - \sin^2 x$

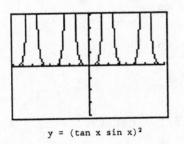

$y = (\tan x \sin x)^2$

The calculator graphs of

$$y = \tan^2 x - \sin^2 x$$

and $y = (\tan x \sin x)^2$

appear to be the same.

$\tan^2 x - \sin^2 x$

$$= \frac{\sin^2 x}{\cos^2 x} - \frac{\sin^2 x \cos^2 x}{\cos^2 x}$$

$$= \frac{\sin^2 x (1 - \cos^2 x)}{\cos^2 x}$$

$$= \frac{\sin^2 x}{\cos^2 x} (1 - \cos^2 x)$$

$$= \tan^2 x \sin^2 x$$

$$= (\tan x \sin x)^2$$

The given statement is an identity.

85. Show that $\sin (\csc s) = 1$ is not an identity.

Let $s = 2$.

$\sin (\csc 2) \approx .89109401$

The given statement is not an identity for all real numbers s.

87. Show that $\csc t = \sqrt{1 + \cot^2 t}$ is not an identity.

Let $t = 1$.

$$\csc 1 \approx 1.1883951$$
$$\sqrt{1 + \cot^2 1} \approx 1.1883951$$

But let $t = 4$.

$$\csc 4 \approx -1.3213487$$
$$\sqrt{1 + \cot^2 4} \approx 1.3213487$$

Recall that $\sqrt{a}$ denotes the positive square root of a.

The statement is not an identity for all real numbers t.

89. Let $\tan \theta = t$ and show that
$\sin \theta \cos \theta = \dfrac{t}{t^2 + 1}$.

Work with the right side.

$$\frac{t}{t^2 + 1} = \frac{\tan \theta}{\tan^2 \theta + 1}$$

$$= \frac{\tan \theta}{\sec^2 \theta}$$

$$= \tan \theta \cos^2 \theta$$

$$= \frac{\sin \theta}{\cos \theta} \cos^2 \theta$$

$$= \sin \theta \cos \theta$$

Section 5.3

3. $\cos 75°$

$$= \cos (30° + 45°)$$

$$= \cos 30° \cos 45° - \sin 30° \sin 45°$$

$$= \frac{\sqrt{3}}{2} \cdot \frac{\sqrt{2}}{2} - \frac{1}{2} \cdot \frac{\sqrt{2}}{2}$$

$$= \frac{\sqrt{6} - \sqrt{2}}{4}$$

5. $\cos (-75°)$

$$= \cos [-30° + (-45°)]$$

$$= \cos (-30°) \cos (-45°)$$
$$\quad - \sin (-30°) \sin (-45°)$$

$$= \cos 30° \cos 45°$$
$$\quad - (-\sin 30°)(-\sin 45°)$$

$$= \frac{\sqrt{3}}{2} \cdot \frac{\sqrt{2}}{2} - \left(-\frac{1}{2}\right)\left(-\frac{\sqrt{2}}{2}\right)$$

$$= \frac{\sqrt{6}}{4} - \frac{\sqrt{2}}{4} = \frac{\sqrt{6} - \sqrt{2}}{4}$$

7. $\cos (-105°)$

$$= \cos [-45° + (-60°)]$$

$$= \cos (-45°) \cos (-60°)$$
$$\quad - \sin (-45°) \sin (-60°)$$

$$= \cos 45° \cos 60°$$
$$\quad - (-\sin 45°)(-\sin 60°)$$

$$= \frac{\sqrt{2}}{2} \cdot \frac{1}{2} - \left(-\frac{\sqrt{2}}{2}\right)\left(-\frac{\sqrt{3}}{2}\right)$$

$$= \frac{\sqrt{2}}{4} - \frac{\sqrt{6}}{4}$$

$$= \frac{\sqrt{2} - \sqrt{6}}{4}$$

9. $\cos \left(-\dfrac{\pi}{12}\right)$

$$= \cos \left(\frac{\pi}{6} - \frac{\pi}{4}\right)$$

$$= \cos \frac{\pi}{6} \cos \frac{\pi}{4} + \sin \frac{\pi}{6} \sin \frac{\pi}{4}$$

$$= \frac{\sqrt{3}}{2} \cdot \frac{\sqrt{2}}{2} + \frac{1}{2} \cdot \frac{\sqrt{2}}{2}$$

$$= \frac{\sqrt{6}}{4} + \frac{\sqrt{2}}{4}$$

$$= \frac{\sqrt{6} + \sqrt{2}}{4}$$

11. $\cos 40° \cos 50° - \sin 40° \sin 50°$

$$= \cos (40° + 50°)$$

$$= \cos 90°$$

$$= 0$$

13. $\cos \dfrac{2\pi}{5} \cos \dfrac{\pi}{10} - \sin \dfrac{2\pi}{5} \sin \dfrac{\pi}{10}$

$$= \cos \left(\frac{2\pi}{5} + \frac{\pi}{10}\right)$$

$$= \cos \frac{5\pi}{10}$$

$$= \cos \frac{\pi}{2}$$

$$= 0$$

15. $\tan 87° = \cot (90° - 87°) = \cot 3°$

17. $\cos \dfrac{\pi}{12} = \sin \left(\dfrac{\pi}{2} - \dfrac{\pi}{12}\right) = \sin \dfrac{5\pi}{12}$

19. $\csc (-14° \ 24')$

$$= \sec [90° - (-14° \ 24')]$$

$$= \sec 104° \ 24'$$

21. $\sin \dfrac{5\pi}{8} = \cos \left(\dfrac{\pi}{2} - \dfrac{5\pi}{8}\right)$

            $= \cos \left(-\dfrac{\pi}{8}\right)$

23. $\sec 146° \ 42' = \csc (90° - 146° \ 42')$

                   $= \csc (-56° \ 42')$

25. $\cot 176.9814°$

      $= \tan (90° - 176.9814°)$

      $= \tan (-86.9814°)$

27. $\cot \dfrac{\pi}{3} =$ _____ $\dfrac{\pi}{6}$

    Since $\dfrac{\pi}{6} = \dfrac{\pi}{2} - \dfrac{\pi}{3}$,

        $\cot \dfrac{\pi}{3} = \tan \dfrac{\pi}{6}$.

    Answer: tan

29. _____ $33° = \sin 57°$

       $\sin 57° = \cos (90° - 57°)$

              $= \cos 33°$

    Answer: cos

31. $\cos 70° = \dfrac{1}{\underline{\quad\quad} 20°}$

       $\cos 70° = \sin (90° - 70°)$

              $= \sin 20°$

              $= \dfrac{1}{\csc 20°}$

    Answer: csc

33. $\tan \theta = \cot (45° + 2\theta)$

    Since $\tan \theta = \cot (90° - \theta)$,

        $90° - \theta = 45° + 2\theta$

            $3\theta = 45°$

              $\theta = 15°$.

35. $\sec \theta = \csc \left(\dfrac{\theta}{2} + 20°\right)$

    Since $\sec \theta = \csc (90° - \theta)$,

       $90° - \theta = \dfrac{\theta}{2} + 20°$.

       $\dfrac{\theta}{2} + 20° = 90° - \theta$

     $\theta + \dfrac{\theta}{2} + 20° = 90°$

    $2\theta + \theta + 40° = 180°$

            $3\theta = 140°$

              $\theta = \dfrac{140°}{3}$

37. $\sin (3\theta - 15°) = \cos (\theta + 25°)$

    Since $\sin \theta = \cos (90° - \theta)$,

    $\sin (3\theta - 15°) = \cos [90° - (3\theta - 15°)]$.

   $90° - (3\theta - 15°) = \theta + 25°$

      $105° - 3\theta = \theta + 25°$

           $80° = 4\theta$

           $20° = \theta$

39. $\cos (0° - \theta)$

    $= \cos 0° \cos \theta + \sin \theta \sin 0°$

    $= 1 \cos \theta + 0 \sin \theta$

    $= \cos \theta$

41. $\cos (180° - \theta)$

    $= \cos 180° \cos \theta + \sin 180° \sin \theta$

    $= (-1) \cos \theta + 0 \sin \theta$

    $= -\cos \theta$

43. $\cos (0° + \theta)$

    $= \cos 0° \cos \theta - \sin 0° \sin \theta$

    $= 1 \cos \theta - 0 \sin \theta$

    $= \cos \theta$

45. $\cos (180° + \theta)$

    $= \cos 180° \cos \theta - \sin 180° \sin \theta$

    $= (-1) \cos \theta - 0 \sin \theta$

    $= -\cos \theta$

47. $\cos s = -\frac{1}{5}$ and $\sin t = \frac{3}{5}$, s and t are in quadrant II.

Solve $\sin^2 s + \cos^2 s = 1$ for $\sin s$.
Since s is in quadrant II, $\sin s > 0$.

$$\sin s = \sqrt{1 - \left(-\frac{1}{5}\right)^2} = \frac{\sqrt{24}}{5}$$

Solve $\sin^2 t + \cos^2 t = 1$ for $\cos t$.
Since t is in quadrant II, $\cos t < 0$.

$$\cos t = -\sqrt{1 - \left(\frac{3}{5}\right)^2} = -\frac{4}{5}$$

$\cos (s + t)$
$$= \cos s \cos t - \sin s \sin t$$
$$= \left(-\frac{1}{5}\right)\left(-\frac{4}{5}\right) - \left(\frac{\sqrt{24}}{5}\right)\left(\frac{3}{5}\right)$$
$$= \frac{4}{25} - \frac{3\sqrt{24}}{25} = \frac{4 - 6\sqrt{6}}{25}$$

$\cos (s - t)$
$$= \cos s \cos t + \sin s \sin t$$
$$= \left(-\frac{1}{5}\right)\left(-\frac{4}{5}\right) + \left(\frac{\sqrt{24}}{5}\right)\left(\frac{3}{5}\right)$$
$$= \frac{4}{25} + \frac{3\sqrt{24}}{25} = \frac{4 + 6\sqrt{6}}{25}$$

49. $\sin s = \frac{3}{5}$ and $\sin t = -\frac{12}{13}$, s is in quadrant I and t is in quadrant III.
$\cos s > 0$ and $\cos t < 0$.

$$\cos s = \sqrt{1 - \left(\frac{3}{5}\right)^2} = \frac{4}{5}$$

$$\cos t = -\sqrt{1 - \left(-\frac{12}{13}\right)^2} = -\frac{5}{13}$$

$\cos (s + t)$
$$= \cos s \cos t - \sin s \sin t$$
$$= \left(\frac{4}{5}\right)\left(-\frac{5}{13}\right) - \left(\frac{3}{5}\right)\left(-\frac{12}{13}\right) = \frac{16}{65}$$

$\cos (s - t)$
$$= \cos s \cos t + \sin s \sin t$$
$$= \left(\frac{4}{5}\right)\left(-\frac{5}{13}\right) + \left(\frac{3}{5}\right)\left(-\frac{12}{13}\right) = -\frac{56}{65}$$

51. $\sin s = \frac{\sqrt{5}}{7}$ and $\sin t = \frac{\sqrt{6}}{8}$, s and t are in quadrant I. $\cos s > 0$ and $\cos t > 0$.

$$\cos s = \sqrt{1 - \left(\frac{\sqrt{5}}{7}\right)^2} = \frac{\sqrt{44}}{7}$$

$$\cos t = \sqrt{1 - \left(\frac{\sqrt{6}}{8}\right)^2} = \frac{\sqrt{58}}{8}$$

$\cos (s + t)$
$$= \cos s \cos t - \sin s \sin t$$
$$= \left(\frac{\sqrt{44}}{7}\right)\left(\frac{\sqrt{58}}{8}\right) - \left(\frac{\sqrt{5}}{7}\right)\left(\frac{\sqrt{6}}{8}\right)$$
$$= \frac{2\sqrt{638} - \sqrt{30}}{56}$$

$\cos (s - t)$
$$= \cos s \cos t + \sin s \sin t$$
$$= \left(\frac{\sqrt{44}}{7}\right)\left(\frac{\sqrt{58}}{8}\right) + \left(\frac{\sqrt{5}}{7}\right)\left(\frac{\sqrt{6}}{8}\right)$$
$$= \frac{2\sqrt{638} + \sqrt{30}}{56}$$

53. $\cos 42° = \cos (30° + 12°)$
Since $42° = 30° + 12°$, the statement is true.

55. $\cos 74° = \cos 60° \cos 14°$
$$+ \sin 60° \sin 14°$$
$\cos (A + B) = \cos A \cos B - \sin A \sin B$
$\cos 74° = \cos (60° + 14°)$
$$= \cos 60° \cos 14°$$
$$- \sin 60° \sin 14°$$

The statement is false.

57. $\cos \frac{\pi}{3} = \cos \frac{\pi}{12} \cos \frac{\pi}{4} - \sin \frac{\pi}{12} \sin \frac{\pi}{4}$

$\cos (A + B) = \cos A \cos B - \sin A \sin B$

$\cos \frac{\pi}{3} = \cos \left(\frac{\pi}{12} + \frac{\pi}{4}\right) = \cos \frac{\pi}{12} \cos \frac{\pi}{4} - \sin \frac{\pi}{12} \sin \frac{\pi}{4}$

The statement is true.

59. $\cos 70° \cos 20° - \sin 70° \sin 20° = 0$

$\cos (A + B) = \cos A \cos B - \sin A \sin B$

$\cos 70° \cos 20° - \sin 70° \sin 20° = \cos (70° + 20°) = \cos 90° = 0$

The statement is true.

61. $\tan \left(\theta - \frac{\pi}{2}\right) = \cot \theta$

Simplify the left side.

$$\tan \left(\theta - \frac{\pi}{2}\right) = \frac{\sin \left(\theta - \frac{\pi}{2}\right)}{\cos \left(\theta - \frac{\pi}{2}\right)}$$

$$= \frac{\sin \theta \cos \frac{\pi}{2} - \cos \theta \sin \frac{\pi}{2}}{\cos \theta \cos \frac{\pi}{2} + \sin \theta \sin \frac{\pi}{2}}$$

$$= \frac{\sin \theta (0) - \cos \theta (1)}{\cos \theta (0) + \sin \theta (1)}$$

$$= \frac{-\cos \theta}{\sin \theta}$$

$$= -\cot \theta$$

The statement is false.

63. Verify $\cos \left(\frac{\pi}{2} + x\right) = -\sin x$.

$\cos \left(\frac{\pi}{2} + x\right) = \cos \frac{\pi}{2} \cos x - \sin \frac{\pi}{2} \sin x = 0 \cos x - 1 \sin x = -\sin x$

65. Verify $\cos 2x = \cos^2 x - \sin^2 x$.

$\cos 2x = \cos (x + x) = \cos x \cos x - \sin x \sin x = \cos^2 x - \sin^2 x$

67. Verify $\cos (\pi + s - t) = -\sin s \sin t - \cos s \cos t$.

$\cos (\pi + s - t)$

$= \cos [\pi + (s - t)] = \cos \pi \cos (s - t) - \sin \pi \sin (s - t)$

$= (-1) \cos (s - t) - 0 \sin (s - t) = (-1)(\cos s \cos t + \sin s \sin t)$

$= -\cos s \cos t - \sin s \sin t$

69. Verify $\cos(\alpha + \beta)\cos(\alpha - \beta) = 1 - \sin^2\alpha - \sin^2\beta$.

Work with the left side.

$\cos(\alpha + \beta)\cos(\alpha - \beta)$

$= (\cos\alpha\cos\beta - \sin\alpha\sin\beta)(\cos\alpha\cos\beta + \sin\alpha\sin\beta)$

$= \cos^2\alpha\cos^2\beta - \sin^2\alpha\sin^2\beta + \cos\alpha\cos\beta\sin\alpha\sin\beta - \sin\alpha\sin\beta\cos\alpha\cos\beta$

$= (1 - \sin^2\beta)\cos^2\alpha - (1 - \cos^2\alpha)\sin^2\beta$

$= \cos^2\alpha - \sin^2\beta\cos^2\alpha - \sin^2\beta + \cos^2\alpha\sin^2\beta = \cos^2\alpha - \sin^2\beta$

Work with the right side.

$1 - \sin^2\alpha - \sin^2\beta$

$= \sin^2\alpha + \cos^2\alpha - \sin^2\alpha - \sin^2\beta = \cos^2\alpha - \sin^2\beta$

71. The cosine of the sum of two angles is equal to the product of the cosines of the angles minus the product of the sines of the angles.

73. $\cos 195° = \cos(180° + 15°)$

$= \cos 180° \cos 15° - \sin 180° \sin 15°$

$= (-1)\cos 15° - (0)\sin 15°$

$= -\cos 15°$

74. $-\cos 15° = -\cos(45° - 30°)$

$= -(\cos 45° \cos 30° + \sin 45° \sin 30°)$

$= -\left(\frac{\sqrt{2}}{2} \cdot \frac{\sqrt{3}}{2} + \frac{\sqrt{2}}{2} \cdot \frac{1}{2}\right)$

$= -\frac{\sqrt{6} + \sqrt{2}}{4}$

75. $\cos 195° = -\cos 15° = -\frac{\sqrt{6} + \sqrt{2}}{4}$

76. (a) $\cos 255°$

$= \cos(180° + 75°)$

$= \cos 180° \cos 75° - \sin 180° \sin 75°$

$= (-1)\cos 75° - (0)\sin 75°$

$= -\cos 75°$

$= -\cos(45° + 30°)$

$= -(\cos 45° \cos 30° - \sin 45° \sin 30°)$

$= -\left(\frac{\sqrt{2}}{2} \cdot \frac{\sqrt{3}}{2} - \frac{\sqrt{2}}{2} \cdot \frac{1}{2}\right)$

$$= -\frac{\sqrt{6} - \sqrt{2}}{4}$$

$$= \frac{\sqrt{2} - \sqrt{6}}{4}$$

(b) $\cos \frac{11\pi}{12}$

$$= \cos \left(\pi - \frac{\pi}{12}\right)$$

$$= \cos \pi \cos \frac{\pi}{12} + \sin \pi \sin \frac{\pi}{12}$$

$$= (-1) \cos \frac{\pi}{12} + (0) \sin \frac{\pi}{12}$$

$$= -\cos \frac{\pi}{12}$$

$$= -\cos \left(\frac{\pi}{3} - \frac{\pi}{4}\right)$$

$$= -\left(\cos \frac{\pi}{3} \cos \frac{\pi}{4} + \sin \frac{\pi}{3} \sin \frac{\pi}{4}\right)$$

$$= -\left(\frac{1}{2} \cdot \frac{\sqrt{2}}{2} + \frac{\sqrt{3}}{2} \cdot \frac{\sqrt{2}}{2}\right)$$

$$= -\frac{\sqrt{2} + \sqrt{6}}{4}$$

77. Verify $\cos A = \sin (90° - A)$.

Start with the identity $\cos (90° - \theta)$
$= \sin \theta$ and replace θ with $90° - A$.

$$\cos [90° - (90° - A)] = \sin (90° - A)$$
$$\cos (90° - 90° + A) = \sin (90° - A)$$
$$\cos A = \sin (90° - A)$$

79. Verify $\dfrac{f(x + h) - f(x)}{h}$

$$= \cos x \left(\frac{\cos h - 1}{h}\right) - \sin x \left(\frac{\sin h}{h}\right).$$

Let $f(x) = \cos x$.

$$f(x + h)$$

$$= \cos (x + h)$$

$$= \cos x \cos h - \sin x \sin h$$

$$\frac{f(x + h) - f(x)}{h}$$

$$= \frac{\cos x \cos h - \sin x \sin h - \cos x}{h}$$

$$= \frac{\cos x \cos h - \cos x - \sin x \sin h}{h}$$

$$= \frac{\cos x \cos h - \cos x}{h} - \frac{\sin x \sin h}{h}$$

$$= \cos x \left(\frac{\cos h - 1}{h}\right) - \sin x \left(\frac{\sin h}{h}\right)$$

81. Since there are 60 cycles per sec, the number of cycles in .05 sec is given by

$$(.05 \text{ sec})(60 \text{ cycles per sec}) = 3 \text{ cycles}.$$

83. **(a)** Graph

$$P = \frac{a}{r} \cos \left[\frac{2\pi r}{\lambda} - ct\right]$$

$$= \frac{.4}{10} \cos \left[\frac{2\pi(10)}{4.9} - 1026t\right]$$

$$= .04 \cos \left(\frac{20\pi}{4.9} - 1026t\right).$$

See the answer graph in the back of the textbook. The pressure P is oscillating.

(b) Graph

$$P = \frac{a}{r} \cos \left[\frac{2\pi r}{\lambda} - ct\right]$$

$$= \frac{3}{r} \cos \left[\frac{2\pi(r)}{4.9} - 1026(10)\right]$$

$$= \frac{3}{r} \cos \left(\frac{2\pi r}{4.9} - 10,260\right).$$

See the answer graph in the back of the textbook. The pressure oscillates, and amplitude decreases as r increases.

(c) $P = \dfrac{a}{r} \cos \left[\dfrac{2\pi r}{\lambda} - ct\right]$

Let $r = n\lambda$.

$$= \frac{a}{n\lambda} \cos \left[\frac{2\pi n\lambda}{\lambda} - ct\right]$$

$$= \frac{a}{n\lambda} \cos [2\pi n - ct]$$

$$= \frac{a}{n\lambda}[\cos{(2\pi n)}\cos{(ct)}$$

$$+ \sin{(2\pi n)}\sin{(ct)}]$$

$$= \frac{a}{n\lambda}\big[(1)\cos{(ct)}$$

$$+ (0)\sin{(ct)}\big]$$

$$= \frac{a}{n\lambda}\cos{(ct)}$$

Section 5.4

3. $\sin 15°$

$$= \sin{(45° - 30°)}$$

$$= \sin 45° \cos 30° - \cos 45° \sin 30°$$

$$= \frac{\sqrt{2}}{2} \cdot \frac{\sqrt{3}}{2} - \frac{\sqrt{2}}{2} \cdot \frac{1}{2}$$

$$= \frac{\sqrt{6}}{4} - \frac{\sqrt{2}}{4}$$

$$= \frac{\sqrt{6} - \sqrt{2}}{4}$$

5. $\tan 15° = \tan{(60° - 45°)}$

$$= \frac{\tan 60° - \tan 45°}{1 + \tan 60° \tan 45°}$$

$$= \frac{\sqrt{3} - 1}{1 + \sqrt{3}(1)} = \frac{\sqrt{3} - 1}{1 + \sqrt{3}} \cdot \frac{1 - \sqrt{3}}{1 - \sqrt{3}}$$

$$= \frac{\sqrt{3} - 3 - 1 + \sqrt{3}}{1 - 3} = \frac{-4 + 2\sqrt{3}}{-2}$$

$$= \frac{-2(2 - \sqrt{3})}{-2} = 2 - \sqrt{3}$$

7. $\sin{(-105°)}$

$$= \sin{(45° - 150°)}$$

$$= \sin 45° \cos 150° - \cos 45° \sin 150°$$

$$= \left(\frac{\sqrt{2}}{2}\right)\left(-\frac{\sqrt{3}}{2}\right) - \left(\frac{\sqrt{2}}{2}\right)\left(\frac{1}{2}\right)$$

$$= \frac{-\sqrt{6} - \sqrt{2}}{4}$$

9. $\sin \frac{5\pi}{12} = \sin{\left(\frac{\pi}{4} + \frac{\pi}{6}\right)}$

$$= \sin \frac{\pi}{4} \cos \frac{\pi}{6} + \cos \frac{\pi}{4} \sin \frac{\pi}{6}$$

$$= \frac{\sqrt{2}}{2} \cdot \frac{\sqrt{3}}{2} + \frac{\sqrt{2}}{2} \cdot \frac{1}{2}$$

$$= \frac{\sqrt{6}}{4} + \frac{\sqrt{2}}{4}$$

$$= \frac{\sqrt{6} + \sqrt{2}}{4}$$

11. $\tan \frac{\pi}{12} = \tan{\left(\frac{\pi}{4} - \frac{\pi}{6}\right)}$

$$= \frac{\tan \frac{\pi}{4} - \tan \frac{\pi}{6}}{1 + \tan \frac{\pi}{4} \tan \frac{\pi}{6}}$$

$$= \frac{1 - \frac{\sqrt{3}}{3}}{1 + (1)\frac{\sqrt{3}}{3}} = \frac{1 - \frac{\sqrt{3}}{3}}{1 + \frac{\sqrt{3}}{3}}$$

$$= \frac{3 - \sqrt{3}}{3 + \sqrt{3}}$$

$$= \frac{3 - \sqrt{3}}{3 + \sqrt{3}} \cdot \frac{3 - \sqrt{3}}{3 - \sqrt{3}}$$

$$= \frac{12 - 6\sqrt{3}}{6}$$

$$= 2 - \sqrt{3}$$

13. $\sin{\left(-\frac{7\pi}{12}\right)}$

$$= \sin{\left(-\frac{\pi}{4} - \frac{\pi}{3}\right)}$$

$$= \sin{\left(-\frac{\pi}{4}\right)} \cos \frac{\pi}{3} - \cos{\left(-\frac{\pi}{4}\right)} \sin \frac{\pi}{3}$$

$$= -\sin \frac{\pi}{4} \cos \frac{\pi}{3} - \cos \frac{\pi}{4} \sin \frac{\pi}{3}$$

$$= -\frac{\sqrt{2}}{2} \cdot \frac{1}{2} - \frac{\sqrt{2}}{2} \cdot \frac{\sqrt{3}}{2} = -\frac{\sqrt{2}}{4} - \frac{\sqrt{6}}{4}$$

$$= \frac{-\sqrt{2} - \sqrt{6}}{4}$$

15. $\sin 76° \cos 31° - \cos 76° \sin 31°$

$= \sin (76° - 31°)$

$= \sin 45°$

$= \dfrac{\sqrt{2}}{2}$

17. $\dfrac{\tan 80° + \tan 55°}{1 - \tan 80° \tan 55°}$

$= \tan (80° + 55°)$

$= \tan 135°$

$= -1$

19. $\dfrac{\tan 100° + \tan 80°}{1 - \tan 100° \tan 80°}$

$= \tan (100° + 80°)$

$= \tan 180°$

$= 0$

21. $\sin \dfrac{\pi}{5} \cos \dfrac{3\pi}{10} + \cos \dfrac{\pi}{5} \sin \dfrac{3\pi}{10}$

$= \sin \left(\dfrac{\pi}{5} + \dfrac{3\pi}{10}\right)$

$= \sin \dfrac{\pi}{2}$

$= 1$

23. $\cos (30° + \theta)$

$= \cos 30° \cos \theta - \sin 30° \sin \theta$

$= \dfrac{\sqrt{3}}{2} \cos \theta - \dfrac{1}{2} \sin \theta$

$= \dfrac{1}{2}(\sqrt{3} \cos \theta - \sin \theta)$

$= \dfrac{\sqrt{3} \cos \theta - \sin \theta}{2}$

25. $\cos (60° + \theta)$

$= \cos 60° \cos \theta - \sin 60° \sin \theta$

$= \dfrac{1}{2} \cos \theta - \dfrac{\sqrt{3}}{2} \sin \theta$

$= \dfrac{1}{2}(\cos \theta - \sqrt{3} \sin \theta)$

$= \dfrac{\cos \theta - \sqrt{3} \sin \theta}{2}$

27. $\cos \left(\dfrac{3\pi}{4} - x\right)$

$= \cos \dfrac{3\pi}{4} \cos x + \sin \dfrac{3\pi}{4} \sin x$

$= -\cos \dfrac{\pi}{4} \cos x + \sin \dfrac{\pi}{4} \sin x$

$= -\dfrac{\sqrt{2}}{2} \cos x + \dfrac{\sqrt{2}}{2} \sin x$

$= \dfrac{\sqrt{2}}{2}(-\cos x + \sin x)$

$= \dfrac{\sqrt{2}(\sin x - \cos x)}{2}$

29. $\tan (\theta + 30°)$

$= \dfrac{\tan \theta + \tan 30°}{1 - \tan \theta \tan 30°}$

$= \dfrac{\tan \theta + \dfrac{1}{\sqrt{3}}}{1 - \dfrac{1}{\sqrt{3}} \tan \theta}$

$= \dfrac{\sqrt{3} \tan \theta + 1}{\sqrt{3} - \tan \theta}$

31. $\sin \left(\dfrac{\pi}{4} + x\right)$

$= \sin \dfrac{\pi}{4} \cos x + \cos \dfrac{\pi}{4} \sin x$

$= \dfrac{\sqrt{2}}{2} \cos x + \dfrac{\sqrt{2}}{2} \sin x$

$= \dfrac{\sqrt{2}(\cos x + \sin x)}{2}$

33. $\sin (270° - \theta)$

$= \sin 270° \cos \theta - \cos 270° \sin \theta$

$= -1(\cos \theta) - 0(\sin \theta)$

$= -\cos \theta$

35. $\tan (360° - \theta)$

$= \dfrac{\tan 360° - \tan \theta}{1 + \tan 360° \tan \theta}$

$= \dfrac{0 - \tan \theta}{1 + 0(\tan \theta)}$

$= -\tan \theta$

37. $\tan (\pi - \theta)$

$$= \frac{\tan \pi - \tan \theta}{1 + \tan \pi \tan \theta} = \frac{0 - \tan \theta}{1 + 0(\tan \theta)}$$

$$= -\tan \theta$$

39. When you try to evaluate

$$\frac{\tan 65.902° + \tan 24.098°}{1 - \tan 65.902° \tan 24.098°},$$

you will get 0 in the denominator, since $(\tan 65.902°)(\tan 24.098°) = 1$. Notice that the given quotient is equal to

$$\tan (65.902° + 24.098°) = \tan 90°,$$

which is undefined.

41. $\cos s = \frac{3}{5}$, $\sin t = \frac{5}{13}$, s and t are in quadrant I.

First find sin s, tan s, cos t, and tan t. Since s and t are in quadrant I, all are positive.

$$\sin s = \sqrt{1 - \left(\frac{3}{5}\right)^2} = \frac{4}{5}$$

$$\tan s = \frac{\cos s}{\sin s} = \frac{\frac{4}{5}}{\frac{3}{5}} = \frac{4}{3}$$

$$\cos t = \sqrt{1 - \left(\frac{5}{13}\right)^2} = \frac{12}{13}$$

$$\tan t = \frac{\frac{5}{13}}{\frac{12}{13}} = \frac{5}{12}$$

$$\sin (s + t)$$

$$= \frac{4}{5} \cdot \frac{12}{13} + \frac{5}{13} \cdot \frac{3}{5} = \frac{63}{65}$$

$$\sin (s - t)$$

$$= \frac{4}{5} \cdot \frac{12}{13} - \frac{5}{13} \cdot \frac{3}{5} = \frac{33}{65}$$

$\tan (s + t)$

$$= \frac{\tan s + \tan t}{1 - \tan s \tan t}$$

$$= \frac{\frac{4}{3} + \frac{5}{12}}{1 - \left(\frac{4}{3}\right)\left(\frac{5}{12}\right)} = \frac{63}{16}$$

$\tan (s - t)$

$$= \frac{\tan s - \tan t}{1 + \tan s \tan t}$$

$$= \frac{\frac{4}{3} - \frac{5}{12}}{1 + \left(\frac{4}{5}\right)\left(\frac{5}{12}\right)} = \frac{33}{56}$$

To find the quadrant of s + t, notice that sin (s + t) > 0, which implies s + t is in quadrant I or II. tan (s + t) > 0, which implies s + t is in quadrant I or III. Therefore, s + t is in quadrant I. Then, sin (s - t) > 0 and tan (s - t) > 0. From the preceding, s - t must also be in quadrant I.

43. $\sin s = \frac{2}{3}$, $\sin t = -\frac{1}{3}$, s is in quadrant II, and t is in quadrant IV. Since s is in quadrant II, cos s < 0.

$$\cos s = -\sqrt{1 - \left(\frac{2}{3}\right)^2} = -\frac{\sqrt{5}}{3}$$

Since t is in quadrant IV, cos t > 0.

$$\cos t = \sqrt{1 - \left(-\frac{1}{3}\right)^2} = \frac{\sqrt{8}}{3} = \frac{2\sqrt{2}}{3}$$

$$\tan s = \frac{\frac{2}{3}}{-\frac{\sqrt{5}}{3}} = -\frac{2}{\sqrt{5}} = -\frac{2\sqrt{5}}{5}$$

$$\tan t = \frac{-\frac{1}{3}}{\frac{2\sqrt{2}}{3}} = -\frac{1}{2\sqrt{2}} = -\frac{\sqrt{2}}{4}$$

sin (s + t)

$$= \frac{2}{3} \cdot \left(\frac{2\sqrt{2}}{3}\right) + \left(-\frac{\sqrt{5}}{3}\right)\left(-\frac{1}{3}\right)$$

$$= \frac{4\sqrt{2}}{9} + \frac{\sqrt{5}}{9} = \frac{4\sqrt{2} + \sqrt{5}}{9}$$

sin (s - t)

$$= \frac{2}{3}\left(\frac{2\sqrt{2}}{3}\right) - \left(-\frac{\sqrt{5}}{3}\right)\left(-\frac{1}{3}\right)$$

$$= \frac{4\sqrt{2}}{9} - \frac{\sqrt{5}}{9} = \frac{4\sqrt{2} - \sqrt{5}}{9}$$

Different forms of tan (s + t) and tan (s - t) will be obtained depending on whether tan s and tan t are written with rationalized denominators.

tan (s + t)

$$= \frac{-\frac{2\sqrt{5}}{5} + \left(-\frac{\sqrt{2}}{4}\right)}{1 - \left(-\frac{2\sqrt{5}}{5}\right)\left(-\frac{\sqrt{2}}{4}\right)}$$

$$= \frac{-8\sqrt{5} - 5\sqrt{2}}{20 - 2\sqrt{10}}$$

or

tan (s + t)

$$= \frac{-\frac{2}{\sqrt{5}} + \left(-\frac{1}{2\sqrt{2}}\right)}{1 - \left(-\frac{2}{\sqrt{5}}\right)\left(-\frac{1}{2\sqrt{2}}\right)}$$

$$= \frac{-4\sqrt{2} - \sqrt{5}}{2\sqrt{10} - 2} = \frac{4\sqrt{2} + \sqrt{5}}{2 - 2\sqrt{10}}$$

tan (s - t)

$$= \frac{-\frac{2\sqrt{5}}{5} - \left(-\frac{\sqrt{2}}{4}\right)}{1 + \left(-\frac{2\sqrt{5}}{5}\right)\left(-\frac{\sqrt{2}}{4}\right)}$$

$$= \frac{-8\sqrt{5} + 5\sqrt{2}}{20 + 2\sqrt{10}}$$

or

tan (s - t)

$$= \frac{-\frac{2}{\sqrt{5}} - \left(-\frac{1}{2\sqrt{2}}\right)}{1 + \left(-\frac{2}{\sqrt{5}}\right)\left(-\frac{2}{2\sqrt{2}}\right)}$$

$$= \frac{-4\sqrt{2} + \sqrt{5}}{2\sqrt{10} + 2}$$

To find the quadrant of s + t, notice that sin (s + t) > 0, which implies s + t is in quadrant I or II. tan (s + t) < 0, which implies s + t is in quadrant II or IV. Therefore, s + t is in quadrant II. Also notice that sin (s - t) > 0, which implies s - t is in quadrant I or II, and tan (s - t) < 0, which implies s - t is in quadrant II or IV. Therefore, s - t is in quadrant II.

45. $\cos s = -\frac{8}{17}$, $\cos t = -\frac{3}{5}$, s and t are in quadrant III.

Since s is in quadrant III, sin s < 0.

$$\sin s = -\sqrt{1 - \left(-\frac{8}{17}\right)^2} = -\frac{15}{17}$$

Since t is in quadrant III, sin t < 0.

$$\sin t = -\sqrt{1 - \left(-\frac{3}{5}\right)^2} = -\frac{4}{5}$$

$$\tan s = \frac{-\frac{15}{17}}{-\frac{8}{17}} = \frac{15}{8}$$

$$\tan t = \frac{-\frac{4}{5}}{-\frac{3}{5}} = \frac{4}{3}$$

sin (s + t)

$$= \left(-\frac{15}{17}\right)\left(-\frac{3}{5}\right) + \left(-\frac{4}{5}\right)\left(-\frac{8}{17}\right) = \frac{77}{85}$$

sin (s − t)

$$= \left(-\frac{15}{17}\right)\left(-\frac{3}{5}\right) - \left(-\frac{4}{5}\right)\left(-\frac{8}{17}\right) = \frac{13}{85}$$

tan (s + t)

$$= \frac{\frac{15}{8} + \frac{4}{3}}{1 - \left(\frac{15}{8}\right)\left(\frac{4}{3}\right)} = \frac{\frac{77}{24}}{-\frac{36}{24}} = -\frac{77}{36}$$

tan (s − t)

$$= \frac{\frac{15}{8} - \frac{4}{3}}{1 + \left(\frac{15}{8}\right)\left(\frac{4}{3}\right)} = \frac{13}{84}$$

To find the quadrant of s + t, notice that sin (s + t) > 0, which implies s + t is in quadrant I or II, and tan (s + t) < 0, which implies s + t is in quadrant II or IV. Therefore, s + t is in quadrant II. Also notice that sin (s − t) > 0, which implies s − t is in quadrant I or II, and tan (s − t) > 0, which implies s − t is in quadrant I or III. Therefore, s − t is in quadrant I.

47. $\sin s = -\frac{4}{5}$, $\cos t = \frac{12}{13}$, s is in quadrant III, t is in quadrant IV.

Since s is in quadrant III, cos s < 0.

$$\cos s = -\sqrt{1 - \left(-\frac{4}{5}\right)^2} = -\frac{3}{5}$$

Since t is in quadrant IV, sin t < 0.

$$\sin t = -\sqrt{1 - \left(\frac{12}{13}\right)^2} = -\frac{5}{13}$$

$$\tan s = \frac{-\frac{4}{5}}{-\frac{3}{5}} = \frac{4}{3}$$

$$\tan t = \frac{-\frac{5}{13}}{\frac{12}{13}} = -\frac{5}{12}$$

sin (s + t)

$$= \left(-\frac{4}{5}\right)\left(\frac{12}{13}\right) + \left(-\frac{5}{13}\right)\left(-\frac{3}{5}\right) = -\frac{33}{65}$$

sin (s − t)

$$= \left(-\frac{4}{5}\right)\left(\frac{12}{13}\right) - \left(-\frac{5}{13}\right)\left(-\frac{3}{5}\right) = -\frac{63}{65}$$

tan (s + t)

$$= \frac{\frac{4}{3} - \frac{5}{12}}{1 - \left(\frac{4}{3}\right)\left(-\frac{5}{12}\right)} = \frac{48 - 15}{36 + 20} = \frac{33}{56}$$

tan (s − t)

$$= \frac{\frac{4}{3} - \left(-\frac{5}{12}\right)}{1 + \left(\frac{4}{3}\right)\left(-\frac{5}{12}\right)} = \frac{48 + 15}{36 - 20} = \frac{63}{16}$$

To find the quadrants of s + t and s − t, notice that sin (s + t) < 0 and sin (s − t) < 0, which implies that s + t and s − t are in quadrant III or IV. Also, notice that tan (s + t) > 0 and tan (s − t) > 0, which implies that both s + t and s − t are in quadrant I or III. Therefore, both s + t and s − t are in quadrant III.

49. $\cos s = -\frac{\sqrt{7}}{4}$, $\sin t = \frac{\sqrt{3}}{5}$, s and t are in quadrant II.

Since s is in quadrant II, sin s > 0.

$$\sin s = \sqrt{1 - \left(-\frac{\sqrt{7}}{4}\right)^2} = \frac{3}{4}$$

Since t is in quadrant II, cos t < 0.

$$\cos t = -\sqrt{1 - \left(\frac{\sqrt{3}}{5}\right)^2} = -\frac{\sqrt{22}}{5}$$

$$\tan s = \frac{\frac{3}{4}}{-\frac{\sqrt{7}}{4}} = -\frac{3}{\sqrt{7}} = -\frac{3\sqrt{7}}{7}$$

$$\tan t = \frac{\frac{\sqrt{3}}{5}}{-\frac{\sqrt{22}}{5}} = -\frac{\sqrt{3}}{\sqrt{22}} = -\frac{\sqrt{66}}{22}$$

$\sin (s + t)$

$$= \left(\frac{3}{4}\right)\left(-\frac{\sqrt{22}}{5}\right) + \left(\frac{\sqrt{3}}{5}\right)\left(-\frac{\sqrt{7}}{4}\right)$$

$$= \frac{-3\sqrt{22} - \sqrt{21}}{20} = \frac{-(3\sqrt{22} + \sqrt{21})}{20}$$

$\sin (s - t)$

$$= \left(\frac{3}{4}\right)\left(-\frac{\sqrt{22}}{5}\right) - \left(\frac{\sqrt{3}}{5}\right)\left(-\frac{\sqrt{7}}{4}\right)$$

$$= \frac{-3\sqrt{22} + \sqrt{21}}{20}$$

$\tan (s + t)$

$$= \frac{\left(-\frac{3\sqrt{7}}{7}\right) + \left(-\frac{\sqrt{66}}{22}\right)}{1 - \left(-\frac{3\sqrt{7}}{7}\right)\left(-\frac{\sqrt{66}}{22}\right)}$$

$$= \frac{-(66\sqrt{7} + 7\sqrt{66})}{154 - 3\sqrt{462}}$$

$\tan (s - t)$

$$= \frac{\left(-\frac{3\sqrt{7}}{7}\right) - \left(-\frac{\sqrt{66}}{22}\right)}{1 + \left(-\frac{3\sqrt{7}}{7}\right)\left(-\frac{\sqrt{66}}{22}\right)}$$

$$= \frac{-66\sqrt{7} + 7\sqrt{66}}{154 + 3\sqrt{462}}$$

To find the quadrants of s + t and s − t, notice that sin (s + t) < 0 and sin (s − t) < 0, which mean s + t and s − t are in quadrant III or IV. Also notice that tan (s + t) < 0 and tan (s − t) < 0, which imply that s + t and s − t are in quadrant II or IV. Therefore, both s + t and s − t are in quadrant IV.

51.

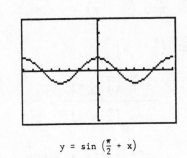

$$y = \sin \left(\frac{\pi}{2} + x\right)$$

The calculator graph of

$$y = \sin \left(\frac{\pi}{2} + x\right)$$

appears to be the same as the graph of

$$y = \cos x.$$

Verify $\sin \left(\frac{\pi}{2} + x\right) = \cos x$.

$\sin \left(\frac{\pi}{2} + x\right)$

$$= \sin \frac{\pi}{2} \cos x + \sin x \cos \frac{\pi}{2}$$

$$= 1 \cdot \cos x + \sin x \cdot 0$$

$$= \cos x$$

53.

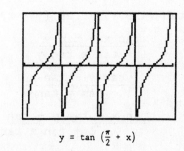

$$y = \tan \left(\frac{\pi}{2} + x\right)$$

The calculator graph of

$$y = \tan \left(\frac{\pi}{2} + x\right)$$

appears to be the same as the graph of

$$y = -\cot x.$$

Verify $\tan \left(\frac{\pi}{2} + x\right) = -\cot x$.

$\tan \left(\frac{\pi}{2} + x\right)$

$$= \frac{\sin \left(\frac{\pi}{2} + x\right)}{\cos \left(\frac{\pi}{2} + x\right)}$$

$$= \frac{\sin \frac{\pi}{2} \cos x + \cos \frac{\pi}{2} \sin x}{\cos \frac{\pi}{2} \cos x - \sin \frac{\pi}{2} \sin x}$$

$$= \frac{(1)(\cos x) + (0)(\sin x)}{(0)(\cos x) - (1)(\sin x)}$$

$$= \frac{\cos x}{-\sin x}$$

$$= -\cot x$$

55. Verify $\sin 2x = 2 \sin x \cos x$.

$\sin 2x$

$= \sin (x + x)$

$= \sin x \cos x + \sin x \cos x$

$= 2 \sin x \cos x$

57. Verify $\tan (x - y) - \tan (y - x)$

$= \dfrac{2(\tan x - \tan y)}{1 + \tan x \tan y}$.

$\tan (x - y) - \tan (y - x)$

$$= \frac{\tan x - \tan y}{1 + \tan x \tan y}$$

$$- \frac{\tan y - \tan x}{1 + \tan y \tan x}$$

$$= \frac{\tan x - \tan y - \tan y + \tan x}{1 + \tan x \tan y}$$

$$= \frac{2 \tan x - 2 \tan y}{1 + \tan x \tan y}$$

$$= \frac{2(\tan x - \tan y)}{1 + \tan x \tan y}$$

59. Verify $\dfrac{\cos (\alpha - \beta)}{\cos \alpha \sin \beta} = \tan \alpha + \cot \beta$.

$\dfrac{\cos (\alpha - \beta)}{\cos \alpha \sin \beta}$

$$= \frac{\cos \alpha \cos \beta + \sin \alpha \sin \beta}{\cos \alpha \sin \beta}$$

$$= \frac{\cos \beta}{\sin \beta} + \frac{\sin \alpha}{\cos \alpha}$$

$$= \cot \beta + \tan \alpha$$

61. Verify $\dfrac{\sin (x - y)}{\sin (x + y)} = \dfrac{\tan x - \tan y}{\tan x + \tan y}$.

Work with the right side.

$\dfrac{\tan x - \tan y}{\tan x + \tan y}$

$$= \frac{\dfrac{\sin x}{\cos x} - \dfrac{\sin y}{\cos y}}{\dfrac{\sin x}{\cos x} + \dfrac{\sin y}{\cos y}}$$

$$= \frac{\sin x \cos y - \cos x \sin y}{\sin x \cos y + \cos x \sin y}$$

$$= \frac{\sin (x - y)}{\sin (x + y)}$$

63. Verify $\dfrac{\sin (s - t)}{\sin t} + \dfrac{\cos (s - t)}{\cos t}$

$= \dfrac{\sin s}{\sin t \cos t}$.

$\dfrac{\sin (s - t)}{\sin t} + \dfrac{\cos (s - t)}{\cos t}$

$$= \frac{\sin s \cos t - \sin t \cos s}{\sin t}$$

$$+ \frac{\cos s \cos t + \sin s \sin t}{\cos t}$$

$$= \frac{\sin s \cos^2 t - \sin t \cos t \cos s}{\sin t \cos t}$$

$$+ \frac{\sin t \cos t \cos s + \sin^2 t \sin s}{\sin t \cos t}$$

$$= \frac{\sin s \cos^2 t + \sin s \sin^2 t}{\sin t \cos t}$$

$$= \frac{\sin s (\cos^2 t + \sin^2 t)}{\sin t \cos t}$$

$$= \frac{\sin s}{\sin t \cos t}$$

65. $\sin 165°$

$= \sin (180° - 15°)$

$= \sin 180° \cos 15° - \cos 180° \sin 15°$

$= \sin 15° = \sin (45° - 30°)$

$= \sin 45° \cos 30° - \cos 45° \sin 30°$

$= \dfrac{\sqrt{2}}{2} \cdot \dfrac{\sqrt{3}}{2} - \dfrac{\sqrt{2}}{2} \cdot \dfrac{1}{2}$

$= \dfrac{\sqrt{6} - \sqrt{2}}{4}$

67. $\tan 255°$

$= \tan (180° + 75°)$

$= \dfrac{\tan 180° + \tan 75°}{1 - \tan 180° \tan 75°}$

$= \tan 75° = \tan (45° + 30°)$

$= \dfrac{\tan 45° + \tan 30°}{1 - \tan 45° \tan 30°}$

$= \dfrac{1 + \dfrac{\sqrt{3}}{3}}{1 - \dfrac{\sqrt{3}}{3}}$

$= \dfrac{3 + \sqrt{3}}{3 - \sqrt{3}} \cdot \dfrac{3 + \sqrt{3}}{3 + \sqrt{3}}$

$= \dfrac{12 + 6\sqrt{3}}{6} = 2 + \sqrt{3}$

69. $\sin 285°$

$= \sin (360° - 75°)$

$= \sin 360° \cos 75° - \cos 360° \sin 75°$

$= -\sin 75° = -\sin (30° + 45°)$

$= -(\sin 30° \cos 45° + \cos 30° \sin 45°)$

$= -\left(\dfrac{1}{2} \cdot \dfrac{\sqrt{2}}{2} + \dfrac{\sqrt{3}}{2} \cdot \dfrac{\sqrt{2}}{2}\right)$

$= \dfrac{-\sqrt{2} - \sqrt{6}}{4}$

71. $\sin \dfrac{11\pi}{12}$

$= \sin \left(\pi - \dfrac{\pi}{12}\right)$

$= \sin \pi \cos \dfrac{\pi}{12} - \cos \pi \sin \dfrac{\pi}{12}$

$= \sin \dfrac{\pi}{12} = \sin \left(\dfrac{\pi}{4} - \dfrac{\pi}{6}\right)$

$= \sin \dfrac{\pi}{4} \cos \dfrac{\pi}{6} - \cos \dfrac{\pi}{4} \sin \dfrac{\pi}{6}$

$= \dfrac{\sqrt{2}}{2} \cdot \dfrac{\sqrt{3}}{2} - \dfrac{\sqrt{2}}{2} \cdot \dfrac{1}{2}$

$= \dfrac{\sqrt{6} - \sqrt{2}}{4}$

73. $\tan \left(-\dfrac{13\pi}{12}\right)$

$= -\tan \left(\pi + \dfrac{\pi}{12}\right)$

$= -\dfrac{\tan \pi + \tan \dfrac{\pi}{12}}{1 - \tan \pi \tan \dfrac{\pi}{12}}$

$= -\tan \dfrac{\pi}{12} = -\tan \left(\dfrac{\pi}{4} - \dfrac{\pi}{6}\right)$

$= -\dfrac{\tan \dfrac{\pi}{4} - \tan \dfrac{\pi}{6}}{1 + \tan \dfrac{\pi}{4} \tan \dfrac{\pi}{6}}$

$= -\dfrac{1 - \dfrac{\sqrt{3}}{3}}{1 + \dfrac{\sqrt{3}}{3}} = -\dfrac{3 - \sqrt{3}}{3 + \sqrt{3}} \cdot \dfrac{3 - \sqrt{3}}{3 - \sqrt{3}}$

$= -\dfrac{12 - 6\sqrt{3}}{6} = -2 + \sqrt{3}$

75. Verify $\tan (A - B) = \dfrac{\tan A - \tan B}{1 + \tan A \tan B}$.

$\tan (A - B)$

$= \tan [A + (-B)]$

$= \dfrac{\tan A + \tan (-B)}{1 - \tan A \tan (-B)}$

(Remember that $\tan (-B) = -\tan B$.)

$= \dfrac{\tan A - \tan B}{1 - \tan A (-\tan B)}$

$= \dfrac{\tan A - \tan B}{1 + \tan A \tan B}$

77. $\sin (A + B + C)$

$= \sin (A + B) \cos C + \sin C \cos (A + B)$

$= (\sin A \cos B + \sin B \cos A) \cdot \cos C$

$\quad + \sin C \cdot (\cos A \cos B - \sin A \sin B)$

$= \sin A \cos B \cos C + \sin B \cos A \cos C$

$\quad + \sin C \cos A \cos B - \sin C \sin A \sin B$

79. Let $f(x) = \sin x$. Show that

$$\frac{f(x+h) - f(x)}{h}$$

$$= \sin x \left(\frac{\cos h - 1}{h}\right) + \cos x \left(\frac{\sin h}{h}\right).$$

$$\frac{f(x+h) - f(x)}{h}$$

$$= \frac{\sin(x+h) - \sin x}{h}$$

$$= \frac{\sin x \cos h + \sin h \cos x - \sin x}{h}$$

$$= \sin x \left(\frac{\cos h - 1}{h}\right) + \cos x \left(\frac{\sin h}{h}\right)$$

80. Since angle β and angle ABC are supplementary, the measure of angle ABC is $180° - \beta$.

81. $\alpha + (180° - \beta) + \theta = 180°$

$$\alpha - \beta + \theta = 0$$

$$\theta = \beta - \alpha$$

82. $\tan \theta = \tan(\beta - \alpha)$

$$= \frac{\tan \beta - \tan \alpha}{1 + \tan \beta \tan \alpha}$$

83. From Exercise 82,

$$\tan \theta = \frac{\tan \beta - \tan \alpha}{1 + \tan \beta \tan \alpha}.$$

Let $\tan \alpha = m_1$ and $\tan \beta = m_2$. Then

$$\tan \theta = \frac{m_2 - m_1}{1 + m_1 m_2}.$$

84. $x + y = 9$, $2x + y = -1$

Change the equations of the given lines to slope-intercept form to find their slopes.

$$x + y = 9$$

$$y = -x + 9$$

$$m_2 = -1$$

$$2x + y = -1$$

$$y = -2x - 1$$

$$m_1 = -2$$

$$\tan \theta = \frac{m_2 - m_1}{1 + m_2 m_1}$$

$$= \frac{-1 - (-2)}{1 + (-1)(-2)}$$

$$= \frac{1}{1 + 2}$$

$$= \frac{1}{3}$$

$$\theta = 18.4°$$

85. $5x - 2y + 4 = 0$, $3x + 5y = 6$

Change the equations of the given lines to slope-intercept form to find their slopes.

$$5x - 2y + 4 = 0$$

$$-2y = -5x - 4$$

$$y = \frac{5}{2}x + 2$$

$$m_1 = \frac{5}{2}$$

$$3x + 5y = 6$$

$$5y = -3x + 6$$

$$y = -\frac{3}{5}x + \frac{6}{5}$$

$$m_2 = -\frac{3}{5}$$

$$\tan \theta = \frac{m_2 - m_1}{1 + m_2 m_1}$$

$$= \frac{-\frac{3}{5} - \frac{5}{2}}{1 + \left(-\frac{3}{5}\right)\left(\frac{5}{2}\right)}$$

$$= \frac{-\frac{31}{10}}{-\frac{5}{10}}$$

$$= \frac{31}{5} = 6.2$$

$$\theta = 80.8°$$

87. (a) $F = \dfrac{.6W \sin(\theta + 90)°}{\sin 12°}$

$= \dfrac{.6(170) \sin(30 + 90)°}{\sin 12°}$

≈ 425 lb

(This is a good reason why people frequently have back problems.)

(b)

$F = \dfrac{.6W \sin(\theta + 90)°}{\sin 12°}$

$= \dfrac{.6W (\sin \theta° \cos 90° + \sin 90° \cos \theta°)}{\sin 12°}$

$= \dfrac{.6}{\sin 12°}W \cos \theta$

$\approx 2.9W \cos \theta$

(c) F will be maximum when $\cos \theta = 1$ or $\theta = 0°$. ($\theta = 0°$ corresponds to the back being horizontal which gives a maximum force on the back muscles. This agrees with intuition since stress on the back increases as one bends farther until the back is parallel with the ground.)

Section 5.5

1. $\sin 2X = 2 \sin X \cos X$
$\sin X \cos X = \frac{1}{2} \sin 2X$

If $\sin 2X = .4$, then

$\sin X \cos X = \frac{1}{2}(.4) = .2.$

Therefore, the screen will display .2 for the final expression.

3. $\cos 2\theta = \frac{3}{5}$, θ is in quadrant I.

$\cos 2\theta = 2 \cos^2 \theta - 1$

$\frac{3}{5} = 2 \cos^2 \theta - 1$

$2 \cos^2 \theta = \frac{3}{5} + 1 = \frac{8}{5}$

$\cos^2 \theta = \frac{8}{10} = \frac{4}{5}$

Since θ is in quadrant I, $\cos \theta > 0$.

$\cos \theta = \dfrac{2}{\sqrt{5}} = \dfrac{2\sqrt{5}}{5}$

Since θ is in quadrant I, $\sin \theta > 0$.

$\sin \theta = \sqrt{1 - \cos^2 \theta}$

$= \sqrt{1 - \left(\frac{2\sqrt{5}}{5}\right)^2} = \sqrt{\frac{1}{5}}$

$= \dfrac{1}{\sqrt{5}} = \dfrac{\sqrt{5}}{5}$

$\tan \theta = \dfrac{\sin \theta}{\cos \theta} = \dfrac{\frac{\sqrt{5}}{5}}{\frac{2\sqrt{5}}{5}} = \dfrac{1}{2}$

$\cot \theta = \dfrac{1}{\tan \theta} = \dfrac{1}{\frac{1}{2}} = 2$

$\sec \theta = \dfrac{1}{\cos \theta} = \dfrac{1}{\frac{2\sqrt{5}}{5}} = \dfrac{5}{2\sqrt{5}} = \dfrac{\sqrt{5}}{2}$

$\csc \theta = \dfrac{1}{\sin \theta} = \dfrac{1}{\frac{\sqrt{5}}{5}} = \sqrt{5}$

5. $\cos 2x = -\frac{5}{12}$, $\frac{\pi}{2} < x < \pi$

$\cos 2x = 2 \cos^2 x - 1$

$2 \cos^2 x = \cos 2x + 1$

$= -\frac{5}{12} + 1 = \frac{7}{12}$

$\cos^2 x = \frac{7}{24}$

Since $\frac{\pi}{2} < x < \pi$, cos x < 0.

$$\cos x = -\sqrt{\frac{7}{24}} = -\frac{\sqrt{7}}{2\sqrt{6}} = -\frac{\sqrt{42}}{12}$$

$$\sin^2 x = 1 - \cos^2 x = 1 - \frac{7}{24} = \frac{17}{24}$$

Since $\frac{\pi}{2} < x < \pi$, sin x > 0.

$$\sin x = \frac{\sqrt{17}}{2\sqrt{6}} = \frac{\sqrt{102}}{12}$$

$$\tan x = \frac{\frac{\sqrt{17}}{2\sqrt{6}}}{-\frac{\sqrt{7}}{2\sqrt{6}}} = -\frac{\sqrt{17}}{\sqrt{7}} = -\frac{\sqrt{119}}{7}$$

$$\cot x = -\frac{\sqrt{7}}{\sqrt{17}} = -\frac{\sqrt{119}}{17}$$

$$\sec x = -\frac{2\sqrt{6}}{\sqrt{7}} = -\frac{2\sqrt{42}}{7}$$

$$\csc x = \frac{2\sqrt{6}}{\sqrt{17}} = \frac{2\sqrt{102}}{17}$$

7. $\sin \theta = \frac{2}{5}$, cos θ < 0

$$\cos 2\theta = 1 - 2\sin^2 \theta = 1 - 2\left(\frac{2}{5}\right)^2$$
$$= 1 - \frac{8}{25} = \frac{17}{25}$$

$$\cos^2 2\theta + \sin^2 2\theta = 1$$

$$\sin^2 2\theta = 1 - \cos^2 2\theta$$
$$= 1 - \left(\frac{17}{25}\right)^2 = \frac{336}{625}$$

Since cos θ < 0, sin 2θ < 0 because
sin 2θ = 2 sin θ cos θ < 0.

$$\sin 2\theta = -\sqrt{\frac{336}{625}} = -\frac{4\sqrt{21}}{25}$$

$$\tan 2\theta = \frac{\sin 2\theta}{\cos 2\theta} = \frac{-\frac{4\sqrt{21}}{25}}{\frac{17}{25}} = -\frac{4\sqrt{21}}{17}$$

$$\cot 2\theta = -\frac{17}{4\sqrt{21}} = -\frac{17\sqrt{21}}{84}$$

$$\sec 2\theta = \frac{25}{17}$$

$$\csc 2\theta = -\frac{25}{4\sqrt{21}} = -\frac{25\sqrt{21}}{84}$$

9. tan x = 2, cos x > 0

$$\tan 2x = \frac{2 \tan x}{1 - \tan^2 x}$$
$$= \frac{2(2)}{1 - (2)^2} = -\frac{4}{3}$$

Since both tan x and cos x are
positive, x must be in quadrant I.
Thus, 2x must be in quadrant II.

$$\sec^2 2x = 1 + \tan^2 2x = 1 + \frac{16}{9}$$
$$= \frac{25}{9}$$

$$\sec 2x = -\frac{5}{3}$$

$$\cos 2x = \frac{1}{\sec 2x} = -\frac{3}{5}$$

$$\cot 2x = \frac{2}{\tan 2x} = -\frac{3}{4}$$

$$\sin 2x = \tan 2x \cos 2x$$
$$= \left(-\frac{4}{3}\right)\left(-\frac{3}{5}\right) = \frac{4}{5}$$

$$\csc 2x = \frac{1}{\sin 2x} = \frac{5}{4}$$

11. $\sin \alpha = -\frac{\sqrt{5}}{7}$, cos α > 0

$$\cos^2 \alpha = 1 - \sin^2 \alpha$$
$$= 1 - \left(-\frac{\sqrt{5}}{7}\right)^2 = \frac{44}{49}$$

Since cos α > 0,

$$\cos \alpha = \sqrt{\frac{44}{49}} = \frac{\sqrt{44}}{7} = \frac{2\sqrt{11}}{7}.$$

$$\cos 2\alpha = 1 - 2 \sin^2 \alpha$$

$$= 1 - 2\left(-\frac{\sqrt{5}}{7}\right)^2 = \frac{39}{49}$$

$$\sin 2\alpha = 2 \sin \alpha \cos \alpha$$

$$= 2\left(-\frac{\sqrt{5}}{7}\right)\left(\frac{2\sqrt{11}}{7}\right) = -\frac{4\sqrt{55}}{49}$$

$$\tan 2\alpha = \frac{\sin 2\alpha}{\cos 2\alpha} = \frac{-\frac{4\sqrt{55}}{49}}{\frac{39}{49}} = -\frac{4\sqrt{55}}{39}$$

$$\cot 2\alpha = \frac{1}{\tan 2\alpha} = -\frac{39\sqrt{55}}{220}$$

$$\sec 2\alpha = \frac{1}{\cos 2\alpha} = \frac{49}{39}$$

$$\csc 2\alpha = \frac{1}{\sin 2\alpha} = -\frac{49\sqrt{55}}{220}$$

13. $2 \cos^2 15° - 1$

Since $\cos 2A = 2 \cos^2 A - 1$,

$$2 \cos^2 15° - 1 = \cos 2(15°)$$

$$= \cos 30° = \frac{\sqrt{3}}{2}.$$

15. $\dfrac{2 \tan 15°}{1 - \tan^2 15°}$

$$= \frac{\tan 15° + \tan 15°}{1 - \tan^2 15°}$$

$$= \tan (15° + 15°)$$

$$= \tan 30° = \frac{1}{\sqrt{3}} = \frac{\sqrt{3}}{3}$$

17. $2 \sin \dfrac{\pi}{3} \cos \dfrac{\pi}{3} = \sin 2\left(\dfrac{\pi}{3}\right)$

$$= \sin \frac{2\pi}{3} = \frac{\sqrt{3}}{2}$$

19. $1 - 2 \sin^2 22\frac{1}{2}°$

$$= \cos 2\left(22\frac{1}{2}°\right) = \cos 45° = \frac{\sqrt{2}}{2}$$

21. $2 \cos^2 67\frac{1}{2}° - 1$

$$= \cos^2 67\frac{1}{2}° - \sin^2 67\frac{1}{2}°$$

$$= \cos 2\left(67\frac{1}{2}°\right) = \cos 135° = -\frac{\sqrt{2}}{2}$$

23. $\sin \dfrac{\pi}{8} \cos \dfrac{\pi}{8}$

Since $\sin 2A = 2 \sin A \cos A$,

$$\frac{1}{2} \sin 2A = \sin A \cos A.$$

$$\sin \frac{\pi}{8} \cos \frac{\pi}{8} = \frac{1}{2} \sin 2\left(\frac{\pi}{8}\right)$$

$$= \frac{1}{2} \sin \frac{\pi}{4}$$

$$= \frac{1}{2} \cdot \frac{\sqrt{2}}{2} = \frac{\sqrt{2}}{4}$$

25. $\dfrac{\tan 51°}{1 - \tan^2 51°}$

Since $\dfrac{2 \tan A}{1 - \tan^2 A} = \tan 2A$,

then

$$\frac{1}{2}\left(\frac{2 \tan A}{1 - \tan^2 A}\right) = \frac{1}{2} \tan 2A$$

$$\frac{\tan A}{1 - \tan^2 A} = \frac{1}{2} \tan 2A.$$

$$\frac{\tan 51°}{1 - \tan^2 51°} = \frac{1}{2} \tan 2(51°)$$

$$= \frac{1}{2} \tan 102°$$

27. $\dfrac{1}{4} - \dfrac{1}{2} \sin^2 47.1° = \dfrac{1}{4}(1 - 2 \sin^2 47.1°)$

Since $\cos 2A = 1 - 2 \sin^2 A$,

$$\frac{1}{4}(1 - 2 \sin^2 47.1°)$$

$$= \frac{1}{4} \cos 2(47.1°)$$

$$= \frac{1}{4} \cos 94.2°.$$

29. $\sin^2 \frac{2\pi}{5} - \cos^2 \frac{2\pi}{5} = -\left(\cos^2 \frac{2\pi}{5} - \sin^2 \frac{2\pi}{5}\right)$

Since $\cos^2 A - \sin^2 A = \cos 2A$,

$\quad -\left(\cos^2 \frac{2\pi}{5} - \sin^2 \frac{2\pi}{5}\right)$

$\quad\quad = -\cos \left(2 \cdot \frac{2\pi}{5}\right) = -\cos \frac{4\pi}{5}.$

31. $2 \sin 5x \cos 5x$

Since $2 \sin A \cos A = \sin 2A$,

$\quad 2 \sin 5x \cos 5x = \sin 2(5x)$

$\quad\quad\quad\quad\quad\quad = \sin 10x.$

35. $\sin 2(45°) = \sin 90° = 1$

$\sin 2(45°) = 2 \sin 45° \cos 45°$

$\quad\quad = 2\left(\frac{\sqrt{2}}{2} \cdot \frac{\sqrt{2}}{2}\right) = 1$

37. $\cos 2(60°) = \cos 120° = -\frac{1}{2}$

$\cos 2(60°) = \cos^2 60° - \sin^2 60°$

$\quad\quad = \left(\frac{1}{2}\right)^2 - \left(\frac{\sqrt{3}}{2}\right)^2 = -\frac{1}{2}$

39. $\cos 2\left(\frac{5\pi}{3}\right) = \cos \frac{10\pi}{3} = \cos \frac{4\pi}{3}$

$\quad\quad = -\cos \frac{\pi}{3}$

$\quad\quad = -\frac{1}{2}$

$\cos 2\left(\frac{5\pi}{3}\right) = \cos^2 \left(\frac{5\pi}{3}\right) - \sin^2 \left(\frac{5\pi}{3}\right)$

$\quad\quad = \cos^2 \left(\frac{\pi}{3}\right) - \left(-\sin \frac{\pi}{3}\right)^2$

$\quad\quad = \left(\frac{1}{2}\right)^2 - \left(-\frac{\sqrt{3}}{2}\right)^2$

$\quad\quad = \frac{1}{4} - \frac{3}{4}$

$\quad\quad = -\frac{2}{4}$

$\quad\quad = -\frac{1}{2}$

41. $\tan 2\left(-\frac{\pi}{3}\right) = \tan \left(-\frac{2\pi}{3}\right)$

$\quad\quad\quad\quad = \sqrt{3}$

$\tan 2\left(-\frac{\pi}{3}\right) = -\tan 2\left(\frac{\pi}{3}\right)$

$\quad\quad = -\frac{2 \tan \frac{\pi}{3}}{1 - \tan^2 \frac{\pi}{3}} = -\frac{2\sqrt{3}}{1 - (\sqrt{3})^2}$

$\quad\quad = -\frac{2\sqrt{3}}{1 - 3} = \sqrt{3}$

43. $\tan 2\left(-\frac{4\pi}{3}\right) = \tan \left(-\frac{8\pi}{3}\right) = \sqrt{3}$

$\tan 2\left(-\frac{4\pi}{3}\right) = -\tan 2\left(\frac{4\pi}{3}\right)$

$\quad\quad = -\frac{2 \tan \frac{4\pi}{3}}{1 - \tan^2 \frac{4\pi}{3}}$

$\quad\quad = -\frac{2\sqrt{3}}{1 - (\sqrt{3})^2} = \frac{2\sqrt{3}}{1 - 3}$

$\quad\quad = \sqrt{3}$

45. $\sin 2\left(-\frac{11\pi}{2}\right) = \sin (-11\pi) = 0$

$\sin 2\left(-\frac{11\pi}{2}\right) = 2 \sin \left(-\frac{11\pi}{2}\right) \cos \left(-\frac{11\pi}{2}\right)$

$\quad\quad = 2(0)(-1) = 0$

47.

$y = \cos^4 x - \sin^4 x$

The calculator graph of

$$y = \frac{\cot^2 x - 1}{2 \cot x}$$

appears to be the same as the graph of

$$y = \cot 2x.$$

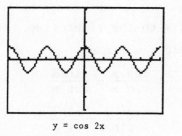

y = cos 2x

The calculator graph of

$$y = \cos^4 x - \sin^4 x$$

appears to be the same as the graph of

$$y = \cos 2x.$$

$\cos^4 x - \sin^4 x$

$= (\cos^2 x + \sin^2 x)(\cos^2 x - \sin^2 x)$

$= 1 \cdot \cos 2x$

$= \cos 2x$

$$\frac{\cot^2 x - 1}{2 \cot x} = \frac{\dfrac{1}{\tan^2 x} - 1}{\dfrac{2}{\tan x}}$$

$$= \frac{\left(\dfrac{1}{\tan^2 x} - 1\right) \tan^2 x}{\left(\dfrac{2}{\tan x}\right) \tan^2 x}$$

$$= \frac{1 - \tan^2 x}{2 \tan x}$$

$$= \frac{1}{\dfrac{2 \tan x}{1 - \tan^2 x}}$$

$$= \frac{1}{\tan 2x}$$

$$= \cot 2x$$

49.

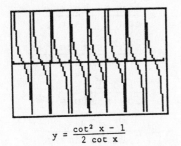

$y = \dfrac{\cot^2 x - 1}{2 \cot x}$

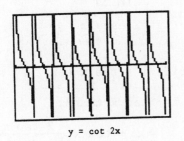

y = cot 2x

51. Verify $(\sin \gamma + \cos \gamma)^2 = \sin 2\gamma + 1$.

$(\sin \gamma + \cos \gamma)^2$

$= \sin^2 \gamma + \cos^2 \gamma + 2 \sin \gamma \cos \gamma$

$= 1 + \sin 2\gamma$

53. Verify $\tan 8k - \tan 8k \tan^2 4k = 2 \tan 4k$.

$\tan 8k - \tan 8k \tan^2 4k$

$= \tan 8k (1 - \tan^2 4k)$

$= \dfrac{2 \tan 4k}{1 - \tan^2 4k}(1 - \tan^2 4k)$

$= 2 \tan 4k$

55. Verify $\cos 2y = \dfrac{2 - \sec^2 y}{\sec^2 y}$.

Work with the right side.

$$\dfrac{2 - \sec^2 y}{\sec^2 y}$$

$$= \dfrac{\left(2 - \dfrac{1}{\cos^2 y}\right) \cos^2 y}{\left(\dfrac{1}{\cos^2 y}\right) \cos^2 y}$$

$$= 2 \cos^2 y - 1$$

$$= \cos 2y$$

57. Verify $\sin 4\alpha = 4 \sin \alpha \cos \alpha \cos 2\alpha$.

Work with the right side.

$$4 \sin \alpha \cos \alpha \cos 2\alpha$$

$$= 2(2 \sin \alpha \cos \alpha) \cos 2\alpha$$

$$= 2 \sin 2\alpha \cos 2\alpha$$

$$= \sin 2(2\alpha)$$

$$= \sin 4\alpha$$

59. Verify $\tan (\theta - 45°) + \tan (\theta + 45°)$
$= 2 \tan 2\theta$.

$$\tan (\theta - 45°) + \tan (\theta + 45°)$$

$$= \dfrac{\tan \theta - \tan 45°}{1 + \tan \theta \tan 45°}$$

$$+ \dfrac{\tan \theta + \tan 45°}{1 - \tan \theta \tan 45°}$$

$$= \dfrac{\tan \theta - 1}{1 + \tan \theta} + \dfrac{\tan \theta + 1}{1 - \tan \theta}$$

$$= \dfrac{\tan \theta - 1}{\tan \theta + 1} - \dfrac{\tan \theta + 1}{\tan \theta - 1}$$

$$= \dfrac{\tan^2 \theta - 2 \tan \theta + 1}{(\tan \theta + 1)(\tan \theta - 1)}$$

$$- \dfrac{\tan^2 \theta + 2 \tan \theta + 1}{(\tan \theta + 1)(\tan \theta - 1)}$$

$$= \dfrac{-4 \tan \theta}{\tan^2 \theta - 1} = \dfrac{4 \tan \theta}{1 - \tan^2 \theta}$$

$$= \dfrac{2(2 \tan \theta)}{1 - \tan^2 \theta}$$

$$= 2 \tan 2\theta$$

61. Verify $\dfrac{2 \cos 2\alpha}{\sin 2\alpha} = \cot \alpha - \tan \alpha$.

Work with the right side.

$$\cot \alpha - \tan \alpha$$

$$= \dfrac{\cos \alpha}{\sin \alpha} - \dfrac{\sin \alpha}{\cos \alpha}$$

$$= \dfrac{\cos^2 \alpha - \sin^2 \alpha}{\sin \alpha \cos \alpha}$$

$$= \dfrac{2(\cos^2 \alpha - \sin^2 \alpha)}{2 \sin \alpha \cos \alpha}$$

$$= \dfrac{2 \cos 2\alpha}{\sin 2\alpha}$$

63. Verify $\sin 2\alpha \cos 2\alpha$
$= \sin 2\alpha - 4 \sin^3 \alpha \cos \alpha$.

$$\sin 2\alpha \cos 2\alpha$$

$$= (2 \sin \alpha \cos \alpha)(1 - 2 \sin^2 \alpha)$$

$$= 2 \sin \alpha \cos \alpha - 4 \sin^3 \alpha \cos \alpha$$

$$= \sin 2\alpha - 4 \sin^3 \alpha \cos \alpha$$

65. Verify $\tan s + \cot s = 2 \csc 2s$.

$$\tan s + \cot s$$

$$= \dfrac{\sin s}{\cos s} + \dfrac{\cos s}{\sin s}$$

$$= \dfrac{\sin^2 s + \cos^2 s}{\cos s \sin s}$$

$$= \dfrac{1}{\cos s \sin s}$$

$$= \dfrac{2}{2 \sin s \cos s}$$

$$= \dfrac{2}{\sin 2s}$$

$$= 2 \csc 2s$$

67. Verify $1 + \tan x \tan 2x = \sec 2x$.

$$1 + \tan x \tan 2x$$

$$= 1 + \tan x \left(\dfrac{2 \tan x}{1 - \tan^2 x}\right)$$

$$= 1 + \dfrac{2 \tan^2 x}{1 - \tan^2 x}$$

$$= \frac{1 - \tan^2 x + 2 \tan^2 x}{1 - \tan^2 x}$$

$$= \frac{1 + \tan^2 x}{1 - \tan^2 x}$$

$$= \frac{\dfrac{\cos^2 x + \sin^2 x}{\cos^2 x}}{\dfrac{\cos^2 x - \sin^2 x}{\cos^2 x}}$$

$$= \frac{1}{\cos^2 x - \sin^2 x}$$

$$= \frac{1}{\cos 2x}$$

$$= \sec 2x$$

In Exercises 69–75, other forms may be possible.

69. $\tan^2 2x = \left(\dfrac{2 \tan x}{1 - \tan^2 x} \right)^2$

$$= \frac{4 \tan^2 x}{1 - 2 \tan^2 x + \tan^4 x}$$

71. cos 3x

$$= \cos (x + 2x)$$

$$= \cos x \cos 2x - \sin x \sin 2x$$

$$= \cos x (2 \cos^2 x - 1)$$

$$\quad - \sin x (2 \sin x \cos x)$$

$$= 2 \cos^3 x - \cos x - 2 \sin^2 x \cos x$$

$$= 2 \cos^3 x - \cos x - 2(1 - \cos^2 x) \cos x$$

$$= 2 \cos^3 x - \cos x - 2 \cos x + 2 \cos^3 x$$

$$= 4 \cos^3 x - 3 \cos x$$

73. tan 3x

$$= \tan (2x + x)$$

$$= \frac{\tan 2x + \tan x}{1 - \tan 2x \tan x}$$

$$= \frac{\dfrac{2 \tan x}{1 - \tan^2 x} + \tan x}{1 - \dfrac{2 \tan x}{1 - \tan^2 x} \tan x}$$

$$= \frac{2 \tan x + \tan x - \tan^3 x}{1 - \tan^2 x - 2 \tan^2 x}$$

$$= \frac{3 \tan x - \tan^3 x}{1 - 3 \tan^2 x}$$

75. tan 4x

$$= \frac{2 \tan 2x}{1 - \tan^2 2x}$$

$$= \frac{2 \left(\dfrac{2 \tan x}{1 - \tan^2 x} \right)}{1 - \left(\dfrac{2 \tan x}{1 - \tan^2 x} \right)^2}$$

$$= \frac{\dfrac{4 \tan x}{1 - \tan^2 x}}{1 - \dfrac{4 \tan^2 x}{1 - 2 \tan^2 x + \tan^4 x}}$$

$$= \frac{4 \tan x(1 - \tan^2 x)}{1 - 2 \tan^2 x + \tan^4 x - 4 \tan^2 x}$$

$$= \frac{4 \tan x - 4 \tan^3 x}{1 - 6 \tan^2 x + \tan^4 x}$$

$$= \frac{4(\tan x - \tan^3 x)}{1 - 6 \tan^2 x + \tan^4 x}$$

77. $\sin^2 2x + \cos^2 2x$

Use the Pythagorean identity

$\sin^2 \theta + \cos^2 \theta = 1$, and let $\theta = 2x$.

$$\sin^2 2x + \cos^2 2x = 1.$$

79. $\cot^2 3r + 1$

Use the Pythagorean identity

$1 + \cot^2 \theta = \csc^2 \theta$, with $\theta = 3r$.

$$\cot^2 3r + 1 = \csc^2 3r$$

81. $\phi = 47° \ 12'$, $h = 387.0$ ft

Use a calculator and substitute these

values in

$g = 978.0524(1 + .005297 \sin^2 \phi$

$\quad - .0000059 \sin^2 2\phi) - .000094h.$

$g = 978.0524[1 + .005297 \sin^2 47° 12'$

$\quad - .0000059 \sin^2 (2 \cdot 47° 12')]$

$\quad - .000094(387.0)$

$= 980.799$ cm per sec^2

83.   cos 45° sin 25°

$$= \frac{1}{2}[\sin(45° + 25°) - \sin(45° - 25°)]$$

$$= \frac{1}{2}(\sin 70° - \sin 20°)$$

85.   3 cos 5x cos 3x

$$= 3 \cdot \frac{1}{2}[\cos(5x + 3x) + \cos(5x - 3x)]$$

$$= \frac{3}{2}(\cos 8x + \cos 2x)$$

87.   sin (−θ) sin (−3θ)

$$= \frac{1}{2}\Big[\cos[-\theta - (-3\theta)]$$

$$\quad - \cos[-\theta + (-3\theta)]\Big]$$

$$= \frac{1}{2}[\cos(2\theta) - \cos(-4\theta)]$$

$$= \frac{1}{2}(\cos 2\theta - \cos 4\theta)$$

89.   −8 cos 4y cos 5y

$$= -8 \cdot \frac{1}{2}[\cos(4y + 5y) + \cos(4y - 5y)]$$

$$= -4[\cos 9y + \cos(-y)]$$

$$= -4(\cos 9y + \cos y)$$

91.   From Example 6,

$$W = \frac{(163 \sin 120\pi t)^2}{15}$$

$$W \approx 1771.3 (\sin 120\pi t)^2.$$

Thus,

1771.3 (sin 120πt)²

$$= 1771.3 \sin 120\pi t \cdot \sin 120\pi t$$

$$= (1771.3)\left(\frac{1}{2}\right)[\cos(120\pi t - 120\pi t)$$

$$\quad - \cos(120\pi t + 120\pi t)]$$

$$= 885.6(\cos 0 - \cos 240\pi t)$$

$$= 885.6(1 - \cos 240\pi t)$$

$$= -885.6 \cos 240\pi t + 885.6.$$

If we compare this to

W = a cos (ωt) + c,

then a = −885.6, ω = 240π, and c = 885.6. See the answer graph in the back of the textbook.

93.   **(a)** The period is equal to

$$\frac{2\pi}{b} = \frac{2\pi}{2\pi\omega} = \frac{1}{\omega}.$$

(b) W = VI

$$= a \sin(2\pi\omega t) \cdot b \sin(2\pi\omega t)$$

$$= ab \sin^2(2\pi\omega t)$$

$$= \frac{ab}{2}[1 - \cos(4\pi\omega t)],$$

which has a period of

$$\frac{2\pi}{4\pi\omega} = \frac{1}{2\omega},$$

exactly half of 1/ω. The graph of the wattage will oscillate at twice the rate of the graph of the voltage or the amperage. For example, if the voltage oscillates at 60 cycles per sec then the wattage will oscillate at 120 cycles per sec.

Section 5.6

1.   With your graphing calculator in radian mode, find sin⁻¹ .25 and store it in X. Then duplicate the graphing calculator screen given in the textbook. The screen will display

tan (X/2) = .1270166538

for the final expression.

3. Since 195° is in quadrant III and since the sine is negative in quadrant III, use the negative square root.

5. Since 225° is in quadrant III and since the tangent is positive in quadrant III, use the positive square root.

7. $\sin 15° = \sin \left(\dfrac{30°}{2}\right)$

Since 15° is in quadrant I, $\sin 15° > 0$.

$$\sin 15° = \sqrt{\dfrac{1 - \cos 30°}{2}}$$

$$= \sqrt{\dfrac{1 - \dfrac{\sqrt{3}}{2}}{2}}$$

$$= \sqrt{\dfrac{2 - \sqrt{3}}{4}}$$

$$= \dfrac{\sqrt{2 - \sqrt{3}}}{2}$$

9. $\cos \dfrac{\pi}{8} = \cos \left(\dfrac{\dfrac{\pi}{4}}{2}\right)$

Since $\pi/8$ is in quadrant I, $\cos \pi/8 > 0$.

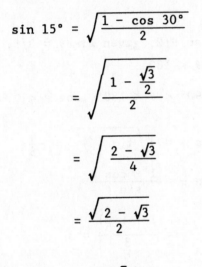

$$\cos \dfrac{\pi}{8} = \sqrt{\dfrac{1 + \cos \left(\dfrac{\pi}{4}\right)}{2}}$$

$$= \sqrt{\dfrac{1 + \dfrac{\sqrt{2}}{2}}{2}}$$

$$= \sqrt{\dfrac{2 + \sqrt{2}}{4}}$$

$$= \dfrac{\sqrt{2 + \sqrt{2}}}{2}$$

11. $\tan 67.5° = \tan \left(\dfrac{135°}{2}\right)$

$$= \dfrac{\sin 135°}{1 + \cos 135°}$$

$$= \dfrac{\sin 45°}{1 + (-\cos 45°)}$$

$$= \dfrac{\dfrac{\sqrt{2}}{2}}{1 - \dfrac{\sqrt{2}}{2}}$$

$$= \dfrac{\sqrt{2}}{2 - \sqrt{2}} \cdot \dfrac{2 + \sqrt{2}}{2 + \sqrt{2}}$$

$$= \dfrac{2\sqrt{2} + 2}{2}$$

$$= \sqrt{2} + 1$$

13. $\sin 67.5° = \sin \left(\dfrac{135°}{2}\right)$

Since 67.5° is in quadrant I, $\sin 67.5° > 0$.

$$\sin 67.5° = \sqrt{\dfrac{1 - \cos 135°}{2}}$$

$$= \sqrt{\dfrac{1 + \cos 45°}{2}}$$

$$= \sqrt{\dfrac{1 + \dfrac{\sqrt{2}}{2}}{2}} = \sqrt{\dfrac{2 + \sqrt{2}}{4}}$$

$$= \dfrac{\sqrt{2 + \sqrt{2}}}{2}$$

15. $\cos 195° = \cos \left(\dfrac{390°}{2}\right)$

Since 195° is in quadrant III, $\cos 195° < 0$.

$$\cos 195° = -\sqrt{\dfrac{1 + \cos 390°}{2}}$$

$$= -\sqrt{\dfrac{1 + \cos 30°}{2}}$$

$$= -\sqrt{\dfrac{1 + \dfrac{\sqrt{3}}{2}}{2}} = -\sqrt{\dfrac{2 + \sqrt{3}}{4}}$$

$$= \dfrac{-\sqrt{2 + \sqrt{3}}}{2}$$

17. $\cos 165° = \cos\left(\dfrac{330°}{2}\right)$

Since 165° is in quadrant II,
$\cos 165° < 0$.

$\cos 165° = -\sqrt{\dfrac{1 + \cos 330°}{2}}$

$= -\sqrt{\dfrac{1 + \cos 30°}{2}}$

$= -\sqrt{\dfrac{1 + \dfrac{\sqrt{3}}{2}}{2}} = -\sqrt{\dfrac{2 + \sqrt{3}}{4}}$

$= \dfrac{-\sqrt{2 + \sqrt{3}}}{2}$

19. To find $\sin 7.5°$, first notice that 7.5° is in quadrant I. Therefore $\sin 7.5° > 0$. Next notice that $7.5° = \dfrac{1}{2} \cdot \dfrac{1}{2} \cdot 30°$.

$\sin 15° = \sin\left(\dfrac{30°}{2}\right)$

$= \sqrt{\dfrac{1 - \cos 30°}{2}}$

$= \sqrt{\dfrac{1 - \dfrac{\sqrt{3}}{2}}{2}}$

$= \dfrac{\sqrt{2 - \sqrt{3}}}{2}$

Since

$\sin 7.5° = \sin\left(\dfrac{15°}{2}\right) = \sqrt{\dfrac{1 - \cos 15°}{2}}$,

we can find $\cos 15°$ by using the identity $\sin^2 \theta + \cos^2 \theta = 1$ and using

$\sin 15° = \dfrac{\sqrt{2 - \sqrt{3}}}{2}$.

21. Find $\cos \theta/2$, given $\cos \theta = 1/4$, with $0 < \theta < \pi/2$.

$0 < \theta < \dfrac{\pi}{2}$

$0 < \dfrac{\theta}{2} < \dfrac{\pi}{4}$

Thus, $\cos \theta/2 > 0$.

$\cos \dfrac{\theta}{2} = \sqrt{\dfrac{1 + \cos \theta}{2}}$

$= \sqrt{\dfrac{1 + \dfrac{1}{4}}{2}} = \sqrt{\dfrac{5}{8}}$

$= \dfrac{\sqrt{10}}{4}$

23. Find $\tan \theta/2$, given $\sin \theta = 3/5$, with $90° < \theta < 180°$.

Since $90° < \theta < 180°$, $\cos \theta < 0$.

$\cos \theta = -\sqrt{1 - \left(\dfrac{3}{5}\right)^2} = -\dfrac{4}{5}$

$\tan \dfrac{\theta}{2} = \dfrac{1 - \cos \theta}{\sin \theta}$

$= \dfrac{1 - \left(-\dfrac{4}{5}\right)}{\dfrac{3}{5}} = 3$

25. Find $\sin \alpha/2$, given $\tan \alpha = 2$, with $0 < \alpha < \pi/2$.

Since α is in quadrant I, $\sec \alpha > 0$.

$\sec^2 \alpha = \tan^2 \alpha + 1$

$= (2)^2 + 1$

$= 5$

$\sec \alpha = \sqrt{5}$

$\cos \alpha = \dfrac{1}{\sec \alpha} = \dfrac{1}{\sqrt{5}} = \dfrac{\sqrt{5}}{5}$

$0 < \alpha < \dfrac{\pi}{2}$

$0 < \dfrac{\alpha}{2} < \dfrac{\pi}{4}$

Thus, $\sin \alpha/2 > 0$.

$$\sin \frac{\alpha}{2} = \sqrt{\frac{1 - \cos \alpha}{2}}$$

$$= \sqrt{\frac{1 - \frac{\sqrt{5}}{5}}{2}} = \frac{\sqrt{50 - 10\sqrt{5}}}{10}$$

27. Find $\tan \beta/2$, given $\tan \beta = \sqrt{7}/3$, with $180° < \beta < 270°$.

$$\sec^2 \beta = \tan^2 \beta + 1$$

$$= \left(\frac{\sqrt{7}}{3}\right)^2 + 1 = \frac{16}{9}$$

Since β is in quadrant III, $\sec \beta < 0$ and $\sin \beta < 0$.

$$\sec \beta = -\sqrt{\frac{16}{9}} = -\frac{4}{3}$$

$$\cos \beta = \frac{1}{\sec \beta} = \frac{1}{-\frac{4}{3}} = -\frac{3}{4}$$

$$\sin \beta = -\sqrt{1 - \cos^2 \beta}$$

$$= -\sqrt{1 - \frac{9}{16}} = -\frac{\sqrt{7}}{4}$$

$$\tan \frac{\beta}{2} = \frac{\sin \beta}{1 + \cos \beta}$$

$$= \frac{-\frac{\sqrt{7}}{4}}{1 + \left(-\frac{3}{4}\right)} = -\sqrt{7}$$

29. Find $\sin \theta$, given $\cos 2\theta = 3/5$, θ is in quadrant I.

Since θ is in quadrant I, $\sin \theta > 0$.

$$\sin \theta = \sin \frac{1}{2}(2\theta)$$

$$= \sqrt{\frac{1 - \cos 2\theta}{2}}$$

$$= \sqrt{\frac{1 - \frac{3}{5}}{2}} = \frac{\sqrt{5}}{5}$$

31. Find $\cos x$, given $\cos 2x = -5/12$, $\frac{\pi}{2} < x < \pi$.

Since $\frac{\pi}{2} < x < \pi$, $\cos x < 0$.

$$\cos x = \cos \frac{1}{2}(2x)$$

$$= -\sqrt{\frac{1 + \cos 2x}{2}}$$

$$\cos x = -\sqrt{\frac{1 + \left(-\frac{5}{12}\right)}{2}}$$

$$= -\frac{\sqrt{42}}{12}$$

33. $\sqrt{\frac{1 - \cos 40°}{2}} = \sin \frac{40°}{2}$

$$= \sin 20°$$

35. $\sqrt{\frac{1 - \cos 147°}{1 + \cos 147°}} = \tan \frac{147°}{2}$

$$= \tan 73.5°$$

37. $\frac{1 - \cos 59.74°}{\sin 59.74°} = \tan \frac{59.74°}{2}$

$$= \tan 29.87°$$

39. $\pm\sqrt{\frac{1 + \cos 18x}{2}} = \cos \frac{18x}{2}$

$$= \cos 9x$$

41. $\pm\sqrt{\frac{1 - \cos 8\theta}{1 + \cos 8\theta}} = \tan \frac{8\theta}{2}$

$$= \tan 4\theta$$

43. $\pm\sqrt{\frac{1 + \cos \left(\frac{x}{4}\right)}{2}} = \cos \frac{\frac{x}{4}}{2}$

$$= \cos \frac{x}{8}$$

45.

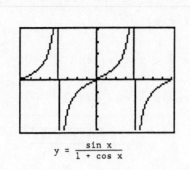

$$y = \frac{\sin x}{1 + \cos x}$$

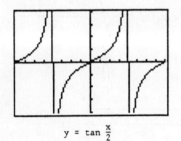

$$y = \tan \frac{x}{2}$$

47.

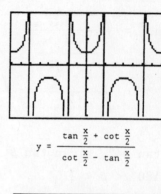

$$y = \frac{\tan \frac{x}{2} + \cot \frac{x}{2}}{\cot \frac{x}{2} - \tan \frac{x}{2}}$$

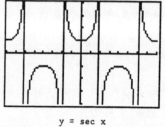

$$y = \sec x$$

The calculator graph of

$$y = \frac{\sin x}{1 + \cos x}$$

appears to be the same as the graph of

$$y = \tan \frac{x}{2}.$$

$$\frac{\sin x}{1 + \cos x} = \frac{\sin 2\left(\frac{x}{2}\right)}{1 + \cos 2\left(\frac{x}{2}\right)}$$

$$= \frac{2 \sin \left(\frac{x}{2}\right) \cos \left(\frac{x}{2}\right)}{1 + \left[2 \cos^2 \left(\frac{x}{2}\right) - 1\right]}$$

$$= \frac{2 \sin \left(\frac{x}{2}\right) \cos \left(\frac{x}{2}\right)}{2 \cos^2 \left(\frac{x}{2}\right)}$$

$$= \frac{\sin \frac{x}{2}}{\cos \frac{x}{2}}$$

$$= \tan \frac{x}{2}$$

The calculator graph of

$$y = \frac{\tan \frac{x}{2} + \cot \frac{x}{2}}{\cot \frac{x}{2} - \tan \frac{x}{2}}$$

appears to be the same as the graph of

$$y = \sec x.$$

$$\frac{\tan \frac{x}{2} + \cot \frac{x}{2}}{\cot \frac{x}{2} - \tan \frac{x}{2}}$$

$$= \frac{\dfrac{\sin \frac{x}{2}}{\cos \frac{x}{2}} + \dfrac{\cos \frac{x}{2}}{\sin \frac{x}{2}}}{\dfrac{\cos \frac{x}{2}}{\sin \frac{x}{2}} - \dfrac{\sin \frac{x}{2}}{\cos \frac{x}{2}}} \cdot \frac{\sin \frac{x}{2} \cos \frac{x}{2}}{\sin \frac{x}{2} \cos \frac{x}{2}}$$

$$= \frac{\sin^2 \frac{x}{2} + \cos^2 \frac{x}{2}}{\cos^2 \frac{x}{2} - \sin^2 \frac{x}{2}}$$

$$= \frac{1}{\cos 2\left(\frac{x}{2}\right)}$$

$$= \frac{1}{\cos x}$$

$$= \sec x$$

49. Verify $\sec^2 \frac{x}{2} = \frac{2}{1 + \cos x}$.

$$\sec^2 \frac{x}{2} = \frac{1}{\cos^2 \frac{x}{2}}$$

$$= \frac{1}{\left(\pm\sqrt{\frac{1 + \cos x}{2}}\right)^2}$$

$$= \frac{1}{\frac{1 + \cos x}{2}}$$

$$= \frac{2}{1 + \cos x}$$

51. Verify $\sin^2 \frac{x}{2} = \frac{\tan x - \sin x}{2 \tan x}$.

Work with the left side.

$$\sin^2 \frac{x}{2} = \left(\pm\sqrt{\frac{1 - \cos x}{2}}\right)^2$$

$$= \frac{1 - \cos x}{2}$$

Work with the right side.

$$\frac{\tan x - \sin x}{2 \tan x}$$

$$= \frac{\frac{\sin x}{\cos x} - \sin x}{2 \frac{\sin x}{\cos x}}$$

$$= \frac{\sin x - \cos x \sin x}{2 \sin x}$$

$$= \frac{1 - \cos x}{2}$$

53. Verify $\frac{2}{1 + \cos x} - \tan^2 \frac{x}{2} = 1$.

$$\frac{2}{1 + \cos x} - \tan^2 \frac{x}{2}$$

$$= \frac{2}{1 + \cos x} - \left(\pm\sqrt{\frac{1 - \cos x}{1 + \cos x}}\right)^2$$

$$= \frac{2}{1 + \cos x} - \frac{1 - \cos x}{1 + \cos x}$$

$$= \frac{2 - 1 + \cos x}{1 + \cos x} = \frac{1 + \cos x}{1 + \cos x}$$

$$= 1$$

55. Verify $1 - \tan^2 \frac{\theta}{2} = \frac{2 \cos \theta}{1 + \cos \theta}$.

$$1 - \tan^2 \frac{\theta}{2}$$

$$= 1 - \left(\frac{\sin \theta}{1 + \cos \theta}\right)^2$$

$$= \frac{(1 + \cos \theta)^2 - \sin^2 \theta}{(1 + \cos \theta)^2}$$

$$= \frac{1 + 2 \cos \theta + \cos^2 \theta - \sin^2 \theta}{(1 + \cos \theta)^2}$$

$$= \frac{1 + 2 \cos \theta + \cos^2 \theta - (1 - \cos^2 \theta)}{(1 + \cos \theta)^2}$$

$$= \frac{1 + 2 \cos^2 \theta - 1 + 2 \cos \theta}{(1 + \cos \theta)^2}$$

$$= \frac{2 \cos \theta (1 + \cos \theta)}{(1 + \cos \theta)^2}$$

$$= \frac{2 \cos \theta}{1 + \cos \theta}$$

57. $\tan \frac{A}{2} = \frac{\sin A}{1 + \cos A}$

$$= \frac{\sin A}{1 + \cos A} \cdot \frac{1 - \cos A}{1 - \cos A}$$

$$= \frac{\sin A (1 - \cos A)}{1 - \cos^2 A}$$

$$= \frac{\sin A (1 - \cos A)}{\sin^2 A}$$

$$= \frac{1 - \cos A}{\sin A}$$

59. $m = \dfrac{5}{4}$

$$\sin \dfrac{\alpha}{2} = \dfrac{1}{m} = \dfrac{4}{5} = .8$$

$$\dfrac{\alpha}{2} \approx 53°$$

$$\alpha \approx 106°$$

61. $m = \dfrac{5}{2}$

$$\sin \dfrac{\alpha}{2} = \dfrac{1}{m} = \dfrac{2}{5} = .4$$

$$\dfrac{\alpha}{2} \approx 23.6°$$

$$\alpha \approx 47°$$

63. $\alpha = 60°$

$$\sin \dfrac{\alpha}{2} = \dfrac{1}{m}$$

$$\dfrac{1}{m} = \sin 30° = \dfrac{1}{2}$$

$$m = 2$$

64. They are both radii of the circle.

65. It is the supplement of a 30° angle.

66. Their sum is 180° − 150° = 30°, and they are equal.

67. Since triangle ACB is a 30°-60°-90° right triangle with AB = 2 and CB = $\sqrt{3}$,

$$DC = BD + BC = 2 + \sqrt{3}.$$

68. In right triangle DAC,

$$(AD)^2 = (AC)^2 + (DC)^2$$
$$= 1^2 + (2 + \sqrt{3})^2$$
$$= 1 + 4 + 4\sqrt{3} + 3$$
$$= 8 + 4\sqrt{3}$$

$$= 6 + 4\sqrt{3} + 2$$
$$= 6 + 2\sqrt{12} + 2$$
$$= (\sqrt{6} + \sqrt{2})^2$$
$$AD = \sqrt{6} + \sqrt{2}.$$

69. In right triangle ADE,

$$\cos D = \dfrac{AD}{DE}$$

$$\cos 15° = \dfrac{\sqrt{6} + \sqrt{2}}{4}.$$

70. In right triangle ADE,

$$(DE)^2 = (AE)^2 + (AD)^2$$
$$4^2 = (AE)^2 + (\sqrt{6} + \sqrt{2})^2$$
$$16 = (AE)^2 + 8 + 4\sqrt{3}$$
$$(AE)^2 = 8 - 4\sqrt{3}$$
$$= 6 - 2\sqrt{12} + 2$$
$$= (\sqrt{6} - \sqrt{2})^2$$
$$AE = \sqrt{6} - \sqrt{2}.$$

Thus,

$$\sin D = \dfrac{AE}{DE}$$

$$\sin 15° = \dfrac{\sqrt{6} - \sqrt{2}}{4}.$$

71. Since angle EAD is a right angle and triangle EAD is a right triangle,

angle AEC = 180° − 90° − 15° = 75°,

implying that angle EAC = 15°. Therefore,

$$\tan 15° = \dfrac{EA}{AC}$$
$$= \dfrac{4 - (2 + \sqrt{3})}{1}$$
$$= 2 - \sqrt{3}.$$

Chapter 5 Review Exercises

1. $\sin(-x) = -\sin x$

 $\tan(-x) = -\tan x$

 $\cot(-x) = -\cot x$

 $\csc(-x) = -\csc x$

The functions sine, tangent, cotangent, and cosecant satisfy the condition $f(-x) = -f(x)$.

3. $\sin(2X) = 2\sin X \cos X$

Since $\sin X = .6$,

 $\cos^2 X = 1 - \sin^2 X$

 $\cos^2 X = 1 - (.6)^2$

 $\cos^2 X = .64$

 $\cos X = \pm.8.$

Since $\pi/2 < X < \pi$, $\cos X = -.8$. Therefore,

 $\sin(2X) = 2(.6)(-.8) = -.96.$

Thus, the screen will display $-.96$ for the final expression.

5. $\cos x = \dfrac{3}{5}$, x is in quadrant IV.

 $\sin^2 x = 1 - \cos^2 x = 1 - \left(\dfrac{3}{5}\right)^2$

 $= \dfrac{16}{25}$

Since x is in quadrant IV, $\sin x < 0$.

 $\sin x = -\sqrt{\dfrac{16}{25}} = -\dfrac{4}{5}$

 $\tan x = \dfrac{\sin x}{\cos x} = \dfrac{-\dfrac{4}{5}}{\dfrac{3}{5}} = -\dfrac{4}{3}$

 $\sec x = \dfrac{1}{\cos x} = \dfrac{1}{\dfrac{3}{5}} = \dfrac{5}{3}$

 $\csc x = \dfrac{1}{\sin x} = \dfrac{1}{-\dfrac{4}{5}} = -\dfrac{5}{4}$

 $\cot x = \dfrac{1}{\tan x} = \dfrac{1}{-\dfrac{4}{3}} = -\dfrac{3}{4}$

7.   (a) $\sin \dfrac{\pi}{12} = \sin\left(\dfrac{\pi}{4} - \dfrac{\pi}{6}\right)$

 $= \sin\dfrac{\pi}{4}\cos\dfrac{\pi}{6} - \cos\dfrac{\pi}{4}\sin\dfrac{\pi}{6}$

 $= \dfrac{\sqrt{2}}{2} \cdot \dfrac{\sqrt{3}}{2} - \dfrac{\sqrt{2}}{2} \cdot \dfrac{1}{2}$

 $= \dfrac{\sqrt{6}}{4} - \dfrac{\sqrt{2}}{4} = \dfrac{\sqrt{6} - \sqrt{2}}{4}$

 $\cos\dfrac{\pi}{12} = \cos\left(\dfrac{\pi}{4} - \dfrac{\pi}{6}\right)$

 $= \cos\dfrac{\pi}{4}\cos\dfrac{\pi}{6} + \sin\dfrac{\pi}{4}\sin\dfrac{\pi}{6}$

 $= \dfrac{\sqrt{2}}{2} \cdot \dfrac{\sqrt{3}}{2} + \dfrac{\sqrt{2}}{2} \cdot \dfrac{1}{2}$

 $= \dfrac{\sqrt{6}}{4} + \dfrac{\sqrt{2}}{4} = \dfrac{\sqrt{6} + \sqrt{2}}{4}$

 $\tan\dfrac{\pi}{12} = \tan\left(\dfrac{\pi}{4} - \dfrac{\pi}{6}\right)$

 $= \dfrac{\tan\dfrac{\pi}{4} - \tan\dfrac{\pi}{6}}{1 + \tan\dfrac{\pi}{4}\tan\dfrac{\pi}{6}}$

 $= \dfrac{1 - \dfrac{\sqrt{3}}{3}}{1 + (1)\dfrac{\sqrt{3}}{3}}$

 $= \dfrac{3 - \sqrt{3}}{3 + \sqrt{3}} \cdot \dfrac{3 - \sqrt{3}}{3 - \sqrt{3}}$

 $= \dfrac{12 - 6\sqrt{3}}{6} = 2 - \sqrt{3}$

(b)   In this exercise $\pi/12$ is in quadrant I, where all circular functions are positive.

$$\sin \frac{\pi}{12} = \sin \left(\frac{\frac{\pi}{6}}{2} \right)$$

$$= \sqrt{\frac{1 - \cos \frac{\pi}{6}}{2}}$$

$$= \sqrt{\frac{1 - \frac{\sqrt{3}}{2}}{2}}$$

$$= \sqrt{\frac{2 - \sqrt{3}}{4}}$$

$$= \frac{\sqrt{2 - \sqrt{3}}}{2}$$

$$\cos \frac{\pi}{12} = \cos \left(\frac{\frac{\pi}{6}}{2} \right)$$

$$= \sqrt{\frac{1 + \cos \frac{\pi}{6}}{2}}$$

$$= \sqrt{\frac{1 + \frac{\sqrt{3}}{2}}{2}}$$

$$= \sqrt{\frac{2 + \sqrt{3}}{4}}$$

$$= \frac{\sqrt{2 + \sqrt{3}}}{2}$$

$$\tan \frac{\pi}{12} = \tan \left(\frac{\frac{\pi}{6}}{2} \right)$$

$$= \frac{\sin \frac{\pi}{6}}{1 + \cos \frac{\pi}{6}}$$

$$= \frac{\frac{1}{2}}{1 + \frac{\sqrt{3}}{2}} = \frac{1}{2 + \sqrt{3}}$$

$$= \frac{1}{2 + \sqrt{3}} \cdot \frac{2 - \sqrt{3}}{2 - \sqrt{3}}$$

$$= 2 - \sqrt{3}$$

9. $\cos 210° = \cos (150° + 60°)$

$$= \cos 150° \cos 60°$$
$$- \sin 150° \sin 60° \quad \text{(e)}$$

11. $\tan (-35°) = \cot [90° - (-35°)]$

$$= \cot 125° \quad \text{(j)}$$

13. $\cos 35° = \cos (-35°) \quad \text{(i)}$

15. $\sin 75° = \sin (15° + 60°)$

$$= \sin 15° \cos 60°$$
$$+ \cos 15° \sin 60° \quad \text{(h)}$$

17. $\cos 300° = \cos 2(150°)$

$$= \cos^2 150° - \sin^2 150° \quad \text{(g)}$$

19. $\csc x = \dfrac{1}{\sin x} \quad \text{(a)}$

21. $\cot x = \dfrac{\cos x}{\sin x} \quad \text{(f)}$

23. $\tan^2 x + 1 = \sec^2 x$

$$= \frac{1}{\cos^2 x} \quad \text{(e)}$$

25. $\sec^2 \theta - \tan^2 \theta$

$$= \frac{1}{\cos^2 \theta} - \frac{\sin^2 \theta}{\cos^2 \theta}$$

$$= \frac{1 - \sin^2 \theta}{\cos^2 \theta}$$

$$= \frac{\cos^2 \theta}{\cos^2 \theta} = 1$$

27. $\tan^2 \theta (1 + \cot^2 \theta)$

$$= \frac{\sin^2 \theta}{\cos^2 \theta} \left(1 + \frac{\cos^2 \theta}{\sin^2 \theta} \right)$$

$$= \frac{\sin^2 \theta}{\cos^2 \theta} \left(\frac{\sin^2 \theta + \cos^2 \theta}{\sin^2 \theta} \right)$$

$$= \frac{\sin^2 \theta}{\cos^2 \theta} \left(\frac{1}{\sin^2 \theta} \right)$$

$$= \frac{1}{\cos^2 \theta}$$

29. $\csc^2 \theta + \sec^2 \theta$

$$= \frac{1}{\sin^2 \theta} + \frac{1}{\cos^2 \theta}$$

$$= \frac{\cos^2 \theta + \sin^2 \theta}{\sin^2 \theta \cos^2 \theta}$$

$$= \frac{1}{\sin^2 \theta \cos^2 \theta}$$

31. Find $\sin (x + y)$, $\cos (x - y)$, and $\tan (x + y)$, given $\sin x = -1/4$, $\cos y = -4/5$, x and y are in quadrant III.

Since x and y are in quadrant III, $\cos x$ and $\cos y$ are negative.

$$\cos x = -\sqrt{1 - \sin^2 x}$$

$$= -\sqrt{1 - \frac{1}{16}}$$

$$= -\sqrt{\frac{15}{16}}$$

$$= -\frac{\sqrt{15}}{4}$$

$$\sin y = -\sqrt{1 - \cos^2 x}$$

$$= -\sqrt{1 - \left(-\frac{4}{5}\right)^2}$$

$$= -\sqrt{\frac{25 - 16}{25}} = -\frac{3}{5}$$

$\sin (x + y)$

$$= \sin x \cos y + \cos x \sin y$$

$$= \left(-\frac{1}{4}\right)\left(-\frac{4}{5}\right) + \left(-\frac{\sqrt{15}}{4}\right)\left(-\frac{3}{5}\right)$$

$$= \frac{4}{20} + \frac{3\sqrt{15}}{20} = \frac{4 + 3\sqrt{15}}{20}$$

$\cos (x - y)$

$$= \cos x \cos y + \sin x \sin y$$

$$= \left(-\frac{\sqrt{15}}{4}\right)\left(-\frac{4}{5}\right) + \left(-\frac{1}{4}\right)\left(-\frac{3}{5}\right)$$

$$= \frac{\sqrt{15}}{5} + \frac{3}{20} = \frac{4\sqrt{15}}{20} + \frac{3}{20}$$

$$= \frac{4\sqrt{15} + 3}{20}$$

$\tan (x + y)$

$$= \frac{\sin (x + y)}{\cos (x + y)}$$

$$= \frac{\dfrac{4 + 3\sqrt{15}}{20}}{\dfrac{4\sqrt{15} - 3}{20}}$$

$$= \frac{4 + 3\sqrt{15}}{4\sqrt{15} - 3} \cdot \frac{4\sqrt{15} + 3}{4\sqrt{15} + 3}$$

$$= \frac{16\sqrt{15} + 12 + 12(15) + 9\sqrt{15}}{16(15) - 9}$$

$$= \frac{192 + 25\sqrt{15}}{231}$$

To find the quadrant of x + y, notice that $\sin (x + y) > 0$, which implies x + y is in quadrant I or II. Also $\tan (x + y) > 0$, which implies that x + y is in quadrant I or III. Therefore, x + y is in quadrant I.

33. Find $\sin (x + y)$, $\cos (x - y)$, and $\tan (x + y)$, given $\sin x = 1/10$, $\cos y = 4/5$, x is in quadrant I, y is in quadrant IV.

Since x is in quadrant I, $\cos x > 0$.

$$\cos x = \sqrt{1 - \sin^2 x}$$

$$= \sqrt{1 - \frac{1}{100}}$$

$$= \sqrt{\frac{99}{100}} = \frac{3\sqrt{11}}{10}$$

Since y is in quadrant IV, $\sin y < 0$.

$$\sin y = -\sqrt{1 - \cos^2 x}$$

$$= -\sqrt{1 - \frac{16}{25}}$$

$$= -\sqrt{\frac{25 - 16}{25}} = -\frac{3}{5}$$

sin (x + y)

= sin x cos y + cos x sin y

$= \left(\frac{1}{10}\right)\left(\frac{4}{5}\right) + \frac{3\sqrt{11}}{10}\left(-\frac{3}{5}\right)$

$= \frac{4}{50} - \frac{9\sqrt{11}}{50} = \frac{4 - 9\sqrt{11}}{50}$

cos (x - y)

= cos x cos y + sin x sin y

$= \left(\frac{3\sqrt{11}}{10}\right)\left(\frac{4}{5}\right) + \left(\frac{1}{10}\right)\left(-\frac{3}{5}\right)$

$= \frac{12\sqrt{11}}{50} - \frac{3}{50} = \frac{12\sqrt{11} - 3}{50}$

tan (x + y)

$= \frac{\sin (x + y)}{\cos (x + y)}$

$= \frac{\dfrac{4 - 9\sqrt{11}}{50}}{\dfrac{12\sqrt{11} + 3}{50}}$

$= \frac{4 - 9\sqrt{11}}{12\sqrt{11} + 3} \cdot \frac{12\sqrt{11} - 3}{12\sqrt{11} - 3}$

$= \frac{48\sqrt{11} - 12 - 11(108) + 27\sqrt{11}}{144(11) - 9}$

$= \frac{75\sqrt{11} - 1200}{1575}$

$= \frac{75(\sqrt{11}) - 75(16)}{75(21)}$

$= \frac{\sqrt{11} - 16}{21}$

To find the quadrant of x + y, notice that sin (x + y) < 0, which implies x + y is in quadrant III or IV. Notice also that tan (x + y) < 0, which implies that x + y is in quadrant II or IV. Therefore, x + y is in quadrant IV.

35. Find sin θ and cos θ, given cos 2θ $= -\frac{3}{4}$, 90° < 2θ < 180°.

Since 2θ is in quadrant II, θ is in quadrant I.

cos 2θ = 1 - 2 sin² θ

$-\frac{3}{4}$ = 1 - 2 sin² θ

$-\frac{7}{4}$ = -2 sin² θ

$\frac{7}{8}$ = sin² θ

Since θ is in quadrant I, sin θ > 0.

$\sqrt{\frac{7}{8}}$ = sin θ

sin θ = $\frac{\sqrt{14}}{4}$

cos θ = $\sqrt{1 - \sin^2 \theta}$

$= \sqrt{1 - \frac{7}{8}} = \sqrt{\frac{1}{8}} = \frac{\sqrt{2}}{4}$

37. Find sin 2x and cos 2x, given tan x = 3, sin x < 0.

If tan x = 3 > 0 and sin x < 0, then x is in quadrant III and 2x is in quadrant I or II.

tan 2x $= \frac{2 \tan x}{1 - \tan^2 x}$

$= \frac{2(3)}{1 - 3^2}$

$= -\frac{6}{8} = -\frac{3}{4}$

Since tan 2x < 0, 2x is in quadrant II. Thus, sec 2x < 0 and sin 2x > 0.

sec 2x $= -\sqrt{1 + \tan^2 2x}$

$= -\sqrt{1 + \left(-\frac{3}{4}\right)^2}$

$= -\sqrt{\frac{25}{16}} = -\frac{5}{4}$

$$\cos 2x = \frac{1}{\sec 2x}$$

$$= -\frac{4}{5}$$

$$\sin 2x = \sqrt{1 - \left(-\frac{4}{5}\right)^2}$$

$$= \frac{3}{5}$$

39. Find $\cos \theta/2$, given $\cos \theta = -1/2$, $90° < \theta < 180°$.

Since θ is in quadrant II, $\theta/2$ is in quadrant I, so $\cos \theta/2 > 0$.

$$\cos \frac{\theta}{2} = \sqrt{\frac{1 + \left(-\frac{1}{2}\right)}{2}}$$

$$= \sqrt{\frac{\frac{1}{2}}{2}} = \sqrt{\frac{1}{4}} = \frac{1}{2}$$

41. Find $\tan x$, given $\tan 2x = 2$, $\pi < x < 3\pi/2$.

$$\tan 2x = \frac{2 \tan x}{1 - \tan^2 x} = 2$$

$$2 \tan x = 2 - 2 \tan^2 x$$

$$\tan^2 x + \tan x - 1 = 0$$

$$\tan x = \frac{-1 \pm \sqrt{1 + 4}}{2}$$

Since x is in quadrant III, $\tan > 0$.

$$\tan x = \frac{-1 + \sqrt{5}}{2}$$

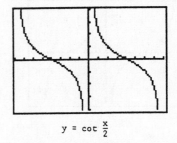

$$y = \cot \frac{x}{2}$$

The calculator graph of

$$y = -\frac{\sin 2x + \sin x}{\cos 2x - \cos x}$$

appears to be the same as the graph of

$$y = \cot \frac{x}{2}.$$

$$-\frac{\sin 2x + \sin x}{\cos 2x - \cos x}$$

$$= -\frac{2 \sin x \cos x + \sin x}{2 \cos^2 x - 1 - \cos x}$$

$$= -\frac{\sin x (2 \cos x + 1)}{(2 \cos x + 1)(\cos x - 1)}$$

$$= -\frac{\sin x}{\cos x - 1}$$

$$= \frac{\sin x}{1 - \cos x}$$

$$= \frac{1}{\frac{1 - \cos x}{\sin x}}$$

$$= \frac{1}{\tan \frac{x}{2}}$$

$$= \cot \frac{x}{2}$$

43.

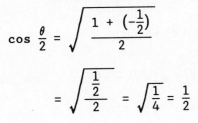

$$y = -\frac{\sin 2x + \sin x}{\cos 2x - \cos x}$$

45.

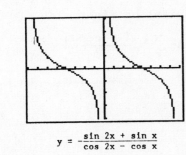

$$y = \frac{\sin x}{1 - \cos x}$$

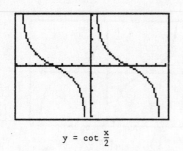

$$y = \cot \frac{x}{2}$$

The calculator graph of

$$y = \frac{\sin x}{1 - \cos x}$$

appears to be the same as the graph of

$$y = \cot \frac{x}{2}.$$

$$\frac{\sin x}{1 - \cos x} = \frac{1}{\dfrac{1 - \cos x}{\sin x}}$$

$$= \frac{1}{\tan \dfrac{x}{2}}$$

$$= \cot \frac{x}{2}$$

47.

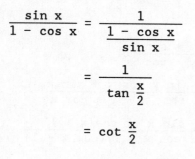

$$y = \frac{2(\sin x - \sin^3 x)}{\cos x}$$

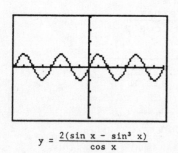

$$y = \sin 2x$$

The calculator graph of

$$y = \frac{2(\sin x - \sin^3 x)}{\cos x}$$

appears to be the same as the graph of

$$y = \sin 2x.$$

$$\frac{2(\sin x - \sin^3 x)}{\cos x}$$

$$= \frac{2 \sin x (1 - \sin^2 x)}{\cos x} \cdot \frac{\cos x}{\cos x}$$

$$= \frac{2 \sin x \cos x (1 - \sin^2 x)}{\cos^2 x}$$

$$= \frac{\sin 2x (1 - \sin^2 x)}{1 - \sin^2 x}$$

$$= \sin 2x$$

49. Verify $\sin^2 x - \sin^2 y = \cos^2 y - \cos^2 x$.

$$\sin^2 x - \sin^2 y$$

$$= (1 - \cos^2 x) - (1 - \cos^2 y)$$

$$= 1 - \cos^2 x - 1 + \cos^2 y$$

$$= \cos^2 y - \cos^2 x$$

51. Verify $\dfrac{\sin^2 x}{2 - 2 \cos x} = \cos^2 \dfrac{x}{2}$.

Work with the left side.

$$\frac{\sin^2 x}{2 - 2 \cos x}$$

$$= \frac{1 - \cos^2 x}{2(1 - \cos x)}$$

$$= \frac{(1 - \cos x)(1 + \cos x)}{2(1 - \cos x)}$$

$$= \frac{1 + \cos x}{2}$$

Work with the right side.

$$\cos^2 \frac{x}{2} = \frac{1 + \cos x}{2}$$

53. Verify $2 \cos A - \sec A$

$= \cos A - \dfrac{\tan A}{\csc A}$.

Work with the right side.

$\cos A - \dfrac{\tan A}{\csc A}$

$= \cos A - \dfrac{\dfrac{\sin A}{\cos A}}{\dfrac{1}{\sin A}}$

$= \cos A - \dfrac{\sin^2 A}{\cos A}$

$= \dfrac{\cos^2 A - \sin^2 A}{\cos A}$

$= \dfrac{\cos^2 A - (1 - \cos^2 A)}{\cos A}$

$= \dfrac{2 \cos^2 A - 1}{\cos A}$

$= 2 \cos A - \dfrac{1}{\cos A}$

$= 2 \cos A - \sec A$

55. Verify $1 + \tan^2 \alpha = 2 \tan \alpha \csc 2\alpha$.

Work with the right side.

$2 \tan \alpha \csc 2\alpha = \dfrac{2 \tan \alpha}{\sin 2\alpha}$

$= \dfrac{2 \dfrac{\sin \alpha}{\cos \alpha}}{2 \sin \alpha \cos \alpha}$

$= \dfrac{2 \sin \alpha}{2 \sin \alpha \cos^2 \alpha}$

$= \dfrac{1}{\cos^2 \alpha} = \sec^2 \alpha$

$= 1 + \tan^2 \alpha$

57. Verify $2 \cos (A + B) \sin (A + B)$

$= \sin 2A \cos 2B + \sin 2B \cos 2A$.

$2 \cos (A + B) \sin (A + B)$

$= \sin [2(A + B)]$

$= \sin (2A + 2B)$

$= \sin 2A \cos 2B + \cos 2A \sin 2B$

$= \sin 2A \cos 2B + \sin 2B \cos 2A$

59. Verify $\tan \theta \sin 2\theta = 2 - 2 \cos^2 \theta$.

$\tan \theta \sin 2\theta$

$= \tan \theta (2 \sin \theta \cos \theta)$

$= \dfrac{\sin \theta}{\cos \theta} (2 \sin \theta \cos \theta)$

$= 2 \sin^2 \theta$

$= 2(1 - \cos^2 \theta)$

$= 2 - 2 \cos^2 \theta$

61. Verify $2 \tan x \csc 2x - \tan^2 x = 1$.

$2 \tan x \csc 2x - \tan^2 x$

$= 2 \tan x \dfrac{1}{\sin 2x} - \tan^2 x$

$= 2 \tan x \dfrac{1}{2 \sin x \cos x} - \tan^2 x$

$= \dfrac{\sin x}{\sin x \cos^2 x} - \dfrac{\sin^2 x}{\cos^2 x}$

$= \dfrac{1}{\cos^2 x} - \dfrac{\sin^2 x}{\cos^2 x} = \dfrac{\cos^2 x}{\cos^2 x} = 1$

63. Verify $\tan \theta \cos^2 \theta$

$= \dfrac{2 \tan \theta \cos^2 \theta - \tan \theta}{1 - \tan^2 \theta}$.

Work with the right side.

$\dfrac{2 \tan \theta \cos^2 \theta - \tan \theta}{1 - \tan^2 \theta}$

$= \dfrac{\tan \theta (2 \cos^2 \theta - 1)}{1 - \dfrac{\sin^2 \theta}{\cos^2 \theta}}$

$= \dfrac{\tan \theta (2 \cos^2 \theta - 1)}{\dfrac{\cos^2 \theta - \sin^2 \theta}{\cos^2 \theta}}$

$= \dfrac{\cos^2 \theta \tan \theta (2 \cos^2 \theta - 1)}{2 \cos^2 \theta - 1}$

$= \cos^2 \theta \tan \theta$

65. Verify $2 \cos^3 x - \cos x$

$$= \frac{\cos^2 x - \sin^2 x}{\sec x}.$$

Work with the right side.

$$\frac{\cos^2 x - \sin^2 x}{\sec x}$$

$$= \frac{2 \cos^2 x - 1}{\dfrac{1}{\cos x}}$$

$$= (2 \cos^2 x - 1) \cos x$$

$$= 2 \cos^3 x - \cos x$$

67. Verify $\cos^4 \theta = \dfrac{3}{8} + \dfrac{1}{2} \cos 2\theta + \dfrac{1}{8} \cos 4\theta$.

Work with the right side.

$$\frac{3}{8} + \frac{1}{2} \cos 2\theta + \frac{1}{8} \cos 4\theta$$

$$= \frac{3}{8} + \frac{2 \cos^2 \theta - 1}{2} + \frac{1}{8}(2 \cos^2 2\theta - 1)$$

$$= \frac{3}{8} + \frac{2 \cos^2 \theta - 1}{2}$$
$$+ \frac{1}{8}[2(2 \cos^2 \theta - 1)^2] - \frac{1}{8}$$

$$= \frac{3}{8} + \frac{2 \cos^2 \theta - 1}{2}$$
$$+ \frac{1}{4}(4 \cos^4 \theta - 4 \cos^2 \theta + 1) - \frac{1}{8}$$

$$= \frac{3}{8} - \frac{1}{2} + \frac{1}{4} - \frac{1}{8} + \cos^2 \theta$$
$$- \cos^2 \theta + \cos^4 \theta$$

$$= \cos^4 \theta$$

69. Verify $\sec^2 \alpha - 1 = \dfrac{\sec 2\alpha - 1}{\sec 2\alpha + 1}.$

Work with the right side.

$$\frac{\sec 2\alpha - 1}{\sec 2\alpha + 1}$$

$$= \frac{\dfrac{1}{\cos 2\alpha} - 1}{\dfrac{1}{\cos 2\alpha} + 1}$$

$$= \frac{\dfrac{1}{\cos^2 \alpha - \sin^2 \alpha} - 1}{\dfrac{1}{\cos^2 \alpha - \sin^2 \alpha} + 1}$$

$$= \frac{1 - \cos^2 \alpha + \sin^2 \alpha}{1 + \cos^2 \alpha - \sin^2 \alpha}$$

$$= \frac{2 \sin^2 \alpha}{2 \cos^2 \alpha} = \tan^2 \alpha$$

$$= \sec^2 \alpha - 1$$

71. Verify $\tan 4\theta = \dfrac{2 \tan 2\theta}{2 - \sec^2 2\theta}.$

$\tan 4\theta$

$$= \frac{2 \tan 2\theta}{2 - \sec^2 2\theta} = \frac{2 \tan 2\theta}{1 - (\sec^2 2\theta - 1)}$$

$$= \frac{2 \tan 2\theta}{1 - \sec^2 2\theta + 1} = \frac{2 \tan 2\theta}{2 - \sec^2 2\theta}$$

73. Verify $\tan \left(\dfrac{x}{2} + \dfrac{\pi}{4}\right) = \sec x + \tan x.$

Work with the left side.

$$\tan \left(\frac{x}{2} + \frac{\pi}{4}\right)$$

$$= \frac{\tan \dfrac{x}{2} + \tan \dfrac{\pi}{4}}{1 - \tan \dfrac{x}{2} \tan \dfrac{\pi}{4}}$$

$$= \frac{\tan \dfrac{x}{2} + 1}{1 - \tan \dfrac{x}{2}}$$

Work with the right side.

$\sec x + \tan x$

$$= \frac{1}{\cos x} + \frac{\sin x}{\cos x}$$

$$= \frac{\cos^2 \dfrac{x}{2} + \sin^2 \dfrac{x}{2} + 2 \sin \dfrac{x}{2} \cos \dfrac{x}{2}}{\cos^2 \dfrac{x}{2} - \sin^2 \dfrac{x}{2}}$$

$$= \frac{\left(\cos \dfrac{x}{2} + \sin \dfrac{x}{2}\right)^2}{\left(\cos \dfrac{x}{2} - \sin \dfrac{x}{2}\right)\left(\cos \dfrac{x}{2} + \sin \dfrac{x}{2}\right)}$$

$$= \frac{\cos \frac{x}{2} + \sin \frac{x}{2}}{\cos \frac{x}{2} - \sin \frac{x}{2}}$$

$$= \frac{\dfrac{\cos \frac{x}{2}}{\cos \frac{x}{2}} + \dfrac{\sin \frac{x}{2}}{\cos \frac{x}{2}}}{\dfrac{\cos \frac{x}{2}}{\cos \frac{x}{2}} - \dfrac{\sin \frac{x}{2}}{\cos \frac{x}{2}}} = \frac{1 + \tan \frac{x}{2}}{1 - \tan \frac{x}{2}}$$

CHAPTER 5 TEST

[5.1] Let tan x = -2/3, with x in quadrant II. Find each of the following.

    1. sin x   **2.** cot x   **3.** sec x

Let sin s = 1/4 with s in quadrant II and let cos t = -2/3 with t in quadrant III. Find each of the following.

[5.3] **4.** cos (s - t)   **[5.4]** **5.** sin (s + t)

[5.5] **6.** sin 2s   **[5.6]** **7.** cos $\frac{1}{2}$t

Suppose x = -π/12. Use identities to find the exact value of each of the following.

[5.3] **8.** cos x   **[5.4]** **9.** sin x

Answer *true* or *false* for each equation.

[5.6] **10.** cos 17° = 2 cos² 34° - 1

[5.1] **11.** csc (-57°) = -csc 57°

[5.4] **12.** sin 91° = sin 34° cos 57° + cos 34° sin 57°

[5.4] **13.** tan 23° = $\dfrac{\tan 70° + \tan 47°}{1 - \tan 70° \tan 47°}$

[5.2] **14.** A student claims that

$$\frac{\cos \theta}{\sin \theta \cot \theta} = 1$$

cannot be verified as an identity, since by letting $\theta = 90°$ we get $0/0$, an undefined expression. Comment on this student's reasoning.

[5.2] Use identities to write each expression in terms of $\sin \theta$ and $\cos \theta$, and simplify.

15. $\dfrac{\tan \theta}{\sec \theta}$ **16.** $\tan^2 \theta \ (1 + \cot^2 \theta)$

Verify that the equation is an identity.

[5.2] **17.** $\dfrac{\cos^2 \beta}{\sin \beta} = \csc \beta - \sin \beta$ [5.5] **18.** $\dfrac{\sin 2\alpha}{\tan \alpha} = 2 - \dfrac{2}{\csc^2 \alpha}$

[5.6] **19.** $\dfrac{\sec A}{\csc A} \tan \dfrac{A}{2} = \sec A - 1$ [5.5] **20.** $\dfrac{\cos 4\theta}{\cos^4 \theta} = \sec^4 \theta - 8 \tan^2 \theta$

CHAPTER 5 TEST ANSWERS

1. $\dfrac{2\sqrt{13}}{13}$ 2. $-\dfrac{3}{2}$ 3. $-\dfrac{\sqrt{13}}{3}$ 4. $\dfrac{2\sqrt{15}-\sqrt{5}}{12}$ 5. $\dfrac{-2+5\sqrt{3}}{12}$

6. $-\dfrac{\sqrt{15}}{8}$ 7. $-\dfrac{\sqrt{6}}{6}$ 8. $\dfrac{\sqrt{2}+\sqrt{6}}{4}$ 9. $\dfrac{\sqrt{2}-\sqrt{6}}{4}$ 10. False

11. True   12. True   13. False

14. Identities are valid for all angles for which both sides are defined. The equation is an identity for all values of θ which are *not* multiples of 90°.

15. $\sin\theta$ 16. $\dfrac{1}{\cos^2\theta}$

17. RHS $= \csc\beta - \sin\beta = \dfrac{1}{\sin\beta} - \sin\beta = \dfrac{1-\sin^2\beta}{\sin\beta} = \dfrac{\cos^2\beta}{\sin\beta} =$ LHS

18. LHS $= \dfrac{\sin 2\alpha}{\tan\alpha} = \dfrac{2\sin\alpha\cos\alpha}{\dfrac{\sin\alpha}{\cos\alpha}} = 2\cos^2\alpha = 2(1-\sin^2\alpha)$

    $= 2 - \dfrac{2}{\csc^2\alpha} =$ RHS

19. LHS $= \dfrac{\sec A}{\csc A}\tan\dfrac{A}{2} = \dfrac{\dfrac{1}{\cos A}}{\dfrac{1}{\sin A}}\left(\dfrac{1-\cos A}{\sin A}\right) = \dfrac{\sin A}{\cos A}\left(\dfrac{1-\cos A}{\sin A}\right)$

    $= \dfrac{1-\cos A}{\cos A} = \dfrac{1}{\cos A} - \dfrac{\cos A}{\cos A} = \sec A - 1 =$ RHS

20. LHS $= \dfrac{\cos 4\theta}{\cos^4\theta} = \dfrac{1-2\sin^2 2\theta}{\cos^4\theta} = \dfrac{1-2(2\sin\theta\cos\theta)^2}{\cos^4\theta}$

    $= \dfrac{1-8\sin^2\theta\cos^2\theta}{\cos^4\theta} = \dfrac{1}{\cos^4\theta} - \dfrac{8\sin^2\theta}{\cos^2\theta}$

    $= \sec^4\theta - 8\tan^2\theta =$ RHS

CHAPTER 6   INVERSE TRIGONOMETRIC FUNCTIONS
           AND TRIGONOMETRIC EQUATIONS

Section 6.1

1. Since

$$\sin\left(-\frac{\pi}{6}\right) = -\frac{1}{2},$$

the exact value of Y is $-\pi/6$.

3. Suppose $y = \sec^{-1}(a)$. Then

$$\sec y = a$$
$$\cos y = \frac{1}{a}$$
$$y = \cos^{-1}\left(\frac{1}{a}\right)$$

Therefore, $\sec^{-1}(a)$ is calculated as $\cos^{-1}(1/a)$.

5. $y = \arcsin\left(-\frac{1}{2}\right)$

$\sin y = -\frac{1}{2}, -\frac{\pi}{2} \le y \le \frac{\pi}{2}$

y is in quadrant IV. The reference angle is $\pi/6$.

$y = -\frac{\pi}{6}$

7. $y = \tan^{-1} 1$

$\tan y = 1, -\frac{\pi}{2} < y < \frac{\pi}{2}$

y is an quadrant I.

$y = \frac{\pi}{4}$

9. $y = \cos^{-1}(-1)$

$\cos y = -1, 0 \le y \le \pi$

$y = \pi$

11. $y = \sin^{-1}\left(-\frac{\sqrt{3}}{2}\right)$

$\sin y = -\frac{\sqrt{3}}{2}, -\frac{\pi}{2} \le y \le \frac{\pi}{2}$

y is in quadrant IV. The reference angle is $\pi/3$.

$y = -\frac{\pi}{3}$

13. $y = \arctan 0$

$\tan y = 0, -\frac{\pi}{2} < y < \frac{\pi}{2}$

Since $\tan 0 = 0$, $y = 0$.

15. $y = \arccos 0$

$\cos y = 0, 0 \le y \le \pi$

$y = \frac{\pi}{2}$

17. $y = \sin^{-1}\left(\frac{\sqrt{2}}{2}\right)$

$\sin y = \frac{\sqrt{2}}{2}, -\frac{\pi}{2} \le y \le \frac{\pi}{2}$

Since $\sin\frac{\pi}{4} = \frac{\sqrt{2}}{2}$, $y = \frac{\pi}{4}$.

19. $y = \arcos\left(-\frac{\sqrt{3}}{2}\right)$

$\cos y = -\frac{\sqrt{3}}{2}, 0 \le y \le \pi$

Since $\cos\frac{5\pi}{6} = -\frac{\sqrt{3}}{2}$, $y = \frac{5\pi}{6}$.

21. $y = \cot^{-1}(-1)$

$\cot y = -1, 0 < y < \pi$

y is in quadrant II. The reference angle is $\pi/4$.

$y = \frac{3\pi}{4}$

23. $y = \csc^{-1}(-2)$

$\csc y = -2,\ -\dfrac{\pi}{2} \leq y \leq \dfrac{\pi}{2}$

y is in quadrant IV. The reference
angle is $\pi/6$.

$y = -\dfrac{\pi}{6}$

25. $y = \text{arcsec}\left(\dfrac{2\sqrt{3}}{3}\right)$

$\sec y = \dfrac{2\sqrt{3}}{3},\ 0 \leq y \leq \pi,\ y \neq \dfrac{\pi}{2}$

Since $\sec \dfrac{\pi}{6} = \dfrac{2\sqrt{3}}{3},\ y = \dfrac{\pi}{6}$.

27. $y = \text{arccot}\left(\dfrac{\sqrt{3}}{3}\right)$

$\cot y = \dfrac{\sqrt{3}}{3},\ -\dfrac{\pi}{2} \leq y \leq \dfrac{\pi}{2},\ y \neq 0$

Since $\cot \dfrac{\pi}{3} = \dfrac{\sqrt{3}}{3},\ y = \dfrac{\pi}{3}$.

29. $\theta = \arctan(-1)$
    $\tan \theta = -1,\ -90° < \theta < 90°$

θ is in quadrant IV. The reference
angle is 45°.
Thus, $\theta = -45°$.

31. $\theta = \arcsin\left(-\dfrac{\sqrt{3}}{2}\right)$

$\sin \theta = -\dfrac{\sqrt{3}}{2},\ -90° \leq \theta \leq 90°$

θ is in quadrant IV. The reference
angle is 60°.
$\theta = -60°$

33. $\theta = \cot^{-1}\left(-\dfrac{\sqrt{3}}{3}\right)$

$\cot \theta = -\dfrac{\sqrt{3}}{3}$ and $0° < \theta < 180°$

θ is in quadrant II. The reference
angle is 60°.
$\theta = 180° - 60° = 120°$

35. $\theta = \csc^{-1}(-2)$

$\csc \theta = -2$ and $-90° < \theta < 90°$,
$\theta \neq 0°$

θ is in quadrant IV. The reference
angle is 30°.
$\theta = -30°$.

For Exercises 37–41, be sure that your
calculator is in radian mode. Keystroke
sequences may vary based on the type and/
or model of calculator being used.

37. arctan 1.1111111

$\tan \theta = 1.1111111,\ -\dfrac{\pi}{2} < \theta < \dfrac{\pi}{2}$

θ is in quadrant I.
With a calculator:
Enter: 1.1111111 $\boxed{\text{INV}}$ $\boxed{\text{TAN}}$

or $\boxed{\text{2nd}}$ $\boxed{\text{TAN}}$ 1.1111111 $\boxed{\text{ENTER}}$

Display: .83798122

arctan 1.1111111 = .83798122

39. $\cot^{-1}(-.92170128)$
    $\cot \theta = -.92170128,\ 0 < \theta < \pi$

θ is in quadrant II.

With a calculator:
Enter: .92170128

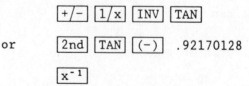

         $\boxed{+/-}$ $\boxed{1/x}$ $\boxed{\text{INV}}$ $\boxed{\text{TAN}}$

or $\boxed{\text{2nd}}$ $\boxed{\text{TAN}}$ $\boxed{(-)}$ .92170128

        $\boxed{x^{-1}}$

The absolute value of this result is
the reference angle of θ. To find θ
in quadrant II, subtract the result
from π. Since the value is negative, we can simply add π.

Continuing with calculator:

Enter: $\boxed{+}$ $\boxed{\pi}$

or $\boxed{+}$ $\boxed{\text{2nd}}$ $\boxed{\pi}$ $\boxed{\text{ENTER}}$

Display: 2.3154725

$\cot^{-1}(-.92170128) = 2.3154725$

41. arcsin .92837781

$\sin\theta = .92837781, \ -\dfrac{\pi}{2} \le \theta \le \dfrac{\pi}{2}$

θ is in quadrant I.

With a calculator:

Enter: .92837781 $\boxed{\text{INV}}$ $\boxed{\text{SIN}}$

or $\boxed{\text{2nd}}$ $\boxed{\text{SIN}}$ .92837781 $\boxed{\text{ENTER}}$

Display: 1.1900238

arcsin .92837781 = 1.1900238

For Exercises 43–47, be sure that your calculator is in degree mode. Keystroke sequences may vary based on the type and/or model of calculator being used.

43. $\theta = \sin^{-1}(-.13349122)$

$\sin\theta = -.13349122, \ -\dfrac{\pi}{2} \le \theta \le \dfrac{\pi}{2}$

In degrees, the range is

$$-90° \le \theta \le 90°.$$

θ is in quadrant IV.

With a calculator:

Enter: .13349122 $\boxed{+/-}$ $\boxed{\text{INV}}$ $\boxed{\text{SIN}}$

or $\boxed{\text{2nd}}$ $\boxed{\text{SIN}}$ $\boxed{(-)}$ .13349122

$\boxed{\text{ENTER}}$

Display: -7.6713835

$\theta = -7.6713835°$

45. $\theta = \arccos(-.39876459)$

$\cos\theta = -.39876459, \ 0 < \theta < \pi$, or

$$0° < \theta < 180°.$$

θ is in quadrant II.

With a calculator:

Enter: .39876459 $\boxed{+/-}$ $\boxed{\text{INV}}$ $\boxed{\text{COS}}$

or $\boxed{\text{2nd}}$ $\boxed{\text{COS}}$ $\boxed{(-)}$ .39876459

$\boxed{\text{ENTER}}$

Display: 113.50097

arccos -.39876459 = 113.50097°

47. $\theta = \csc^{-1} 1.9422833$

$\csc\theta = 1.9422833, \ -\dfrac{\pi}{2} \le \theta \le \dfrac{\pi}{2}$, or

$$0° < \theta < 180°$$

θ is in quadrant I.

With a calculator:

Enter: 1.9422833 $\boxed{1/x}$ $\boxed{\text{INV}}$ $\boxed{\text{SIN}}$

or $\boxed{\text{2nd}}$ $\boxed{\text{SIN}}$ 1.9422833 $\boxed{x^{-1}}$

$\boxed{\text{ENTER}}$

Display: 30.987961

$\csc^{-1} 1.9422833 = 30.987961°$

49. $y = \cot^{-1} x$

$y = \cot^{-1} x$ means $\cot y = x$.

Domain: $(-\infty, \infty)$

Range: $(0, \pi)$

See the answer graph in the back of the textbook.

51. $y = \text{arcsec } x$

$y = \text{arcsec } x$ means $\sec y = x$.

Domain: $(-\infty, -1] \cup [1, \infty)$

Range: $\left[0, \dfrac{\pi}{2}\right) \cup \left(\dfrac{\pi}{2}, \pi\right]$

See the answer graph in the back of the textbook.

53. Be sure that your calculator is in radian mode. $\sin^{-1} 1.003$ produces an error message because 1.003 is not in the domain of the arcsin func- tion, which is $[-1, 1]$.

57. $\tan \left(\arccos \frac{3}{4}\right)$

Let $\omega = \arccos \frac{3}{4}$, $0 \le \omega \le \pi$.

$\cos \omega = \frac{3}{4}$; ω is in quadrant I.

Sketch ω and label a triangle with the side opposite ω equal to

$$\sqrt{4^2 - 3^2} = \sqrt{7}.$$

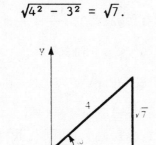

$$\tan \left(\arccos \frac{3}{4}\right) = \tan \omega = \frac{\sqrt{7}}{3}$$

59. $\cos (\tan^{-1} (-2))$

Let $\omega = \tan^{-1} (-2)$, $-\frac{\pi}{2} < \omega < \frac{\pi}{2}$.

$\tan \omega = -2$; ω is in quadrant IV.

Sketch ω and label a triangle with the hypotenuse equal to

$$\sqrt{(-2)^2 + 1} = \sqrt{5}.$$

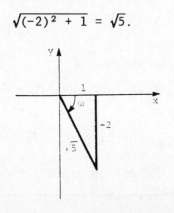

$\cos (\tan^{-1} (-2)) = \cos \omega$

$$= \frac{1}{\sqrt{5}} = \frac{\sqrt{5}}{5}$$

61. $\cot \left(\arcsin \left(-\frac{2}{3}\right)\right)$

Let $\omega = \arcsin \left(-\frac{2}{3}\right)$, $-\frac{\pi}{2} \le \omega \le \frac{\pi}{2}$.

$\sin \omega = -\frac{2}{3}$, ω is in quadrant IV.

Sketch ω and label a triangle with the side adjacent to ω equal to

$$\sqrt{3^2 - (-2)^2} = \sqrt{5}.$$

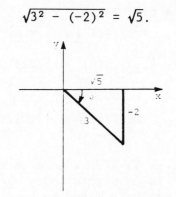

$\cot \left(\arcsin \left(-\frac{2}{3}\right)\right) = \cot \omega$

$$= -\frac{\sqrt{5}}{2}$$

63. $\sec (\sec^{-1} 2)$

Let $\theta = \sec^{-1} 2$, $0 \le \theta \le \pi$.

$\sec \theta = 2$, θ is in quadrant I.

$$\sec (\sec^{-1} 2) = \sec \theta = 2$$

65. $\arccos \left(\cos \frac{\pi}{4}\right)$

Let $\arccos \left(\cos \frac{\pi}{4}\right) = \theta$.

$\cos \theta = \cos \frac{\pi}{4}$, and $0 \le \theta \le \pi$.

$$\theta = \frac{\pi}{4}$$

67. $\arcsin\left(\sin\frac{\pi}{3}\right)$

Let $\arcsin\left(\sin\frac{\pi}{3}\right) = \theta$.

$\sin\theta = \sin\frac{\pi}{3}$ and $-\frac{\pi}{2} \leq \theta \leq \frac{\pi}{2}$.

$\quad\quad\theta = \frac{\pi}{3}$

69. $\sin\left(2\tan^{-1}\frac{12}{5}\right)$

Let $\omega = \tan^{-1}\frac{12}{5}$, $-\frac{\pi}{2} < \omega < \frac{\pi}{2}$.

$\tan\omega = \frac{12}{5}$

Sketch ω in quadrant I and a right
triangle with a hypotenuse equal to
$\sqrt{12^2 + 5^2} = 13$.

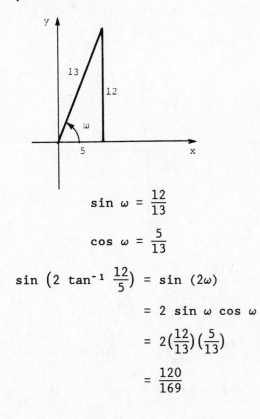

$$\sin\omega = \frac{12}{13}$$

$$\cos\omega = \frac{5}{13}$$

$\sin\left(2\tan^{-1}\frac{12}{5}\right) = \sin(2\omega)$

$\quad\quad = 2\sin\omega\cos\omega$

$\quad\quad = 2\left(\frac{12}{13}\right)\left(\frac{5}{13}\right)$

$\quad\quad = \frac{120}{169}$

71. $\cos\left(2\arctan\frac{4}{3}\right)$

Let $\omega = \arctan\frac{4}{3}$, $-\frac{\pi}{2} < \omega < \frac{\pi}{2}$.

$\tan\omega = \frac{4}{3}$, ω is in quadrant I.

Sketch ω in quadrant I and a right
triangle with a hypotenuse equal to
$\sqrt{4^2 + 3^2} = 5$.

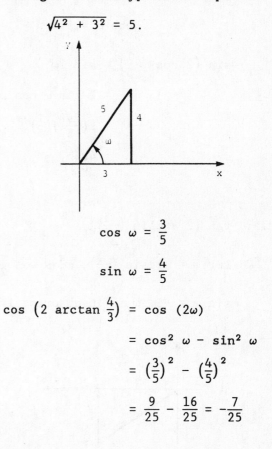

$$\cos\omega = \frac{3}{5}$$

$$\sin\omega = \frac{4}{5}$$

$\cos\left(2\arctan\frac{4}{3}\right) = \cos(2\omega)$

$\quad\quad = \cos^2\omega - \sin^2\omega$

$\quad\quad = \left(\frac{3}{5}\right)^2 - \left(\frac{4}{5}\right)^2$

$\quad\quad = \frac{9}{25} - \frac{16}{25} = -\frac{7}{25}$

73. $\sin\left(2\cos^{-1}\frac{1}{5}\right)$

Let $\omega = \cos^{-1}\frac{1}{5}$, $0 \leq \omega \leq \pi$.

$\cos\omega = \frac{1}{5}$, ω is in quadrant I.

The side opposite ω is

$$\sqrt{5^2 - 1^2} = \sqrt{24} = 2\sqrt{6}.$$

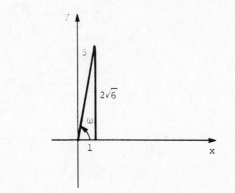

$$\sin \omega = \frac{2\sqrt{6}}{5}$$

$$\cos \omega = \frac{1}{5}$$

$$\sin \left(2 \cos^{-1} \frac{1}{5}\right) = \sin 2\omega$$

$$= 2 \sin \omega \cos \omega$$

$$= 2\left(\frac{2\sqrt{6}}{5}\right)\left(\frac{1}{5}\right)$$

$$= \frac{4\sqrt{6}}{25}$$

75. $\tan \left(2 \arcsin \left(-\frac{3}{5}\right)\right)$

Let $\omega = \arcsin \left(-\frac{3}{5}\right)$, $-\frac{\pi}{2} \leq \omega \leq \frac{\pi}{2}$.

$\sin \omega = -\frac{3}{5}$, ω is in quadrant IV.

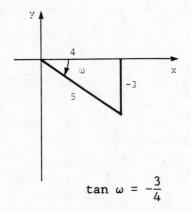

$$\tan \omega = -\frac{3}{4}$$

$$\tan \left(2 \arcsin \left(-\frac{3}{5}\right)\right) = \tan (2\omega)$$

$$= \frac{2 \tan \omega}{1 - \tan^2 \omega}$$

$$= \frac{2\left(-\frac{3}{4}\right)}{1 - \left(-\frac{3}{4}\right)^2}$$

$$= \frac{-\frac{3}{2}}{1 - \frac{9}{16}} \cdot \frac{16}{16}$$

$$= \frac{-24}{16 - 9} = -\frac{24}{7}$$

77. $\sin \left(\sin^{-1} \frac{1}{2} + \tan^{-1} (-3)\right)$

Let $\omega_1 = \sin^{-1} \frac{1}{2}$, $-\frac{\pi}{2} \leq \omega_1 \leq \frac{\pi}{2}$.

$\sin \omega_1 = \frac{1}{2}$; ω_1 is in quadrant I.

$$\cos \omega_1 = \frac{\sqrt{3}}{2}, \ \sin \omega_1 = \frac{1}{2}$$

Let $\omega_2 = \tan^{-1} (-3)$, $-\frac{\pi}{2} < \omega_2 < \frac{\pi}{2}$.

$\tan \omega_2 = -3$

$$= -\frac{3}{1}, \ \omega_2 \text{ is in quadrant IV.}$$

$$\cos \omega_2 = \frac{1}{\sqrt{10}} = \frac{\sqrt{10}}{10}$$

$$\sin \omega_2 = \frac{-3}{\sqrt{10}} = -\frac{3\sqrt{10}}{10}$$

$\sin \left(\sin^{-1} \frac{1}{2} + \tan^{-1} (-3)\right)$

$$= \sin (\omega_1 + \omega_2)$$

$$= \sin \omega_1 \cos \omega_2 + \cos \omega_1 \sin \omega_2$$

$$= \left(\frac{1}{2}\right)\left(\frac{\sqrt{10}}{10}\right) + \left(\frac{\sqrt{3}}{2}\right)\left(\frac{-3\sqrt{10}}{10}\right)$$

$$= \frac{\sqrt{10}}{20} - \frac{3\sqrt{30}}{20}$$

$$= \frac{\sqrt{10} - 3\sqrt{30}}{20}$$

79. $\cos\left(\arcsin\dfrac{3}{5} + \arccos\dfrac{5}{13}\right)$

Let $\omega_1 = \arcsin\dfrac{3}{5}$, $-\dfrac{\pi}{2} \le \omega_1 \le \dfrac{\pi}{2}$.

$\sin\omega_1 = \dfrac{3}{5}$, ω_1 is in quadrant I.

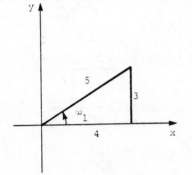

$$\cos\omega_1 = \dfrac{x}{5} = \dfrac{4}{5},$$

$$\sin\omega_1 = \dfrac{3}{5}$$

Let $\omega_2 = \arccos\dfrac{5}{13}$, $0 \le \omega \le \pi$.

$\cos\omega_2 = \dfrac{5}{12}$, ω_2 is in quadrant I.

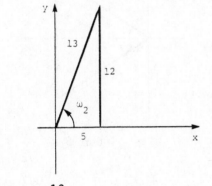

$$\sin\omega_1 = \dfrac{12}{13}$$

$$\cos\omega_2 = \dfrac{5}{13}$$

$\cos\left(\arcsin\dfrac{3}{5} + \arccos\dfrac{5}{13}\right)$

$= \cos(\omega_1 + \omega_2)$

$= \cos\omega_1 \cos\omega_2 - \sin\omega_1 \sin\omega_2$

$= \left(\dfrac{4}{5}\right)\left(\dfrac{5}{13}\right) - \left(\dfrac{3}{5}\right)\left(\dfrac{12}{13}\right)$

$= \dfrac{20}{65} - \dfrac{36}{65} = -\dfrac{16}{65}$

For Exercises 81 and 83, be sure that your calculator is in degree mode. Keystroke sequences may vary based on the type and/or model of calculator being used.

81. $\cos(\tan^{-1} .5)$

Enter: .5 $\boxed{\text{INV}}$ $\boxed{\text{TAN}}$ $\boxed{\text{COS}}$

or $\boxed{\text{COS}}$ $\boxed{\text{2nd}}$ $\boxed{\text{TAN}}$ .5 $\boxed{\text{ENTER}}$

$\cos(\tan^{-1} .5) = .89442719$

83. $\tan(\arcsin .12251014)$

Enter: .12251014 $\boxed{\text{INV}}$ $\boxed{\text{SIN}}$ $\boxed{\text{TAN}}$

or $\boxed{\text{TAN}}$ $\boxed{\text{2nd}}$ $\boxed{\text{SIN}}$ .12251014

$\boxed{\text{ENTER}}$

$\tan(\arcsin .12251014) = .12343998$

85. $\sin(\arccos u)$

Let $\omega = \arccos u$, $0 < \omega < \pi$.

$\cos\omega = u = \dfrac{u}{1}$

If $u > 0$, then ω is in quadrant I.

If $u < 0$, then ω is in quadrant II.

In either quadrant, $\sin\omega > 0$.

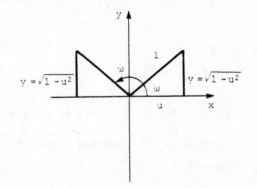

Since $y > 0$, from the Pythagorean theorem,

$$y = \sqrt{1 - u^2}.$$

$$\sin\omega = \dfrac{\sqrt{1 - u^2}}{1}$$

$$= \sqrt{1 - u^2}$$

Then sin (arccos u) = sin ω

$$= \sqrt{1 - u^2}.$$

87. cot (arcsin u)

Let ω = arcsin u, $-\dfrac{\pi}{2} \le \omega \le \dfrac{\pi}{2}$.

Then sin ω = u = $\dfrac{u}{1}$.

If u > 0, then ω is in quadrant I
and cot ω > 0.

If u < 0, then ω is in quadrant IV
and cot ω < 0.

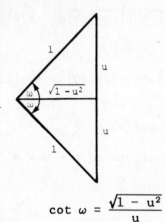

$$\cot \omega = \frac{\sqrt{1 - u^2}}{u}$$

Then

cot (arcsin u) = cot ω

$$= \frac{\sqrt{1 - u^2}}{u}.$$

89. sin $\left(\sec^{-1} \dfrac{u}{2}\right)$

Let ω = $\sec^{-1} \dfrac{u}{2}$, $0 \le \omega \le \pi$, $\omega \ne \dfrac{\pi}{2}$.

Since ω is in quadrant I or II,
sin ω > 0.

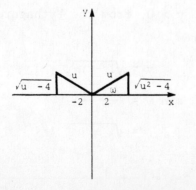

$$\sin \omega = \frac{\sqrt{u^2 - 4}}{u}$$

Then

$$\sin \left(\sec^{-1} \frac{u}{2}\right) = \sin \omega$$

$$= \frac{\sqrt{u^2 - 4}}{|u|}.$$

91. tan $\left(\text{arcsin} \dfrac{u}{\sqrt{u^2 + 2}}\right)$

Let ω = arcsin $\dfrac{u}{\sqrt{u^2 + 2}}$, $-\dfrac{\pi}{2} \le \omega \le \dfrac{\pi}{2}$.

Then sin ω = $\dfrac{u}{\sqrt{u^2 + 2}}$.

If u > 0, then ω is in quadrant I
and tan ω > 0.

If u < 0, then ω is in quadrant IV
and ω < 0.

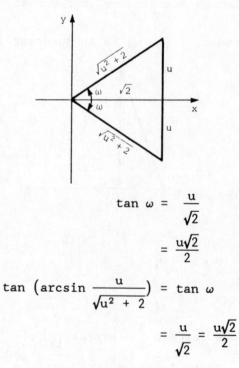

$$\tan \omega = \frac{u}{\sqrt{2}}$$

$$= \frac{u\sqrt{2}}{2}$$

$$\tan \left(\text{arcsin} \frac{u}{\sqrt{u^2 + 2}}\right) = \tan \omega$$

$$= \frac{u}{\sqrt{2}} = \frac{u\sqrt{2}}{2}$$

95. α = 2 arcsin $\dfrac{1}{m}$

(a) m = 1.2

α = 2 arcsin $\left(\dfrac{1}{m}\right)$

α = 2 arcsin $\left(\dfrac{1}{1.2}\right)$ = 113°

(b) m = 1.5

$\alpha = 2 \arcsin\left(\dfrac{1}{1.5}\right) = 84°$

(c) m = 2

$\alpha = 2 \arcsin\left(\dfrac{1}{2}\right) = 60°$

(d) m = 2.5

$\alpha = 2 \arcsin\left(\dfrac{1}{2.5}\right) = 47°$

Section 6.2

1. The equation sin x − b = 0 is equivalent to sin x = b. If we consider the graph of y = sin x in the interval [0, 2π) and the graph of y = b, we see that the graph of y = b is a horizontal line which will intersect the graph of y = sin x in two points if −1 < b < 1, one point if b = ±1, and no points if b > 1 or b < −1.
   Therefore, sin x − b = 0 has two solutions if −1 < b < 1, one solution if b = ±1, and no solutions if b > 1 or b < −1.

3. Since x = a is a solution of tan x − b = 0 and since the period of y = tan x is π, all solutions are given by the expression

   a + nπ, where n is an integer.

5. If 2 cos X − 1 = 0, then

   $\cos X = \dfrac{1}{2}.$

   In the interval [0, 2π),

   $X = \dfrac{\pi}{3} \text{ or } \dfrac{5\pi}{3}.$

7. 2 cot x + 1 = −1

   2 cot x = −2

   cot x = −1

   $x = \dfrac{3\pi}{4}, \dfrac{7\pi}{4}$

9. 2 sin x + 3 = 4

   2 sin x = 1

   $\sin x = \dfrac{1}{2}$

   $x = \dfrac{\pi}{6}, \dfrac{5\pi}{6}$

11. $\tan^2 x + 3 = 0$

    $\tan^2 x = -3$

    $\tan x = \pm\sqrt{-3}$

    The square root of a negative number is not real.
    No solution

13. $(\cot x - 1)(\sqrt{3} \cot x + 1) = 0$

    cot x − 1 = 0

    cot x = 1

    $x = \dfrac{\pi}{4}, \dfrac{5\pi}{4}$

    or

    $\sqrt{3} \cot x + 1 = 0$

    $\cot x = \dfrac{-1}{\sqrt{3}} = -\dfrac{\sqrt{3}}{3}$

    $x = \dfrac{2\pi}{3}, \dfrac{5\pi}{3}$

    $x = \dfrac{\pi}{4}, \dfrac{2\pi}{3}, \dfrac{5\pi}{4}, \dfrac{5\pi}{3}$

15. $\cos^2 x + 2 \cos x + 1 = 0$

    $(\cos x + 1)^2 = 0$

    cos x + 1 = 0

    cos x = −1

    x = π

17. $-2 \sin^2 x = 3 \sin x + 1$

$\qquad 2 \sin^2 x + 3 \sin x + 1 = 0$

$\qquad (2 \sin x + 1)(\sin x + 1) = 0$

$\qquad\qquad\qquad 2 \sin x + 1 = 0$

$\qquad\qquad\qquad\qquad\qquad \sin x = -\dfrac{1}{2}$

$x = \dfrac{7\pi}{6}, \dfrac{11\pi}{6}$

or

$\sin x + 1 = 0$

$\qquad \sin x = -1$

$\qquad\qquad x = \dfrac{3\pi}{2}$

$x = \dfrac{3\pi}{2}$

$x = \dfrac{7\pi}{6}, \dfrac{3\pi}{2}, \dfrac{11\pi}{6}$

19. $\qquad\qquad 2 \cos^4 x = \cos^2 x$

$\qquad\qquad 2 \cos^4 x - \cos^2 x = 0$

$\cos^2 x (2 \cos^2 x - 1) = 0$

$\cos^2 x = 0$

$\ \cos x = 0$

$\qquad\qquad x = \dfrac{\pi}{2}, \dfrac{3\pi}{2}$

or

$2 \cos^2 x - 1 = 0$

$\qquad \cos^2 x = \dfrac{1}{2}$

$\qquad\qquad \cos x = \pm\dfrac{\sqrt{2}}{2}$

$\qquad\qquad\qquad x = \dfrac{\pi}{4}, \dfrac{3\pi}{4}, \dfrac{5\pi}{4}, \dfrac{7\pi}{4}$

$x = \dfrac{\pi}{4}, \dfrac{\pi}{2}, \dfrac{3\pi}{4}, \dfrac{5\pi}{4}, \dfrac{3\pi}{2}, \dfrac{7\pi}{4}$

21. $\qquad 2 \sin^2 x - \sin x - 1 = 0$

$\quad (2 \sin x + 1)(\sin x - 1) = 0$

$2 \sin x + 1 = 0$

$\qquad \sin x = -\dfrac{1}{2}$

$x = \dfrac{7\pi}{6} + 2n\pi, \dfrac{11\pi}{6} + 2n\pi,$ where n is

an integer.

or

$\sin x - 1 = 0$

$\qquad \sin x = 1$

$x = \dfrac{\pi}{2} + 2n\pi,$ where n is an integer.

$x = \dfrac{\pi}{2} + 2n\pi, \dfrac{7\pi}{6} + 2n\pi, \dfrac{11\pi}{6} + 2n\pi,$

where n is an integer.

23. $4 \cos^2 x - 1 = 0$

$\qquad \cos^2 x = \dfrac{1}{4}$

$\qquad\qquad \cos x = \pm\dfrac{1}{2}$

$x = \dfrac{\pi}{3} + 2n\pi, \dfrac{2\pi}{3} + 2n\pi, \dfrac{4\pi}{3} + 2n\pi,$

$\dfrac{5\pi}{3} + 2n\pi,$ where n is an integer.

25. $\qquad \cos^2 x + \cos x - 6 = 0$

$(\cos x + 3)(\cos x - 2) = 0$

$\cos x + 3 = 0 \quad$ or $\quad \cos x - 2 = 0$

$\qquad \cos x = -3 \ $ or $\qquad \cos x = 2$

There is no solution because

$-1 \le \cos x \le 1$ for every value of x.

27. $(\cot \theta - \sqrt{3})(2 \sin \theta + \sqrt{3}) = 0$

$\cot \theta - \sqrt{3} = 0$

$\qquad \cot \theta = \sqrt{3}$

$\theta = 30°, 210°$

or

$$2 \sin \theta + \sqrt{3} = 0$$

$$\sin \theta = -\frac{\sqrt{3}}{2}$$

$$\theta = 240°, \ 300°$$

$$\theta = 30°, \ 210°, \ 240°, \ 300°$$

29. $2 \sin \theta - 1 = \csc \theta$

Use the identity

$$\csc \theta = \frac{1}{\sin \theta}, \ \sin \theta \neq 0.$$

$$2 \sin \theta - 1 = \frac{1}{\sin \theta}$$

$$2 \sin^2 \theta - \sin \theta = 1$$

$$2 \sin^2 \theta - \sin \theta - 1 = 0$$

$$(2 \sin \theta + 1)(\sin \theta - 1) = 0$$

$$2 \sin \theta + 1 = 0$$

$$\sin \theta = -\frac{1}{2}$$

$$\theta = 210°, \ 330°$$

or

$$\sin \theta - 1 = 0$$

$$\sin \theta = 1$$

$$\theta = 90°$$

$$\theta = 90°, \ 210°, \ 330°$$

31. $\tan \theta - \cot \theta = 0$

Use the identity

$$\cot \theta = \frac{1}{\tan \theta}, \ \tan \theta \neq 0.$$

$$\tan \theta - \frac{1}{\tan \theta} = 0$$

$$\tan^2 \theta - 1 = 0$$

$$\tan^2 \theta = 1$$

$$\tan \theta = \pm 1$$

If $\tan \theta = 1$,

$$\theta = 45°, \ 225°.$$

If $\tan \theta = -1$,

$$\theta = 135°, \ 315°.$$

$$\theta = 45°, \ 135°, \ 225°, \ 315°$$

33. $\csc^2 \theta - 2 \cot \theta = 0$

$$(1 + \cot^2 \theta) - 2 \cot \theta = 0$$

$$\cot^2 \theta - 2 \cot \theta + 1 = 0$$

$$(\cot \theta - 1)^2 = 0$$

$$\cot \theta = 1$$

$$\theta = 45°, \ 225°$$

35. $2 \tan^2 \theta \sin \theta - \tan^2 \theta = 0$

$$\tan^2 \theta (2 \sin \theta - 1) = 0$$

$$\tan^2 \theta = 0$$

$$\tan \theta = 0$$

$$\theta = 0°, \ 180°$$

or

$$2 \sin \theta - 1 = 0$$

$$\sin \theta = \frac{1}{2}$$

$$\theta = 30°, \ 150°$$

$$\theta = 0°, \ 30°, \ 150°, \ 180°$$

37. $\sec^2 \theta \tan \theta = 2 \tan \theta$

$$\sec^2 \theta \tan \theta - 2 \tan \theta = 0$$

$$\tan \theta (\sec^2 \theta - 2) = 0$$

$$\tan \theta = 0 \ \text{ or } \ \sec^2 \theta = 2$$

$$\sec \theta = \pm\sqrt{2}$$

$$\theta = 0°, \ 180°$$

or

$$\theta = 45°, \ 135°, \ 225°, \ 315°$$

$$\theta = 0°, \ 45°, \ 135°, \ 180°, \ 225°, \ 315°$$

39.
$$\sin \theta + \cos \theta = 1$$
$$(\sin \theta + \cos \theta)^2 = 1^2$$
$$\sin^2 \theta + 2 \sin \theta \cos \theta + \cos^2 \theta = 1$$
$$1 + 2 \sin \theta \cos \theta = 1$$
$$2 \sin \theta \cos \theta = 0$$
$$\sin \theta \cos \theta = 0$$

$$\sin \theta = 0$$
$$\theta = 0°, 180°$$

or

$$\cos \theta = 0$$
$$\theta = 90°, 270°$$

Check these solutions since both
sides were squared.

$\sin 180° + \cos 180° = -1$ and
$\sin 270° + \cos 270° = -1$ so

180° and 270° are not solutions.

$$\theta = 0°, 90°$$

41.
$$3 \sin^2 x - \sin x = 2$$
$$3 \sin^2 x - \sin x - 2 = 0$$
$$(3 \sin x + 2)(\sin x - 1) = 0$$

$$\sin x = -\frac{2}{3} \quad \text{or} \quad \sin x = 1$$

$$x = 221.8°, 318.2° \quad x = 90°$$
$$x = 90°, 221.8°, 318.2°$$

43.
$$\sec^2 \theta = 2 \tan \theta + 4$$
$$\tan^2 \theta + 1 = 2 \tan \theta + 4$$
$$\tan^2 \theta - 2 \tan \theta - 3 = 0$$
$$(\tan \theta - 3)(\tan \theta + 1) = 0$$
$$\tan \theta = 3 \quad \text{or} \quad \tan \theta = -1$$

$\theta = 71.6°, 251.6°$ or $\theta = 135°, 315°$
$\theta = 71.6°, 135°, 251.6°, 315°$

45.
$$3 \cot^2 \theta = \cot \theta$$
$$3 \cot^2 \theta - \cot \theta = 0$$
$$\cot \theta (3 \cot \theta - 1) = 0$$
$$\cot \theta = 0$$
$$\theta = 90°, 270°$$

or

$$\cot \theta = \frac{1}{3}$$
$$\theta = 71.6°, 251.6°$$

$$\theta = 71.6°, 90°, 251.6°, 270°$$

47.
$$9 \sin^2 x - 6 \sin x = 1$$
$$9 \sin^2 x - 6 \sin x - 1 = 0$$

Use the quadratic formula with
$a = 9$, $b = -6$, $c = -1$.

$$\sin x = \frac{6 \pm \sqrt{36 - [4 \cdot 9(-1)]}}{2 \cdot 9}$$

$$= \frac{6 \pm \sqrt{72}}{18}$$

$$= \frac{1 \pm \sqrt{2}}{3}$$

If $\sin x = \dfrac{1 + \sqrt{2}}{3}$

$$= \frac{1 + 1.4142136}{3}$$

$$= .80473787,$$

$$x = 53.6°, 126.4°.$$

If $\sin x = \dfrac{1 - \sqrt{2}}{3}$

$$= \frac{1 - 1.4142136}{3}$$

$$= -.13807119,$$

$$x = -7.9°.$$

Since this solution is not in the
interval [0°, 360°), we must use it
as a reference angle to find angles
in the interval.

$$x = 180° + (-7.9°) = 187.9°$$
$$\text{or } (-7.9°) + 360° = 352.1°$$

$$x = 53.6°, \ 126.4°, \ 187.9°, \ 352.1°$$

49. $\tan^2 x + 4 \tan x + 2 = 0$

$a = 1, \ b = 4, \ c = 2$

$$\tan x = \frac{-4 \pm \sqrt{16 - 4 \cdot 1 \cdot 2}}{2 \cdot 1}$$

$$= \frac{-4 \pm \sqrt{8}}{2} = \frac{-4 \pm 2\sqrt{2}}{2}$$

$$= -2 \pm \sqrt{2}$$

If $\tan x = -2 + \sqrt{2}$

$$= -2 + 1.4142136$$

$$= -.58578644,$$

$x = 149.6°, \ 329.6°.$

If $\tan x = -2 - \sqrt{2}$

$$= -2 - 1.4142136$$

$$= -3.4142136,$$

$$x = 106.3°, \ 286.3°.$$

$x = 106.3°, \ 149.6°, \ 286.3°, \ 329.6°$

51. $\sin^2 x - 2 \sin x + 3 = 0$

$a = 1, \ b = -2, \ c = 3$

$$\sin x = \frac{3 \pm \sqrt{4 - (4 \cdot 1 \cdot 3)}}{2 \cdot 1}$$

$$= \frac{2 \pm \sqrt{-8}}{2 \cdot 1}$$

Since $\sqrt{-8}$ is not a real number, the equation has no solution.

53. $\cot x + 2 \csc x = 3$

$$\frac{\cos x}{\sin x} + \frac{2}{\sin x} = 3$$

Multiply both sides by $\sin x$.

$$\cos x + 2 = 3 \sin x$$

Square both sides. Squaring both sides may introduce extraneous solutions, so all solutions must be checked in the original equation.

$$(\cos x + 2)^2 = (3 \sin x)^2$$
$$\cos^2 x + 4 \cos x + 4 = 9 \sin^2 x$$
$$\cos^2 x + 4 \cos x + 4 = 9(1 - \cos^2 x)$$
$$\cos^2 x + 4 \cos x + 4 = 9 - 9 \cos^2 x$$
$$10 \cos^2 x + 4 \cos x - 5 = 0$$

Use the quadratic formula with $a = 10, \ b = 4,$ and $c = -5.$

$$\cos x = \frac{-4 \pm \sqrt{4^2 - 4(10)(-5)}}{2(10)}$$

$$\cos x = \frac{-4 \pm \sqrt{216}}{20}$$

$$\cos x = \frac{-4 \pm 6\sqrt{6}}{20}$$

$$\cos x = \frac{-2 \pm 3\sqrt{6}}{10}$$

$\cos x = \dfrac{-2 + 3\sqrt{6}}{10}$ or $\cos x = \dfrac{-2 - 3\sqrt{6}}{10}$

$\cos x = .53484692$ or $\cos x = -.93484692$

$x = 57.7°$ or $302.3°$ or $x = 159.2°$ or $200.8°$

These four angles must be checked in the original equation, $\cot x + 2 \csc x = 3$. We see that $57.7°$ and $159.2°$ check, but that $302.3°$ and $200.8°$ do not check.

55. From Example 1,

$$\cos \theta = -\frac{1}{2} \text{ or } \cos \theta = 1$$

$\theta = 120° + 360° \cdot n$ or $240° + 360° \cdot n$ or $\theta = 0° + 360° \cdot n$, where n is an integer.

Thus,

$\theta = 360° \cdot n$, $120° + 360° \cdot n$,
$240° + 360° \cdot n$, where n is an
integer.

57. The error occurs in going from the
step

$$\sin^2 x - \sin x = 0$$

to the step

$$\sin x - 1 = 0.$$

We cannot divide both sides by sin x
because sin x may equal zero and
division by zero is undefined.
The equation should be solved by
factoring.

$$\sin^2 x - \sin x = 0$$
$$\sin x (\sin x - 1) = 0$$
$$\sin x = 0 \quad \text{or} \quad \sin x = 1$$
$$x = 0 \quad \text{or} \quad x = \frac{\pi}{2}$$

59. $V = \cos 2\pi t$, $0 \le t \le \frac{1}{2}$

(a) $V = 0$, $\cos 2\pi t = 0$
$$2\pi t = \cos^{-1} 0$$
$$2\pi t = \frac{\pi}{2}$$
$$t = \frac{\frac{\pi}{2}}{2\pi}$$
$$= \frac{1}{4} \text{ sec}$$

(b) $V = .5$, $\cos 2\pi t = .5$
$$2\pi t = \cos^{-1} (.5)$$
$$2\pi t = \frac{\pi}{3}$$
$$t = \frac{\frac{\pi}{3}}{2\pi}$$
$$= \frac{1}{6} \text{ sec}$$

(c) $V = .25$, $\cos 2\pi t = .25$
$$2\pi t = \cos^{-1} (.25)$$
$$2\pi t = 1.3181161$$
$$t = \frac{1.3181161}{2\pi}$$
$$= .21 \text{ sec}$$

61. $.342D \cos \theta + h \cos^2 \theta = \frac{16D^2}{V_0^2}$

$V_0 = 60$, $D = 80$, $h = 2$
$.342(80)\cos \theta + 2 \cos^2 \theta$
$$= \frac{16(80)^2}{60^2}$$

$2 \cos^2 \theta + 27.36 \cos \theta - \frac{256}{9} = 0$

$\cos^2 \theta + 13.68 \cos \theta - \frac{128}{9} = 0$

$a = 1$, $b = 13.68$, $c = -128/9$

$\cos \theta = \frac{-13.68 \pm \sqrt{(13.68)^2 - 4(-128/9)}}{2}$

$$= \frac{-13.68 \pm 15.6215}{2}$$

$\cos \theta = .97075$ or $\cos \theta = -14.65075$

-14.65075 does not lead to a solution because it is not in the range
of cos θ.
If $\cos \theta = .97075$,

$$\theta = 14°.$$

63. Since f measures cycles per second
and T measures seconds per cycle,

$$f = \frac{1}{T} \quad \text{or} \quad T = \frac{1}{f}.$$

For example, if f = 5 cycles per
sec, then T = 1/5 sec per cycle.
Thus, T is the time for one complete
cycle, which is the period.

65. $s(t) = \sin t + 2 \cos t$

Let $s(t) = \dfrac{2 + \sqrt{3}}{2}$.

$\sin t + 2 \cos t = \dfrac{2 + \sqrt{3}}{2}$

$2 \sin t + 4 \cos t = 2 + \sqrt{3}$

$\qquad 4 \cos t = (2 + \sqrt{3}) - 2 \sin t$

$\qquad (4 \cos t)^2 = [(2 + \sqrt{3}) - 2 \sin t]^2$

$\qquad 16 \cos^2 t = (7 + 4\sqrt{3})$
$\qquad\qquad\qquad - 4(2 + \sqrt{3}) \sin t$
$\qquad\qquad\qquad + 4 \sin^2 t$

$16(1 - \sin^2 t) = (7 + 4\sqrt{3})$
$\qquad\qquad\qquad - 4(2 + \sqrt{3}) \sin t$
$\qquad\qquad\qquad + 4 \sin^2 t$

$16 - 16 \sin^2 t = (7 + 4\sqrt{3})$
$\qquad\qquad\qquad - 4(2 + \sqrt{3}) \sin t$
$\qquad\qquad\qquad + 4 \sin^2 t$

$0 = 20 \sin^2 t - 4(2 + \sqrt{3}) \sin t$
$\qquad + (-9 + 4\sqrt{3})$

$0 = (2 \sin t - \sqrt{3})[10 \sin t - (4 - 3\sqrt{3})]$

$\sin t = \dfrac{\sqrt{3}}{2}$ or $\sin t = \dfrac{4 - 3\sqrt{3}}{10}$

If $\sin t = \sqrt{3}/2$, then $t = \pi/3$.

One value of t is $\pi/3$.

Section 6.3

1. Estimates of the x-values where the two graphs intersect are .5, 3.1, and 5.8.

3. If $\sin \left(\dfrac{X}{2}\right) = \sin X$ in the interval $[0, 2\pi)$, then X could be 0 or X could be $2\pi/3$ since

$$\sin \dfrac{2\pi}{3} = \sin \dfrac{\pi}{3}.$$

The possible values of X are 0 and $2\pi/3$.

Alternately, if $\sin \left(\dfrac{X}{2}\right) = \sin X$, then

$$\sin \dfrac{X}{2} = 2 \sin \dfrac{X}{2} \cos \dfrac{X}{2}$$

$\sin \dfrac{X}{2} - 2 \sin \dfrac{X}{2} \cos \dfrac{X}{2} = 0$

$\sin \dfrac{X}{2} \left(1 - 2 \cos \dfrac{X}{2}\right) = 0$

$\sin \dfrac{X}{2} = 0$ or $\cos \dfrac{X}{2} = \dfrac{1}{2}$

$\qquad X = 0$ or $\dfrac{2\pi}{3}$.

5. $\cos 2x = \dfrac{\sqrt{3}}{2}$

Since $0 \le x < 2\pi$,
$\qquad 0 \le 2x < 4\pi$.

$2x = \dfrac{\pi}{6}, \dfrac{11\pi}{6}, \dfrac{13\pi}{6}, \dfrac{23\pi}{6}$

$x = \dfrac{\pi}{12}, \dfrac{11\pi}{12}, \dfrac{13\pi}{12}, \dfrac{23\pi}{12}$

7. $\sin 3x = -1$,

Since $0 \le x < 2\pi$,
$\qquad 0 \le 3x < 6\pi$.

$3x = \dfrac{3\pi}{2}, \dfrac{7\pi}{2}, \dfrac{11\pi}{2}$

$x = \dfrac{\pi}{2}, \dfrac{7\pi}{6}, \dfrac{11\pi}{6}$

9. $3 \tan 3x = \sqrt{3}$, $0 \le 3x < 6\pi$

$\tan 3x = \dfrac{\sqrt{3}}{3}$

$3x = \dfrac{\pi}{6}, \dfrac{7\pi}{6}, \dfrac{13\pi}{6}, \dfrac{19\pi}{6}, \dfrac{25\pi}{6}, \dfrac{31\pi}{6}$

$x = \dfrac{\pi}{18}, \dfrac{7\pi}{18}, \dfrac{13\pi}{18}, \dfrac{19\pi}{18}, \dfrac{25\pi}{18}, \dfrac{31\pi}{18}$

11. $\sqrt{2} \cos 2x = -1, \; 0 \le 2x < 4\pi$

$$\cos 2x = \frac{-1}{\sqrt{2}} = -\frac{\sqrt{2}}{2}$$

$$2x = \frac{3\pi}{4}, \frac{5\pi}{4}, \frac{11\pi}{4}, \frac{13\pi}{4}$$

$$x = \frac{3\pi}{8}, \frac{5\pi}{8}, \frac{11\pi}{8}, \frac{13\pi}{8}$$

13. $\sin \frac{x}{2} = \sqrt{2} - \sin \frac{x}{2}, \; 0 \le \frac{x}{2} < \pi$

$$\sin \frac{x}{2} + \sin \frac{x}{2} = \sqrt{2}$$

$$2 \sin \frac{x}{2} = \sqrt{2}$$

$$\sin \frac{x}{2} = \frac{\sqrt{2}}{2}$$

$$\frac{x}{2} = \frac{\pi}{4}, \frac{3\pi}{4}$$

$$x = \frac{\pi}{2}, \frac{3\pi}{2}$$

15. $\tan 4x = 0, \; 0 \le 4x < 8\pi$

$$4x = 0, \; \pi, \; 2\pi, \; 3\pi, \; 4\pi, \; 5\pi,$$
$$6\pi, \; 7\pi$$

$$x = 0, \frac{\pi}{4}, \frac{\pi}{2}, \frac{3\pi}{4}, \pi, \frac{5\pi}{4}, \frac{3\pi}{2}, \frac{7\pi}{4}$$

17. $8 \sec^2 \frac{x}{2} = 4$

$$\sec^2 \frac{x}{2} = \frac{1}{2}$$

$$\sec \frac{x}{2} = \pm \frac{\sqrt{2}}{2}$$

There is no solution since $\sec x \ge 1$
or $\sec x \le -1$.

19. $\sin \frac{x}{2} = \cos \frac{x}{2}, \; 0 \le \frac{x}{2} < \pi$

$$\sin^2 \frac{x}{2} = \cos^2 \frac{x}{2}$$

Use the identity

$$\sin^2 \theta + \cos^2 \theta = 1$$
$$\cos^2 \theta = 1 - \sin^2 \theta.$$

$$\sin^2 \frac{x}{2} = 1 - \sin^2 \frac{x}{2}$$

$$2 \sin^2 \frac{x}{2} = 1$$

$$\sin^2 \frac{x}{2} = \frac{1}{2}$$

$$\sin \frac{x}{2} = \pm \sqrt{\frac{1}{2}} = \pm \frac{\sqrt{2}}{2}$$

If $\sin \frac{x}{2} = \frac{\sqrt{2}}{2}$,

$$\frac{x}{2} = \frac{\pi}{4}, \frac{3\pi}{4}.$$

$\sin \frac{x}{2} = -\frac{\sqrt{2}}{2}$ has no solution in the
interval $[0, \pi)$.

$$x = \frac{\pi}{2}, \frac{3\pi}{2}$$

Check solutions, since both sides
were squared.

$$x = \frac{\pi}{2}:$$

$$\sin \frac{x}{2} = \sin \frac{\pi}{4} = \frac{\sqrt{2}}{2}$$

$$\cos \frac{x}{2} = \cos \frac{\pi}{2} = \frac{\sqrt{2}}{2}$$

$$x = \frac{3\pi}{2}:$$

$$\sin \frac{x}{2} = \sin \frac{3\pi}{4} = \frac{\sqrt{2}}{2}$$

$$\cos \frac{x}{2} = \cos \frac{3\pi}{4} = -\frac{\sqrt{2}}{2}$$

The only solution is $\pi/2$.

21. $\cos 2x + \cos x = 0$, $0 \le x < 2\pi$

Use the identity

$\cos 2x = 2 \cos^2 x - 1$.

$2 \cos^2 x - 1 + \cos x = 0$

$2 \cos^2 x + \cos x - 1 = 0$

$(2 \cos x - 1)(\cos x + 1) = 0$

$2 \cos x - 1 = 0$

$\cos x = \dfrac{1}{2}$

$x = \dfrac{\pi}{3}, \dfrac{5\pi}{3}$

or

$\cos x + 1 = 0$

$\cos x = -1$

$x = \pi$

$x = \dfrac{\pi}{3}, \pi, \dfrac{5\pi}{3}$

23. $\sqrt{2} \sin 3\theta - 1 = 0$

$0° \le 3\theta < 3(360°) = 1080°$

$\sqrt{2} \sin 3\theta = 1$

$\sin 3\theta = \dfrac{1}{\sqrt{2}} = \dfrac{\sqrt{2}}{2}$

In quadrants I and II, sine is positive.

Therefore,

$3\theta = 45°, 135°, 405°, 495°, 765°,$

   $855°$

$\theta = 15°, 45°, 135°, 165°, 255°,$

   $285°$.

25. $\cos \dfrac{\theta}{2} = 1$, $0 \le \dfrac{\theta}{2} < 180°$

$\dfrac{\theta}{2} = 0°$

$\theta = 0°$

27. $2\sqrt{3} \sin \dfrac{\theta}{2} = 3$, $0° \le \dfrac{\theta}{2} < 180°$

$\sin \dfrac{\theta}{2} = \dfrac{3}{2\sqrt{3}}$

$= \dfrac{3\sqrt{3}}{6} = \dfrac{\sqrt{3}}{2}$

$\dfrac{\theta}{2} = 60°, 120°$

$\theta = 120°, 240°$

29. $2 \sin \theta = 2 \cos 2\theta$

$\sin \theta = \cos 2\theta$

Use the identity

$\cos 2\theta = 1 - 2 \sin^2 \theta$.

$\sin \theta = 1 - 2 \sin^2 \theta$

$2 \sin^2 \theta + \sin \theta - 1 = 0$

$(2 \sin \theta - 1)(\sin \theta + 1) = 0$

$2 \sin \theta - 1 = 0$

$\sin \theta = \dfrac{1}{2}$

$\theta = 30°, 150°$

or

$\sin \theta + 1 = 0$

$\sin \theta = -1$

$\theta = 270°$

$\theta = 30°, 150°, 270°$

31. $1 - \sin \theta = \cos 2\theta$

Use the identity

$\cos 2\theta = 1 - 2 \sin^2 \theta$.

$1 - \sin \theta = 1 - 2 \sin^2 \theta$

$2 \sin^2 \theta - \sin \theta = 0$

$\sin \theta (2 \sin \theta - 1) = 0$

$\sin \theta = 0$

$\theta = 0°, 180°$

or

$$2 \sin \theta - 1 = 0$$

$$\sin \theta = \frac{1}{2}$$

$$\theta = 30°, 150°$$

$$\theta = 0°, 30°, 150°, 180°$$

33. $\csc^2 \frac{\theta}{2} = 2 \sec \theta$

Use the identities

$$\csc x = \frac{1}{\sin x} \text{ and } \sec x = \frac{1}{\cos x}.$$

$$\frac{1}{\sin^2 \frac{\theta}{2}} = \frac{2}{\cos \theta}$$

$$2 \sin^2 \frac{\theta}{2} = \cos \theta$$

Use the identity

$$\sin^2 \frac{x}{2} = \frac{1 - \cos x}{2}, \text{ and substitute.}$$

$$2\left(\frac{1 - \cos \theta}{2}\right) = \cos \theta$$

$$1 - \cos \theta = \cos \theta$$

$$1 = 2 \cos \theta$$

$$\frac{1}{2} = \cos \theta$$

$$\theta = 60°, 300°$$

35. $2 - \sin 2\theta = 4 \sin 2\theta,$
     $0° \le 2\theta < 720°$

$$2 = 5 \sin 2\theta$$

$$\sin 2\theta = \frac{2}{5} = .4$$

$$2\theta = 23.6°, 156.4°, 383.6°, 516.4°$$

$$\theta = 11.8°, 78.2°, 191.8°, 258.2°$$

37. $2 \cos^2 2\theta = 1 - \cos 2\theta,$
     $0° \le 2\theta < 720°$

$$2 \cos^2 2\theta + \cos 2\theta - 1 = 0$$

$$(2 \cos 2\theta - 1)(\cos 2\theta + 1) = 0$$

$$2 \cos 2\theta - 1 = 0$$

$$\cos 2\theta = \frac{1}{2}$$

$$2\theta = 60°, 300°, 420°, 660°$$

$$\theta = 30°, 150°, 210°, 330°$$

or

$$\cos 2\theta + 1 = 0$$

$$\cos 2\theta = -1$$

$$2\theta = 180°, 540°$$

$$\theta = 90°, 270°$$

$$\theta = 30°, 90°, 150°, 210°, 270°, 330°$$

39. If $\sin \frac{1}{2}x = -\frac{1}{2},$

we must write

$$\frac{1}{2}x = 210° + n \cdot 360°$$

or

$$\frac{1}{2}x = 330° + n \cdot 360°,$$

where n is any integer. Then multi-plying by 2, we have

$$x = 420° + n \cdot 720°$$

or

$$x = 660° + n \cdot 720°.$$

Since $n \in \{\ldots, -3, -2, -1, 0, 1, 2, 3, \ldots\}$, there is no value of n which will give values of x in the interval $[0°, 360°)$.

41. $h = \frac{35}{3} + \frac{7}{3} \sin \frac{2\pi x}{365}$

(a) Find x such that h = 14.

$$14 = \frac{35}{3} + \frac{7}{3} \sin \frac{2\pi x}{365}$$

$$14 - \frac{35}{3} = \frac{7}{3} \sin \frac{2\pi x}{365}$$

$$\frac{7}{3} = \frac{7}{3} \sin \frac{2\pi x}{365}$$

$$\sin \frac{2\pi x}{365} = 1$$

$$\frac{2\pi x}{365} = \frac{\pi}{2} + 2\pi n, \text{ n is an}$$

integer.

$$x = \left(\frac{\pi}{2} + 2\pi n\right)\left(\frac{365}{2\pi}\right)$$

$$= \frac{365}{4} \pm 365n$$

$$= 91.25 \pm 365n$$

x = 91.25 means about 91.3 days after March 21, on June 20.

(b) h assumes its least value when $\sin \frac{2\pi x}{365}$ takes on its least value, which is -1.

$$\sin \frac{2\pi x}{365} = -1$$

$$\frac{2\pi x}{365} = \frac{3\pi}{2} \pm 2\pi n, \text{ n is an}$$

integer.

$$x = \left(\frac{3\pi}{2} \pm 2\pi n\right)\left(\frac{365}{2\pi}\right)$$

$$= \frac{3(365)}{4} \pm 365n$$

$$= 273.75 \pm 365$$

x = 273.75 means about 273.8 days after March 21, on December 19.

(c) Let h = 10.

$$10 = \frac{35}{3} + \frac{7}{3} \sin \frac{2\pi x}{365}$$

$$30 = 35 + 7 \sin \frac{2\pi x}{365}$$

$$-5 = 7 \sin \frac{2\pi x}{365}$$

$$-\frac{5}{7} = \sin \frac{2\pi x}{365}$$

$$-.71428571 = \sin \left(\frac{2\pi x}{365}\right)$$

In quadrants III and IV, sine is negative.

In quadrant III,

$$\frac{2\pi x}{365} = \pi + .79560295$$

$$= 3.9371956$$

$$x = \frac{365}{2\pi}(3.9371956)$$

$$= 228.7.$$

x = 228.7 means 228.7 days after March 21, on November 4.

In quadrant IV,

$$\frac{2\pi x}{365} = 2\pi - .79560295$$

$$= 5.4875823$$

$$x = \frac{365}{2\pi}(5.4878263)$$

$$= 318.8.$$

x = 318.8 means about 318.8 days after March 21, on February 2.

43. $i = I_{max} \sin 2\pi ft$

Let i = 40, I_{max} = 100, f = 60.

$$40 = 100 \sin 2\pi(60)(t)$$

$$40 = 100 \sin 120\pi t$$

$$.4 = \sin 120\pi t$$

From a calculator,

$$120\pi t = .4115168$$

$$t = \frac{.4115168}{120\pi}$$

$$t = .0010916$$

$$t \approx .001 \text{ sec.}$$

45. $i = I_{max} \sin 2\pi ft$

Let $i = I_{max}$, $f = 60$.

$$I_{max} = I_{max} \sin 2\pi(60)t$$

$$1 = \sin 120\pi t$$

$$120\pi t = \frac{\pi}{2}$$

$$120t = \frac{1}{2}$$

$$t = \frac{1}{240} \approx .004 \text{ sec}$$

47. **(a)** The final graph is

$$P = P_1 + P_2 + P_3 + P_4 + P_5$$

$$= .002 \sin 880\pi t + \frac{.002}{2} \sin 1760\pi t$$

$$+ \frac{.002}{3} \sin 2640\pi t$$

$$+ \frac{.002}{4} \sin 3520\pi t$$

$$+ \frac{.002}{5} \sin 4400\pi t.$$

See the answer graph in the back of the textbook.

(b) The graph approximates a saw-tooth shape.

(c) From the graph we see that the maximum of $P = P_1 + P_2 + P_3 + P_4 + P_5$ is approximately .00317. This maximum occurs five times on $[0, .01]$ when $x \approx .000188$, .00246, .00474, .00701, and .00928.

Section 6.4

1. If $\arcsin x = \arccos x$, we must have $x = \sqrt{2}/2$. The coordinates of the point of intersection are $(\sqrt{2}/2, \pi/4)$. The solution of $\arcsin x - \arccos x = 0$ is $x = \sqrt{2}/2$.

3. If $2 \cos^{-1} X = \pi$, then $\cos^{-1} X = \frac{\pi}{2}$. This means $\cos \frac{\pi}{2} = X$, so $X = 0$. The possible value for X in $[0, 1]$ is 0.

5. $y = 5 \cos x$

$$\frac{y}{5} = \cos x$$

$$x = \arccos \frac{y}{5}$$

7. $2y = \cot 3x$

$$3x = \text{arccot } 2y$$

$$x = \frac{1}{3} \text{arccot } 2y$$

9. $y = 3 \tan 2x$

$$\frac{y}{3} = \tan 2x$$

$$2x = \arctan \frac{y}{3}$$

$$x = \frac{1}{2} \arctan \frac{y}{3}$$

11. $y = 6 \cos \frac{x}{4}$

$$\frac{y}{6} = \cos \frac{x}{4}$$

$$\frac{x}{4} = \arccos \frac{y}{6}$$

$$x = 4 \arccos \frac{y}{6}$$

13. $y = -2 \cos 5x$

$-\dfrac{y}{2} = \cos 5x$

$5x = \arccos \left(-\dfrac{y}{2}\right)$

$x = \dfrac{1}{5} \arccos \left(-\dfrac{y}{2}\right)$

15. $y = \cos (x + 3)$

$x + 3 = \arccos y$

$x = -3 + \arccos y$

17. $y = \sin x - 2$

$y + 2 = \sin x$

$x = \arcsin (y + 2)$

19. $y = 2 \sin x - 4$

$y + 4 = 2 \sin x$

$\dfrac{y + 4}{2} = \sin x$

$x = \arcsin \left(\dfrac{y + 4}{2}\right)$

21.   The expression $\sin x - 2$ means $(\sin x) - 2$ which is not equivalent to the expression $\sin (x - 2)$.

23. $\dfrac{4}{3} \cos^{-1} \dfrac{y}{4} = \pi$

$\cos^{-1} \dfrac{y}{4} = \dfrac{3\pi}{4}$

$\dfrac{y}{4} = \cos \dfrac{3\pi}{4}$

$\dfrac{y}{4} = -\dfrac{\sqrt{2}}{2}$

$y = -2\sqrt{2}$

25. $2 \arccos \left(\dfrac{y - \pi}{3}\right) = 2\pi$

$\arccos \left(\dfrac{y - \pi}{3}\right) = \pi$

$\dfrac{y - \pi}{3} = \text{Cos } \pi$

$\dfrac{y - \pi}{3} = -1$

$y - \pi = -3$

$y = \pi - 3$

27. $\arcsin x = \arctan \dfrac{3}{4}$

Let $\arctan \dfrac{3}{4} = u$, so $\tan u = \dfrac{3}{4}$, u is in quadrant I.

Sketch a triangle and label it. The hypotenuse is $\sqrt{3^2 + 4^2} = 5$.

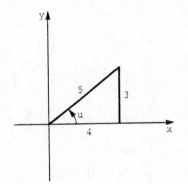

$\sin u = \dfrac{3}{r} = \dfrac{3}{5}$

The equation becomes

$\arcsin x = u$, or $x = \sin u$.

$x = \dfrac{3}{5}$

29. $\cos^{-1} x = \sin^{-1} \dfrac{3}{5}$

Let $\sin^{-1} \dfrac{3}{5} = u$, so $\sin u = \dfrac{3}{5}$.

u is in quadrant I.

194   Chapter 6   **Inverse Trigonometric Functions and Trigonometric Equations**

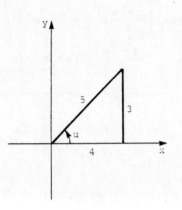

$$\cos u = \frac{4}{5}$$

The equation becomes

$$\cos^{-1} x = u, \quad \text{or} \quad x = \cos u.$$

$$x = \frac{4}{5}$$

31. $\sin^{-1} x - \tan^{-1} 1 = -\frac{\pi}{4}$

$$\sin^{-1} x = \tan^{-1} 1 - \frac{\pi}{4}$$

$$\sin^{-1} x = \frac{\pi}{4} - \frac{\pi}{4}$$

$$\sin^{-1} x = 0$$

$$\sin 0 = x$$

$$x = 0$$

33. $\arccos x + 2 \arcsin \frac{\sqrt{3}}{2} = \pi$

$$\arccos x = \pi - 2 \arcsin \frac{\sqrt{3}}{2}$$

$$\arccos x = \pi - 2\left(\frac{\pi}{3}\right)$$

$$\arccos x = \pi - \frac{2\pi}{3}$$

$$\arccos x = \frac{\pi}{3}$$

$$x = \cos \frac{\pi}{3}$$

$$x = \frac{1}{2}$$

35. $\arcsin 2x + \arccos x = \frac{\pi}{6}$

$$\arcsin 2x = \frac{\pi}{6} - \arccos x$$

$$2x = \sin\left(\frac{\pi}{6} - \arccos x\right)$$

Use identity for sin (A − B).

$$2x = \sin \frac{\pi}{6} \cos (\arccos x)$$

$$- \cos \frac{\pi}{6} \sin (\arccos x)$$

Let u = arccos x.

$$\cos u = x = \frac{x}{1}$$

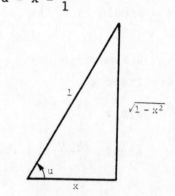

$$\sin u = \sqrt{1 - x^2}$$

$$2x = \sin \frac{\pi}{6} \cdot \cos u - \cos \frac{\pi}{6} \sin u$$

$$2x = \frac{1}{2}x - \frac{\sqrt{3}}{2}(\sqrt{1 - x^2})$$

$$4x = x - \sqrt{3}\sqrt{1 - x^2}$$

$$3x = -\sqrt{3}\sqrt{1 - x^2}$$

$$(3x)^2 = (-\sqrt{3}\sqrt{1 - x^2})^2$$

$$9x^2 = 3(1 - x^2)$$

$$9x^2 = 3 - 3x^2$$

$$12x^2 = 3$$

$$x^2 = \frac{3}{12} = \frac{1}{4}$$

$$x = \pm\frac{1}{2}$$

Check these solutions since both sides were squared.

$x = \frac{1}{2}$:

$\arcsin\left(2 \cdot \frac{1}{2}\right) + \arccos\left(\frac{1}{2}\right)$

$= \frac{\pi}{2} + \frac{\pi}{3} \neq \frac{\pi}{6}$

so 1/2 is not a solution.

$x = -\frac{1}{2}$:

$\arcsin\left(2\left(-\frac{1}{2}\right)\right) + \arccos\left(-\frac{1}{2}\right)$

$= -\frac{\pi}{2} + \frac{2\pi}{3} = \frac{\pi}{6}$

The only solution is -1/2.

37. $\cos^{-1} x + \tan^{-1} x = \frac{\pi}{2}$

$\cos^{-1} x = \frac{\pi}{2} - \tan^{-1} x$

$x = \cos\left(\frac{\pi}{2} - \tan^{-1} x\right)$

Use identity for cos (A − B).

$x = \cos\frac{\pi}{2}\cos\left(\tan^{-1} x\right)$

$\quad + \sin\frac{\pi}{2}\sin\left(\tan^{-1} x\right)$

Let $u = \tan^{-1} x$.

$\tan u = x$

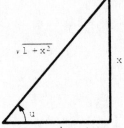

$\cos u = \frac{1}{\sqrt{1 + x^2}}$

$\sin u = \frac{x}{\sqrt{1 + x^2}}$

$x = \cos\frac{\pi}{2}\cdot\cos u + \sin\frac{\pi}{2}\cdot\sin u$

$x = 0\left(\frac{1}{\sqrt{1 + x^2}}\right) + 1\cdot\frac{x}{\sqrt{1 + x^2}}$

$x = \frac{x}{\sqrt{1 + x^2}}$

$x - \frac{x}{\sqrt{1 + x^2}} = 0$

$x\left(1 - \frac{1}{\sqrt{1 + x^2}}\right) = 0$

$x = 0$

or

$1 - \frac{1}{\sqrt{1 + x^2}} = 0$

$\frac{1}{\sqrt{1 + x^2}} = 1$

$\sqrt{1 + x^2} = 1$

$1 + x^2 = 1$

$x^2 = 0$

$x = 0$

39. $d = 550 + 450\cos\frac{\pi}{50}t$

$d - 550 = 450\cos\frac{\pi}{50}t$

$\frac{d - 550}{450} = \cos\frac{\pi}{50}t$

$\frac{\pi}{50}\cdot t = \arccos\left(\frac{d - 550}{450}\right)$

$t = \frac{50}{\pi}\arccos\left(\frac{d - 550}{450}\right)$

41. **(a)** $e = E_{max} \sin 2\pi ft$

$$\frac{e}{E_{max}} = \sin 2\pi ft$$

$$2\pi ft = \arcsin \frac{e}{E_{max}}$$

$$t = \frac{1}{2\pi f} \arcsin \frac{e}{E_{max}}$$

(b) Let $E_{max} = 12$, $e = 5$, and $f = 100$.

$$t = \frac{1}{2\pi(100)} \arcsin \frac{5}{12}$$

$$t = \frac{1}{200\pi} \arcsin .41666667 = .00068$$

43. **(a)** $u = \arcsin x$

$$x = \sin u, \quad -\frac{\pi}{2} \le u \le \frac{\pi}{2}$$

(b) See the answer graph at the back of the textbook.

(c) $\tan u = \dfrac{x}{\sqrt{1 - x^2}} = \dfrac{x\sqrt{1 - x^2}}{1 - x^2}$

(d) $u = \arctan \left(\dfrac{x\sqrt{1 - x^2}}{1 - x^2}\right)$

45. Let $A_1 = .0025$, $\phi_1 = \pi/7$, $A_2 = .001$, $\phi_2 = \pi/6$, and $f = 300$.

(a) $A = \sqrt{(A_1 \cos \phi_1 + A_2 \cos \phi_2)^2 + (A_1 \sin \phi_1 + A_2 \sin \phi_2)^2}$

$\quad = \sqrt{[.0025 \cos (\pi/7) + .001 \cos (\pi/6)]^2 + [.0025 \sin (\pi/7) + .001 \sin (\pi/6)]^2}$

$\quad \approx .0035$

$$\phi = \arctan \left[\frac{A_1 \sin \phi_1 + A_2 \sin \phi_2}{A_1 \cos \phi_1 + A_2 \cos \phi_2}\right]$$

$$= \arctan \left[\frac{.0025 \sin (\pi/7) + .001 \sin (\pi/6)}{.0025 \cos (\pi/7) + .001 \cos (\pi/6)}\right]$$

$$\approx .470$$

Thus, $P = A \sin (2\pi f + \phi)$

$\quad\quad P = .0035 \sin (600\pi t + .47)$.

(b) Graph $P = .0035 \sin (600\pi t + .47)$ and

$$P_1 + P_2 = .0025 \sin \left(600\pi t + \frac{\pi}{7}\right) + .001 \sin \left(600\pi t + \frac{\pi}{6}\right).$$

See the answer graph at the back of the textbook. Their graphs are the same.

47. $\theta = \tan^{-1} \dfrac{4}{x} - \tan^{-1} \dfrac{1}{x}$

(a) Let $\theta = \pi/6$.

$$\frac{\pi}{6} = \tan^{-1} \frac{4}{x} - \tan^{-1} \frac{1}{x}$$

$$\tan \frac{\pi}{6} = \tan \left(\tan^{-1} \frac{4}{x} - \tan^{-1} \frac{1}{x} \right)$$

$$\frac{\sqrt{3}}{3} = \frac{\dfrac{4}{x} - \dfrac{1}{x}}{1 + \dfrac{4}{x} \cdot \dfrac{1}{x}}$$

$$\frac{\sqrt{3}}{3} = \frac{\dfrac{3}{x}}{1 + \dfrac{4}{x^2}}$$

$$\frac{\sqrt{3}}{3} = \frac{\dfrac{3}{x}}{\dfrac{x^2 + 4}{x^2}}$$

$$\frac{\sqrt{3}}{3} = \frac{3x}{x^2 + 4}$$

$$\sqrt{3}x^2 + 4\sqrt{3} = 9x$$

$$\sqrt{3}x^2 - 9x + 4\sqrt{3} = 0$$

Divide by $\sqrt{3}$.

$$x^2 - 3\sqrt{3}x + 4 = 0$$

$$x = \frac{3\sqrt{3} \pm \sqrt{(-3\sqrt{3})^2 - 4(1)(4)}}{2}$$

$$x = \frac{3\sqrt{3} \pm \sqrt{11}}{2}$$

$$x \approx 4.26 \text{ ft or } .94 \text{ ft}$$

(b) Let $\theta = \pi/8$.

$$\frac{\pi}{8} = \tan^{-1} \frac{4}{x} - \tan^{-1} \frac{1}{x}$$

$$\tan \frac{\pi}{8} = \tan \left(\tan^{-1} \frac{4}{x} - \tan^{-1} \frac{1}{x} \right)$$

From part (a),

$$\tan \frac{\pi}{8} = \frac{3x}{x^2 + 4}.$$

$$.414 = \frac{3x}{x^2 + 4}$$

$$.414x^2 + 1.656 = 3x$$

$$.414x^2 - 3x + 1.656 = 0$$

$$x = \frac{3 \pm \sqrt{9 - 4(.414)(1.656)}}{2(.414)}$$

$$x \approx \frac{3 \pm 2.502}{.828}$$

$$x \approx 6.64 \text{ ft or } .60 \text{ ft}$$

Chapter 6   Review Exercises

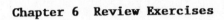

5. $y = \sin^{-1} \left(\dfrac{\sqrt{2}}{2} \right)$

$\sin y = \dfrac{\sqrt{2}}{2}, \ -\dfrac{\pi}{2} \le y \le \dfrac{\pi}{2}$

$y = \dfrac{\pi}{4}$

7. $y = \tan^{-1} (-\sqrt{3})$

$\tan y = -\sqrt{3}, \ -\dfrac{\pi}{2} \le y \le \dfrac{\pi}{2}$

$y = -\dfrac{\pi}{3}$

9. $y = \cos^{-1} \left(-\dfrac{\sqrt{2}}{2} \right)$

$\cos y = -\dfrac{\sqrt{2}}{2}, \ 0 \le y \le \pi$

$y = \dfrac{3\pi}{4}$

11. $y = \sec^{-1} (-2)$

$\sec y = -2, \ 0 \le y \le \pi, \ y \ne \dfrac{\pi}{2}$

$y = \dfrac{2\pi}{3}$

13. $y = \text{arccot}\ (-1)$

$\cot y = -1,\ 0 < y < \pi$

$y = \dfrac{3\pi}{4}$

15. $\theta = \arcsin\left(-\dfrac{\sqrt{3}}{2}\right)$

$\sin \theta = -\dfrac{\sqrt{3}}{2},\ -90° \le \theta \le 90°$

$\theta = -60°$

For Exercises 17–21, be sure your calculator is in degree mode. Keystroke sequences may vary based on the type and/or model of calculator being used.

17. $\theta = \arctan 1.7804675$

Enter: 1.7804675 [INV] [TAN]

or   [2nd] [TAN] 1.7804675 [ENTER]

Display: 60.679245

$\theta = 60.679245°$

19. $\theta = \cos^{-1}\ .80396577$

Enter: .80396577 [INV] [COS]

or   [2nd] [COS] .80396577 [ENTER]

Display: 36.489508

$\theta = 36.489508°$

21. $\theta = \text{arcsec}\ 3.4723155$

Enter: 3.4723155 [1/x] [INV] [COS]

or   [2nd] [COS] 3.4723155 [x⁻¹]

        [ENTER]

Display: 73.262206

$\theta = 73.262206°$

23. If $\theta = \sin^{-1} 3$, then $\sin \theta = 3$ and $-90° \le \theta \le 90°$. But the sine function can never have a value greater than 1, so $\sin \theta = 3$ is impossible and $\sin^{-1} 3$ cannot be defined.

25. The domain of the arccot function is $(-\infty, \infty)$ since the range of the cotangent function is $(-\infty, \infty)$.

27. $\sin\left(\sin^{-1} \dfrac{1}{2}\right)$

Let $\omega = \sin^{-1} \dfrac{1}{2}$.

Then

$\sin \omega = \dfrac{1}{2},\ -\dfrac{\pi}{2} \le \omega \le \dfrac{\pi}{2}.$

$\sin\left(\sin^{-1} \dfrac{1}{2}\right) = \sin \omega = \dfrac{1}{2}$

29. $\cos\ (\text{arccos}\ (-1))$

Let $\omega = \text{arccos}\ (-1)$

Then $\cos \omega = -1,\ 0 \le \omega \le \pi.$

$\cos\ (\text{arccos}\ (-1)) = \cos \omega = -1$

31. $\text{arccos}\left(\cos \dfrac{3\pi}{4}\right) = x$

$\cos x = \cos \dfrac{3\pi}{4},\ 0 \le x \le \pi,$

so $x = \dfrac{3\pi}{4}.$

$\text{arccos}\left(\cos \dfrac{3\pi}{4}\right) = \dfrac{3\pi}{4}$

33. $\tan^{-1}\left(\tan \dfrac{\pi}{4}\right) = x$

$\tan x = \tan \dfrac{\pi}{4},\ -\dfrac{\pi}{2} < x < \dfrac{\pi}{2},$

so $x = \dfrac{\pi}{4}.$

$\tan^{-1}\left(\tan \dfrac{\pi}{4}\right) = \dfrac{\pi}{4}$

35. $\sin \left(\arccos \dfrac{3}{4}\right)$

Let $\omega = \arccos \dfrac{3}{4}$, so $\cos \omega = \dfrac{3}{4}$,

$0 \leq \omega \leq \pi$.

Sketch a triangle and label it using $\cos \omega = 3/4$. The side opposite

$\omega = \sqrt{4^2 - 3^2} = \sqrt{7}$.

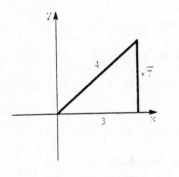

$$\sin \omega = \dfrac{\sqrt{7}}{4}$$

$$\sin \left(\arccos \dfrac{3}{4}\right) = \sin \omega = \dfrac{\sqrt{7}}{4}$$

37. $\cos \left(\csc^{-1} (-2)\right)$

Let $\omega = \csc^{-1} (-2)$, $-\dfrac{\pi}{2} \leq \omega < \dfrac{\pi}{2}$,

$\omega \neq 0$,

$\csc \omega = -2$, ω is in quadrant IV.

The side adjacent to ω is

$$\sqrt{2^2 - (-1)^2} = \sqrt{3}.$$

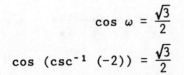

$$\cos \omega = \dfrac{\sqrt{3}}{2}$$

$$\cos \left(\csc^{-1} (-2)\right) = \dfrac{\sqrt{3}}{2}$$

39. $\tan \left(\arcsin \dfrac{3}{5} + \arccos \dfrac{5}{7}\right)$

Let $\omega_1 = \arcsin \dfrac{3}{5}$, $-\dfrac{\pi}{2} \leq \omega_1 \leq \dfrac{\pi}{2}$.

$\sin \omega_1 = \dfrac{3}{5}$ The side adjacent to ω_1

is

$$\sqrt{5^2 - 3^2} = 4.$$

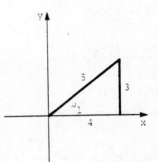

$$\tan \omega_1 = \dfrac{3}{4}$$

Let $\omega_2 = \arccos \dfrac{5}{7}$, $0 \leq \omega_2 \leq \pi$.

$\cos \omega_2 = \dfrac{5}{7}$ The side opposite ω_2 is

$$\sqrt{7^2 - 5^2} = \sqrt{24} = 2\sqrt{6}.$$

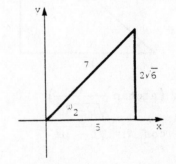

$$\tan \omega_2 = \dfrac{2\sqrt{6}}{5}$$

Use $\tan (\omega_1 + \omega_2)$

$$= \dfrac{\tan \omega_1 + \tan \omega_2}{1 - \tan \omega_1 \tan \omega_2}.$$

$\tan \left(\arcsin \frac{3}{5} + \arccos \frac{5}{7} \right)$

$= \dfrac{\frac{3}{4} + \frac{2\sqrt{6}}{5}}{1 - \left(\frac{3}{4}\right)\left(\frac{2\sqrt{6}}{5}\right)} = \dfrac{\frac{15 + 8\sqrt{6}}{20}}{\frac{20 - 6\sqrt{6}}{20}}$

$= \dfrac{15 + 8\sqrt{6}}{20 - 6\sqrt{6}} \cdot \dfrac{20 + 6\sqrt{6}}{20 + 6\sqrt{6}}$

$= \dfrac{588 + 250\sqrt{6}}{184}$

$= \dfrac{294 + 125\sqrt{6}}{92}$

41. $\cos \left(\arctan \dfrac{u}{\sqrt{1 - u^2}} \right)$

Let $\omega = \arctan \dfrac{u}{\sqrt{1 - u^2}}, \ -\dfrac{\pi}{2} < \omega < \dfrac{\pi}{2}$,

$\tan \omega = \dfrac{u}{\sqrt{1 - u^2}}$ The hypotenuse is

$\sqrt{(1 - u^2) + u^2} = 1.$

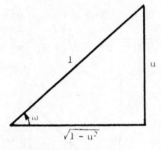

$\cos \left(\arctan \dfrac{u}{\sqrt{1 - u^2}} \right)$

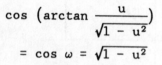

$= \cos \omega = \sqrt{1 - u^2}$

43. $y = \sin^{-1} x$

Domain: $[-1, 1]$

Range: $\left[-\dfrac{\pi}{2}, \dfrac{\pi}{2} \right]$

See the answer graph at the back of
the textbook.

45. $y = \text{arccot } x$

Domain: $(-\infty, \infty)$

Range: $(0, \pi)$

See the answer graph at the back of
the textbook.

47. $2 \tan x - 1 = 0$

$2 \tan x = 1$

$\tan x = \dfrac{1}{2}$

$x = .46364761$ or $x = 3.6052403$

49. $\tan x = \cot x, \ 0 \le x < 2\pi$

Use the identity $\cot x = \dfrac{1}{\tan x}$.

$\tan x = \dfrac{1}{\tan x}, \ \tan \ne 0$

$\tan^2 x = 1$

$\tan x = \pm 1$

If $\tan x = 1$,

$x = \dfrac{\pi}{4}, \dfrac{5\pi}{4}$.

If $\tan x = -1$,

$x = \dfrac{3\pi}{4}, \dfrac{7\pi}{4}$.

$x = \dfrac{\pi}{4}, \dfrac{3\pi}{4}, \dfrac{5\pi}{4}, \dfrac{7\pi}{4}$

51. $\tan^2 2x - 1 = 0, \ 0 \le 2x < 4\pi$

$\tan^2 2x = 1$

$\tan 2x = \pm 1$

If $\tan 2x = 1$,

$2x = \dfrac{\pi}{4}, \dfrac{5\pi}{4}, \dfrac{9\pi}{4}, \dfrac{13\pi}{4}$

$x = \dfrac{\pi}{8}, \dfrac{5\pi}{8}, \dfrac{9\pi}{8}, \dfrac{13\pi}{8}.$

If tan $2x = -1$,

$$2x = \frac{3\pi}{4}, \frac{7\pi}{4}, \frac{11\pi}{4}, \frac{15\pi}{4}$$

$$x = \frac{3\pi}{8}, \frac{7\pi}{8}, \frac{11\pi}{8}, \frac{15\pi}{8}.$$

$$x = \frac{\pi}{8}, \frac{3\pi}{8}, \frac{5\pi}{8}, \frac{7\pi}{8}, \frac{9\pi}{8}, \frac{11\pi}{8}, \frac{13\pi}{8}, \frac{15\pi}{8}.$$

53. $\cos 2x + \cos x = 0$, $0 \le x < 2\pi$

Use the identity

$\cos 2x = 2\cos^2 x - 1.$

$$2\cos^2 x - 1 + \cos x = 0$$

$$2\cos^2 x + \cos x - 1 = 0$$

$$(2\cos x - 1)(\cos x + 1) = 0$$

$$2\cos x - 1 = 0$$

$$\cos x = \frac{1}{2}$$

$$x = \frac{\pi}{3}, \frac{5\pi}{3}$$

or

$\cos x + 1 = 0$

$$\cos x = -1$$

$$x = \pi$$

$$x = \frac{\pi}{3}, \pi, \frac{5\pi}{3}$$

55. $\sin^2 \theta + 3\sin\theta + 2 = 0$,
$0° \le \theta < 360°$

$$(\sin\theta + 2)(\sin\theta + 1) = 0$$

$\sin\theta + 2 = 0$

$\sin\theta = -2$

-2 is not in the range of $\sin\theta$ so there is no solution.

On the other hand,

$\sin\theta + 1 = 0$

$\sin\theta = -1$

$\theta = 270°.$

57. $\sin 2\theta = \cos 2\theta + 1$, $0° \le 2\theta < 720°$

$$(\sin 2\theta)^2 = (\cos 2\theta + 1)^2$$

$$\sin^2 2\theta = \cos^2 2\theta + 2\cos 2\theta + 1$$

Use the identity

$\sin^2 x + \cos^2 x = 1$

$$\sin^2 x = 1 - \cos^2 x.$$

$$1 - \cos^2 2\theta = \cos^2 2\theta$$
$$+ 2\cos 2\theta + 1$$

$$2\cos^2 2\theta + 2\cos 2\theta = 0$$

$$\cos^2 2\theta + \cos 2\theta = 0$$

$$\cos 2\theta(\cos 2\theta + 1) = 0$$

$$\cos 2\theta = 0$$

$$2\theta = 90°, 270°, 450°, 630°$$

or

$\cos 2\theta + 1 = 0$

$$\cos 2\theta = -1$$

$$2\theta = 180°, 540°$$

Possible values for θ are

$\theta = 45°, 135°, 225°, 315°, 90°, 270°.$

Check these solutions since both sides were squared. A value for θ will be a solution if

$\sin 2\theta - \cos 2\theta = 1.$

$\theta = 45°$, $2\theta = 90°$

$\sin 90° - \cos 90° = 1 - 0$

$= 1$

$\theta = 90°$, $2\theta = 180°$

$\sin 180° - \cos 180° = 0 - (-1)$

$= 1$

$\theta = 135°$, $2\theta = 270°$

$\sin 270° - \cos 270° = -1 - 0$

$\ne 1$

$\theta = 225°$, $2\theta = 450°$

$\sin 450° - \cos 450° = 1 - 0$

$= 1$

$\theta = 270°$, $2\theta = 540°$

$\sin 540° - \cos 540° = 0 - (-1)$

$\qquad\qquad\qquad\quad = 1$

$\theta = 315°$, $2\theta = 630°$

$\sin 630° - \cos 630° = -1 - 0$

$\qquad\qquad\qquad\quad \neq 1$

Solutions are 45°, 90°, 225°, 270°.

59. $3 \cos^2 \theta + 2 \cos \theta - 1 = 0$

$(3 \cos \theta - 1)(\cos \theta + 1) = 0$

$3 \cos \theta - 1 = 0$ or $\cos \theta + 1 = 0$

$\qquad \cos \theta = \dfrac{1}{3}$ or $\qquad \cos \theta = -1$

$\theta = 70.5°, \ 289.5°$, or $\theta = 180°$

61. $\sin 2\theta + \sin 4\theta = 0$,

$0° \leq 2\theta < 720°$

$\sin 2\theta + \sin [2(2\theta)] = 0$

Use the identity

$\qquad\qquad \sin 2x = 2 \sin x \cos x$.

$\sin 2\theta + 2 \sin 2\theta \cos 2\theta = 0$

$\sin 2\theta(1 + 2 \cos 2\theta) = 0$

$\sin 2\theta = 0$

$2\theta = 0°, \ 180°, \ 360°, \ 540°$

$\theta = 0°, \ 90°, \ 180°, \ 270°$

or

$1 + 2 \cos 2\theta = 0$

$\qquad \cos 2\theta = -\dfrac{1}{2}$

$2\theta = 120°, \ 240°, \ 480°, \ 600°$

$\theta = 60°, \ 120°, \ 240°, \ 300°$

$\theta = 0°, \ 60°, \ 90°, \ 120°, \ 180°, \ 240°,$

$\qquad 270°, \ 300°$

63. $4y = 2 \sin x$

$2y = \sin x$

$x = \arcsin 2y$

65. $2y = \tan (3x + 2)$

$3x + 2 = \arctan 2y$

$3x = \arctan 2y - 2$

$x = \dfrac{1}{3}(\arctan 2y - 2)$

$x = \left(\dfrac{1}{3} \arctan 2y\right) - \dfrac{2}{3}$

67. $\dfrac{4}{3} \arctan \dfrac{x}{2} = \pi$

$\arctan \dfrac{x}{2} = \dfrac{3\pi}{4}$

But, by definition, the range of arctan is $(-\pi/2, \ \pi/2)$. So this equation has no solution.

69. $\arccos x + \arctan 1 = \dfrac{11\pi}{12}$

$\arccos x = \dfrac{11\pi}{12} - \arctan 1$

$\arccos x = \dfrac{11\pi}{12} - \dfrac{\pi}{4}$

$\arccos x = \dfrac{11\pi}{12} - \dfrac{3\pi}{12} = \dfrac{8\pi}{12}$

$\arccos x = \dfrac{2\pi}{3}$

$\cos \dfrac{2\pi}{3} = x$

$x = -\dfrac{1}{2}$

71. $\dfrac{c_1}{c_2} = \dfrac{\sin \theta_1}{\sin \theta_2}$

$.752 = \dfrac{\sin \theta_1}{\sin \theta_2}$

If $\theta_2 = 90°$,

$.752 = \dfrac{\sin \theta_1}{\sin 90°}$

$\qquad = \dfrac{\sin \theta_1}{1}$

$\qquad = \sin \theta_1$

$\theta_1 = 48.8°$.

73. See the answer graph at the back of the textbook.

75. **(a)** Graph the sum

    $P = .005 \sin 440\pi t + .005 \sin 446\pi t$

    for $.15 \le t \le 1.15$. There appears to be 3 beats per sec. See the answer graph at the back of the textbook.

    (b) Graph the sum

    $P = .005 \sin 440\pi t + .005 \sin 432\pi t$

    for $.15 \le t \le 1.15$. There appears to be 4 beats per sec. See the answer graph at the back of the textbook.

    (c) From part (a), $223 - 220 = 3$. From part (b), $220 - 216 = 4$.

    Therefore, the number of beats appears to be the absolute value of the difference in the frequencies of the two tones.

CHAPTER 6   TEST

Do not use a calculator for this test.

[6.1]   Give the exact real number value of y.

1. $y = \arccos \dfrac{1}{2}$ 2. $y = \csc^{-1}(-1)$

3. $y = \arctan(-\sqrt{3})$ 4. $y = \sec^{-1}\sqrt{2}$

[6.1]   Find each of the following.

5. $\cos\left(\tan^{-1}\dfrac{3}{4}\right)$ 6. $\sin^{-1}\left(\cos\dfrac{\pi}{6}\right)$

7. $\sin\left(\arcsin\dfrac{2}{3} - \arccos\dfrac{1}{4}\right)$

8. $\cot^{-1}\left[\cot\left(-\dfrac{\pi}{3}\right)\right]$ Explain why the answer is *not* $-\pi/3$.

[6.2]   Solve the equations for solutions in the interval [0°, 360°).

9. $\sec\theta - \csc\theta = 0$ 10. $\tan^2\theta = \tan\theta$

11. $\sec^4\theta = 2\sec^2\theta$ 12. $\sin^2\theta = 2\sin\theta - 1$

Solve the equations for solutions in the interval [0, 2π).

[6.2]   13. $\sec^2 x = 2\sqrt{3}\tan x - 2$ [6.3]   14. $2\cos 2x = 1$

[6.3]   15. $\tan\dfrac{\theta}{2} = 1$ [6.3]   16. $\sin 2x = \cos x$

[6.4]   Solve each equation for x.

17. $y = 3 \cos 2x$ **18.** $3y = 2 \tan x - 1$

[6.4]   Solve each equation.

19. $\csc^{-1} 2x = -\dfrac{\pi}{6}$ **20.** $\arcsin x = \arctan \dfrac{4}{3}$

CHAPTER 6   TEST ANSWERS

1. $\pi/3$ **2.** $-\pi/2$ **3.** $-\pi/3$ **4.** $\pi/4$ **5.** $4/5$ **6.** $\pi/3$

7. $\dfrac{2 - 5\sqrt{3}}{12}$ **8.** $2\pi/3$; $-\pi/3$ is not in the range of $\cot^{-1} x$.

9. $45°, 225°$ **10.** $0°, 45°, 180°, 225°$ **11.** $45°, 135°, 225°, 315°$

12. $90°$ **13.** $\pi/3, 4\pi/3$ **14.** $\pi/6, 5\pi/6, 7\pi/6, 11\pi/6$ **15.** $\pi/2$

16. $\pi/6, \pi/2, 5\pi/6, 3\pi/2$ **17.** $x = \dfrac{1}{2} \arccos \dfrac{y}{3}$ **18.** $x = \arctan \dfrac{3y + 1}{2}$

19. -1 **20.** $4/5$

CUMULATIVE REVIEW EXERCISES (Chapters 1–6)

1. How far is the point $(\sqrt{5}, 2)$ from the origin? The x-axis? The y-axis?

2. A merry-go-round makes 2 revolutions per min. Through how many degrees will a horse travel in 3 sec?

3. Use properties of similar triangles to find the value of x in the figure.

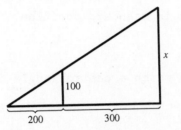

4. The graph of $24x - 7y = 0$, $x \le 0$, is the terminal side of an angle θ in standard position. Find the sine, cosine, and tangent of θ.

5. Without using a calculator, decide which is greater, $\sin \frac{\pi}{10}$ or $\sin^2 \frac{\pi}{10}$.

6. Find a value of θ satisfying the equation $\cos (2\theta + 10°) = \sin (16\theta - 10°)$.

7. Find the exact values of the six trigonometric functions for 240°.

8. Use a calculator to find $\cos 67° 24'$.

9. Air-traffic controllers must know the *cloud ceiling*, the altitude of the lowest point of the clouds. To determine the cloud ceiling, a spotlight located a known distance from the airport is pointed straight up and the angle of elevation from the airport to the circle of light on the bottom of the cloud is determined. (See the figure.) Find the cloud ceiling if the spotlight is located 2.0 mi from the airport and the angle of elevation is 30°.

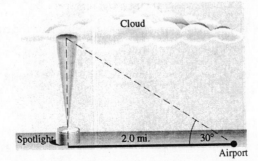

10. A ship leaves port and sails on a bearing of 28.2°. Another ship leaves the same port at the same time and sails on a bearing of 118.2°. If the first ship sails at 20.0 mph and the second sails at 24.0 mph, find the distance between the two ships after 5 hr.

11. The airline distance from Philadelphia to Syracuse is 260 mi, on a bearing of 335°. The distance from Philadelphia to Cincinnati is 510 mi, on a bearing of 245°. Find the bearing from Cincinnati to Syracuse.

12. Consider an angle of 4 radians in standard position. In what quadrant does its terminal side lie?

13. Find the radius of a circle in which a central angle of 5 radians intercepts an arc of length 7 ft.

14. Find the exact value of s in the interval $[\pi, 3\pi/2]$ for which cos s = $-1/2$.

15. A person whirling a rock in a 1-meter-long sling wants the rock to leave the sling at 30 meters per sec. How fast (in revolutions per minute) must the sling whirl to accomplish this?

16. Graph the function y = −cos 6x over a two-period interval.

17. Give the equation of a cosine function having the graph below.

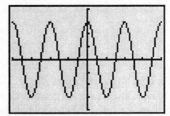

```
WINDOW FORMAT
 Xmin=-4
 Xmax=4
 Xscl=1
 Ymin=-4
 Ymax=4
 Yscl=1
```

18. Graph the function $y = -\sin\left(2x + \dfrac{\pi}{2}\right)$ over a one-period interval.

19. Graph the function y = 1 + 3 sin πx over a two-period interval.

20. Give the equation of a cosine function having the same graph as the function in Exercise 19.

21. Graph the function $y = 2 \csc\left(x + \dfrac{\pi}{4}\right)$ over a one-period interval.

22. Graph the function $y = -3 \cot \frac{1}{4}x$ over a one-period interval.

23. Give the equation of a function having the graph shown.

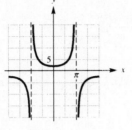

24. Suppose θ is in quadrant III and $\tan \theta = 4$. Use the fundamental identities to find the remaining five trigonometric functions of θ.

25. Use the fundamental identities to find an equivalent expression for $\sec \theta + \tan \theta$ involving only sines and cosines and then simplify it.

26. Rewrite $\sin (-3x)$ without the negative sign inside the parentheses.

27. Factor the trigonometric expression $4 \sec^2 x + 3 \sec x - 1$.

28. Verify the trigonometric identity $(\sin s + \cos s)^2 \csc s = 2 \cos s + \frac{1}{\sin s}$.

29. Use the sum identity for cosine to find the exact value of $\cos 165°$. (*Hint*: $165° = 120° + 45°$.)

30. Verify the identity $\dfrac{\cos (\alpha - \theta) - \cos (\alpha + \theta)}{\cos (\alpha - \theta) + \cos (\alpha + \theta)} = \tan \theta \tan \alpha$.

31. Write $\sin (\theta - 30°)$ as a function of θ.

32. Use the difference identity for tangent to find the exact value of $\tan 15°$.

33. Use a double-angle identity to simplify $\sin 22.5° \cos 22.5°$.

34. Verify the identity $\dfrac{\sin 2\theta}{\sin \theta} - \dfrac{\cos 2\theta}{\cos \theta} = \sec \theta$.

35. Given $\cos 2\theta = 3/5$ and θ terminates in quadrant IV, find the values of the six trigonometric functions of θ.

36. Use a half-angle identity to find tan 165°.

37. Given $\sin \beta = -\sqrt{7}/4$, with β in quadrant III, find $\cos \beta/2$, $\sin \beta/2$, and $\tan \beta/2$.

38. Verify the identity $\cos^2 \dfrac{\alpha}{2} = \dfrac{1 + \sec \alpha}{2 \sec \alpha}$.

39. Suppose $a = \arcsin b$. Which of the following are also solutions to $\sin x = b$: $a - \pi$, $\pi - a$, $a + \pi$?

40. Without using a calculator, determine the value of $\sec \left(\text{arccot } \dfrac{3}{5} \right)$.

41. Write $\sec (\cot^{-1} u)$ as an expression in u.

42. Find all solutions to $\sin^2 \theta + \sin \theta = \cos^2 \theta$ in the interval [0°, 360°).

43. Find all solutions to $\sin 2x \cos x + \cos 2x \sin x = 1$ in the interval $[0, 2\pi)$.

44. Find all solutions to $\sin x = \dfrac{1}{2}\sqrt{3} \tan x$ in the interval $[0, 2\pi)$.

45. Find all solutions to $\cos 2x = 2 \sin x \cos x$ in the interval $[0, 2\pi)$.

46. Find all solutions to $\tan 4\theta - \sec 4\theta = 1$ in the interval [0°, 360°).

47. Solve $y = 4 \sin (3x + 5)$ for x.

48. Solve $\arccos x = \arctan \dfrac{5}{12}$.

SOLUTIONS TO CUMULATIVE REVIEW EXERCISES (Chapters 1–6)

1. The distance from $(\sqrt{5}, 2)$ to the origin is given by

$$\sqrt{(\sqrt{5} - 0)^2 + (2 - 0)^2} = \sqrt{5 + 4}$$
$$= \sqrt{9} = 3.$$

The distance from (x, y) to the x-axis is $|y|$, and the distance from (x, y) to the y-axis is $|x|$. Thus, the distance from $(\sqrt{5}, 2)$ to the x-axis is $|2|$ or 2. The distance to the y-axis is $|\sqrt{5}|$ or $\sqrt{5}$.

2. $\dfrac{2 \text{ revolutions}}{\text{min}} \cdot \dfrac{1 \text{ min}}{60 \text{ sec}} \cdot \dfrac{360°}{1 \text{ revolution}}$

$= 12°$ per sec

In 3 sec, a horse will travel through $3(12°)$ or $36°$.

3. $\dfrac{x}{500} = \dfrac{100}{200}$

$\dfrac{x}{500} = \dfrac{1}{2}$

$2x = 500$

$x = 250$

4. $24x - 7y = 0,\ x \le 0$

Find a point on the terminal side of θ. Let $x = -7$ (since $x \le 0$).

$24(-7) - 7y = 0$

Divide by -7.

$24 + y = 0$

$y = -24$

A point on the terminal side is $(-7, -24)$.

$r = \sqrt{(-7)^2 + (-24)^2}$
$= \sqrt{49 + 576}$
$= \sqrt{625}$
$= 25$

$\sin \theta = \dfrac{y}{r} = \dfrac{-24}{25} = -\dfrac{24}{25}$

$\cos \theta = \dfrac{x}{r} = \dfrac{-7}{25} = -\dfrac{7}{25}$

$\tan \theta = \dfrac{y}{x} = \dfrac{-24}{-7} = \dfrac{24}{7}$

5. If $0 < x < 1$, multiplication by x, x nonnegative, gives $0 < x^2 < x$. Thus, for $0 < x < 1$, $x > x^2$.

Since $0 < \sin \dfrac{\pi}{10} < 1$,

$$\sin \dfrac{\pi}{10} > \sin^2 \dfrac{\pi}{10}.$$

Therefore, $\sin \dfrac{\pi}{10}$ is greater.

6. Since cosine and sine are co-functions, $\cos (2\theta + 10°) = \sin (16\theta - 10°)$ and

$(2\theta + 10°) + (16\theta - 10°) = 90°$

$18\theta = 90°$

$\theta = 5°.$

7. $\sin 240° = -\sin 60° = -\dfrac{\sqrt{3}}{2}$

$\cos 240° = -\cos 60° = -\dfrac{1}{2}$

$\tan 240° = \tan 60° = \sqrt{3}$

$\cot 240° = \dfrac{1}{\sqrt{3}} = \dfrac{\sqrt{3}}{3}$

$\sec 240° = -2$

$\csc 240° = -\dfrac{2}{\sqrt{3}} = -\dfrac{2\sqrt{3}}{3}$

8. $\cos 67° 24' = \cos 67.4° = .38429532$

9.   Let x be the cloud ceiling.

$$\tan 30° = \frac{x}{2.0}$$

$$x = 2.0 \tan 30°$$

$$x \approx 1.154701$$

The cloud ceiling is about 1.2 mi.

10. After 5 hr the ships have sailed 5(20.0) mi or 100 mi and 5(24.0) mi or 120 mi. The difference between their bearings is

$$118.2° - 28.2° = 90°.$$

This tells us that their paths form a right angle. The distance between the ships is the length of the hypotenuse of a right triangle having legs of lengths 100 mi and 120 mi. The length of the hypotenuse is given by

$$\sqrt{100^2 + 120^2} = \sqrt{10,000 + 14,400}$$
$$= \sqrt{24,400}$$
$$\approx 156.20.$$

The distance between them is about 156 mi.

11. In the figure, let P represent Philadelphia, S represent Syracuse, and C represent Cincinnati.

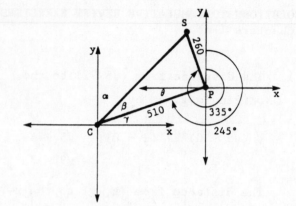

The bearing from Cincinnati to Syracuse will be angle α. To find α, use the fact that

$$\alpha + \beta + \gamma = 90°.$$

Note that $\gamma = \theta$ and $\theta = 270° - 245° = 25°$.

To find β, use the fact that triangle SPC is a right triangle with the right angle at P.

$$\tan \beta = \frac{260}{510}$$

$$\beta = 27.0°$$

$$\alpha + 27° + 25° = 90°$$
$$\alpha + 52° = 90°$$
$$\alpha = 38°$$

The bearing from Cincinnati to Syracuse is 38°.

12. π radians ≈ 3.14 radians

$\frac{3\pi}{2}$ radians ≈ 4.71 radians

Therefore, $\pi < 4 < \frac{3\pi}{2}$, and an angle of 4 radians in standard position is in quadrant III.

13. $\theta = 5$ radians, $s = 7$

$s = r\theta$

$7 = r(5)$

$r = \dfrac{7}{5}$

$r = 1.4$

The radius is 1.4 ft.

14. $\cos \dfrac{\pi}{3} = \dfrac{1}{2}$

If s is in the interval $[\pi, 3\pi/2]$ and if $\cos s = -1/2$, then

$$s = \pi + \frac{\pi}{3} = \frac{4\pi}{3}.$$

15. $v = 30$ m/sec, $r = 1$ m

$v = r\omega$

30 m/sec $= (1$ m$)\omega$

$\omega = 30$ radians/sec

$\dfrac{30 \text{ radians}}{\text{sec}} \cdot \dfrac{60 \text{ sec}}{1 \text{ min}} \cdot \dfrac{1 \text{ revolution}}{2\pi \text{ radians}}$

$= \dfrac{900}{\pi}$ revolutions per min

16. $y = -\cos 6x$

The period is $2\pi/6$ or $\pi/3$.

The amplitude is $|-1|$ which is 1.

To graph over a two-period interval, graph the cosine curve with a period of $\pi/3$ and an amplitude of 1 from $x = 0$ to $x = 2\pi/3$. Then reflect the graph about the x-axis because of the negative sign.

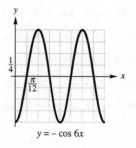

$y = -\cos 6x$

17. From the graph we see that the amplitude is 3 and the period is 2. Since the equation will have the form

$$y = a \cos bx,$$

$a = 3$ and $2\pi/b = 2$. Therefore, $2b = 2\pi$ and $b = \pi$. The equation is

$$y = 3 \cos \pi x.$$

18. $y = -\sin \left(2x + \dfrac{\pi}{2}\right)$

$y = -\sin 2\left(x + \dfrac{\pi}{4}\right)$

The amplitude is $|-1| = 1$.

The period is $2\pi/2 = \pi$.

The phase shift is $|-\pi/4| = \pi/4$ units to the left.

Graph a sine curve with a period of π and an amplitude of 1. Because of the negative sign, reflect the graph about the x-axis. Then translate the graph $\pi/4$ units to the left.

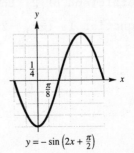

$y = -\sin \left(2x + \dfrac{\pi}{2}\right)$

19. $y = 1 + 3 \sin \pi x$

The amplitude is $|3|$ which is 3.
The period is $2\pi/\pi$ which is 2.
Graph a sine curve with a period
of 2 and an amplitude of 3. Then
translate the graph 1 unit upward.

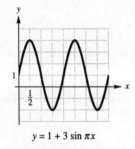

$y = 1 + 3 \sin \pi x$

20.   Recall that for every angle θ;

$$\sin \theta = \cos \left(\frac{\pi}{2} - \theta \right).$$

Therefore, $\sin \pi x = \cos \left(\frac{\pi}{2} - \pi x \right)$.
The graph of

$$y = 1 + 3 \cos \left(\frac{\pi}{2} - \pi x \right)$$

will be the same as the graph of

$$y = 1 + 3 \sin \pi x.$$

21. $y = 2 \csc \left(x + \frac{\pi}{4} \right)$

The range of the cosecant function
is $(-\infty, -1] \cup [1, \infty)$. Because of
the coefficient of 2, the range of
this function is $(-\infty, -2] \cup [2, \infty)$.

Start by graphing a cosecant function with this range. Then translate the graph $\pi/4$ units to the left
since the phase shift is $|-\pi/4| =$
$\pi/4$ to the left.

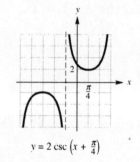

$y = 2 \csc \left(x + \frac{\pi}{4} \right)$

22. $y = -3 \cot \frac{1}{4} x$

The period of the function is $\frac{\pi}{1/4}$
or 4π. Graphing over a one-period
interval, there will be vertical
asymptotes at $x = 0$ and $x = 4\pi$. The
graph intersects the x-axis at
$x = 2\pi$.
The coefficient 3 affects the steep-
ness of the graph, while the negative sign reflects the graph about
the x-axis.
When $x = \pi$,

$$y = -3 \cot \frac{\pi}{4} = -3(1) = -3.$$

When $x = 3\pi$,

$$y = -3 \cot \frac{3\pi}{4} = -3(-1) = 3.$$

Using these two points, the point
where the graph intersects the x-
axis, and the asymptotes, graph the
function.

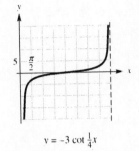

$$y = -3 \cot \tfrac{1}{4} x$$

23. The graph is the graph of a secant
    function with a range of
    $(-\infty, -2] \cup [2, \infty)$, so the coeffi-
    cient is 2.
    The period is 4π, so $2\pi/b = 4\pi$ and
    $b = .5$.
    Therefore, the equation is

    $$y = 2 \sec .5x.$$

24. $\tan \theta = 4$ and θ is in quadrant III.
    In quadrant III all trigonometric
    functions have negative values
    except tangent and cotangent.

    $$\tan^2 \theta + 1 = \sec^2 \theta$$
    $$\sec^2 \theta = 4^2 + 1$$
    $$\sec^2 \theta = 17$$
    $$\sec \theta = \pm\sqrt{17}$$
    $$\sec \theta = -\sqrt{17}$$

    $$\cos \theta = \frac{1}{-\sqrt{17}} = -\frac{\sqrt{17}}{17}$$

    $$\sin^2 \theta = 1 - \cos^2 \theta$$
    $$\sin^2 \theta = 1 - \left(-\frac{1}{\sqrt{17}}\right)^2$$
    $$\sin^2 \theta = 1 - \frac{1}{17}$$
    $$\sin^2 \theta = \frac{16}{17}$$

$$\sin \theta = \pm\frac{4}{\sqrt{17}}$$

$$\sin \theta = -\frac{4\sqrt{17}}{17}$$

$$\csc \theta = \frac{1}{-\dfrac{4}{\sqrt{17}}} = -\frac{\sqrt{17}}{4}$$

$$\cot \theta = \frac{1}{4}$$

To summarize, $\sin \theta = -4\sqrt{17}/17$,
$\cos \theta = -\sqrt{17}/17$, $\cot \theta = 1/4$,
$\sec \theta = -\sqrt{17}$, and $\csc \theta = -\sqrt{17}/4$.

25. $\sec \theta \tan \theta = \dfrac{1}{\cos \theta} + \dfrac{\sin \theta}{\cos \theta}$

    $$= \frac{1 + \sin \theta}{\cos \theta}$$

26. Since $\sin (-\theta) = -\sin \theta$,

    $$\sin (-3x) = -\sin 3x.$$

27. $4 \sec^2 x + 3 \sec x - 1$
    $= (4 \sec x - 1)(\sec x + 1)$

28. Verify $(\sin s + \cos s)^2 \csc s$

    $$= 2 \cos s + \frac{1}{\sin s}.$$

    Work on the left side.

$(\sin s + \cos s)^2 \csc s$

$$= (\sin^2 s + 2 \sin s \cos s + \cos^2 s)\frac{1}{\sin s}$$

$$= (1 + 2 \sin s \cos s)\frac{1}{\sin s}$$

$$= \frac{1}{\sin s} + \frac{2 \sin s \cos s}{\sin s}$$

$$= 2 \cos s + \frac{1}{\sin s}$$

29. $\cos 165° = \cos (120° + 45°)$

$\qquad = \cos 120° \cos 45° - \sin 120° \sin 45°$

$\qquad = \left(-\dfrac{1}{2}\right)\left(\dfrac{\sqrt{2}}{2}\right) - \left(\dfrac{\sqrt{3}}{2}\right)\left(\dfrac{\sqrt{2}}{2}\right)$

$\qquad = -\dfrac{\sqrt{2}}{4} - \dfrac{\sqrt{6}}{4} = -\dfrac{\sqrt{2} + \sqrt{6}}{4}$

30. Verify $\dfrac{\cos (\alpha - \theta) - \cos (\alpha + \theta)}{\cos (\alpha - \theta) + \cos (\alpha + \theta)} = \tan \theta \tan \alpha$.

Work on the left side.

$\dfrac{\cos (\alpha - \theta) - \cos (\alpha + \theta)}{\cos (\alpha - \theta) + \cos (\alpha + \theta)}$

$\quad = \dfrac{(\cos \alpha \cos \theta + \sin \alpha \sin \theta) - (\cos \alpha \cos \theta - \sin \alpha \sin \theta)}{(\cos \alpha \cos \theta + \sin \alpha \sin \theta) + (\cos \alpha \cos \theta - \sin \alpha \sin \theta)}$

$\quad = \dfrac{\cos \alpha \cos \theta + \sin \alpha \sin \theta - \cos \alpha \cos \theta + \sin \alpha \sin \theta}{\cos \alpha \cos \theta + \sin \alpha \sin \theta + \cos \alpha \cos \theta - \sin \alpha \sin \theta}$

$\quad = \dfrac{2 \sin \alpha \sin \theta}{2 \cos \alpha \cos \theta}$

$\quad = \dfrac{\sin \theta}{\cos \theta} \cdot \dfrac{\sin \alpha}{\cos \alpha}$

$\quad = \tan \theta \tan \alpha$

31. $\sin (\theta - 30°) = \sin \theta \cos 30° - \cos \theta \sin 30°$

$\qquad = \sin \theta \left(\dfrac{\sqrt{3}}{2}\right) - \cos \theta \left(\dfrac{1}{2}\right)$

$\qquad = \dfrac{\sqrt{3} \sin \theta - \cos \theta}{2}$

32. $\tan 15° = \tan (45° - 30°)$

$\qquad = \dfrac{\tan 45° - \tan 30°}{1 + \tan 45° \tan 30°}$

$\qquad = \dfrac{1 - \dfrac{\sqrt{3}}{3}}{1 + (1)\left(\dfrac{\sqrt{3}}{3}\right)} \cdot \dfrac{3}{3}$

$\qquad = \dfrac{3 - \sqrt{3}}{3 + \sqrt{3}} = \dfrac{3 - \sqrt{3}}{3 + \sqrt{3}} \cdot \dfrac{3 - \sqrt{3}}{3 - \sqrt{3}}$

$\qquad = \dfrac{9 - 6\sqrt{3} + 3}{9 - 3} = \dfrac{12 - 6\sqrt{3}}{6}$

$\qquad = 2 - \sqrt{3}$

33. Since $\sin 2\theta = 2 \sin \theta \cos \theta$,

$$\sin \theta \cos \theta = \frac{1}{2} \sin 2\theta.$$

Therefore,

$$\sin 22.5° \cos 22.5° = \frac{1}{2} \sin 2(22.5°)$$

$$= \frac{1}{2} \sin 45°$$

$$= \frac{1}{2}\left(\frac{\sqrt{2}}{2}\right)$$

$$= \frac{\sqrt{2}}{4}.$$

34. Verify $\dfrac{\sin 2\theta}{\sin \theta} - \dfrac{\cos 2\theta}{\cos \theta} = \sec \theta$.

Work on the left side.

$$\frac{\sin 2\theta}{\sin \theta} - \frac{\cos 2\theta}{\cos \theta} = \frac{\cos \theta \sin 2\theta - \sin \theta \cos 2\theta}{\sin \theta \cos \theta}$$

$$= \frac{\cos \theta \,(2 \sin \theta \cos \theta) - \sin \theta \,(1 - 2 \sin^2 \theta)}{\sin \theta \cos \theta}$$

$$= \frac{2 \sin \theta \cos^2 \theta - \sin \theta + 2 \sin^3 \theta}{\sin \theta \cos \theta}$$

$$= \frac{2 \sin \theta \,(\cos^2 \theta + \sin^2 \theta) - \sin \theta}{\sin \theta \cos \theta}$$

$$= \frac{2 \sin \theta \,(1) - \sin \theta}{\sin \theta \cos \theta} = \frac{\sin \theta}{\sin \theta \cos \theta}$$

$$= \frac{1}{\cos \theta} = \sec \theta$$

35. $\cos 2\theta = 3/5$ and θ terminates in quadrant IV.

$$\cos 2\theta = 2 \cos^2 \theta - 1$$

$$\frac{3}{5} = 2 \cos^2 \theta - 1$$

$$2 \cos^2 \theta = \frac{8}{5}$$

$$\cos^2 \theta = \frac{4}{5}$$

$$\cos \theta = \pm\frac{2}{\sqrt{5}}$$

$$\cos \theta = \frac{2\sqrt{5}}{5}$$

since θ terminates in quadrant IV.

$$\cos 2\theta = 1 - 2\sin^2 \theta$$

$$\frac{3}{5} = 1 - 2\sin^2 \theta$$

$$2\sin^2 \theta = \frac{2}{5}$$

$$\sin^2 \theta = \frac{1}{5}$$

$$\sin \theta = \pm\frac{1}{\sqrt{5}}$$

$$\sin \theta = -\frac{\sqrt{5}}{5}$$

since θ terminates in quadrant IV.

$$\tan \theta = \frac{-\frac{\sqrt{5}}{5}}{\frac{2\sqrt{5}}{5}} = -\frac{1}{2}$$

$$\cot \theta = \frac{1}{-\frac{1}{2}} = -2$$

$$\sec \theta = \frac{1}{\frac{2}{\sqrt{5}}} = \frac{\sqrt{5}}{2}$$

$$\csc \theta = \frac{1}{-\frac{1}{\sqrt{5}}} = -\sqrt{5}$$

To summarize, $\sin \theta = -\sqrt{5}/5$, $\cos \theta = 2\sqrt{5}/5$, $\tan \theta = -1/2$, $\cot \theta = -2$, $\sec \theta = \sqrt{5}/2$, and $\csc \theta = -\sqrt{5}$.

36. $\tan 165° = \tan \left(\frac{330°}{2}\right)$

$$= \frac{1 - \cos 330°}{\sin 330°}$$

$$= \frac{1 - \frac{\sqrt{3}}{2}}{-\frac{1}{2}}$$

$$= -2 + \sqrt{3}$$

$$= \sqrt{3} - 2$$

37. Since $\sin \beta = -\sqrt{7}/4$ and β is in quadrant III,

$$\pi < \beta < \frac{3\pi}{2} \text{ and } \frac{\pi}{2} < \frac{\beta}{2} < \frac{3\pi}{4}.$$

Thus, $\frac{\beta}{2}$ is in quadrant II.

$$\cos^2 \beta = 1 - \sin^2 \beta$$

$$= 1 - \left(-\frac{\sqrt{7}}{4}\right)^2$$

$$= \frac{16}{16} - \frac{7}{16}$$

$$\cos^2 \beta = \frac{9}{16}$$

$$\cos \beta = \pm\sqrt{\frac{9}{16}}$$

$$\cos \beta = -\frac{3}{4}$$

since β is in quadrant III.

$$\sin \frac{\beta}{2} = \sqrt{\frac{1 - \cos \beta}{2}} = \sqrt{\frac{1 - \left(-\frac{3}{4}\right)}{2}}$$

$$= \sqrt{\frac{7}{8}} = \frac{\sqrt{14}}{4}$$

$$\cos \frac{\beta}{2} = -\sqrt{\frac{1 + \cos \beta}{2}} = -\sqrt{\frac{1 + \left(-\frac{3}{4}\right)}{2}}$$

$$= -\sqrt{\frac{1}{8}} = -\frac{\sqrt{2}}{4}$$

$$\tan \frac{\beta}{2} = \frac{\sin \beta}{1 + \cos \beta} = \frac{-\frac{\sqrt{7}}{4}}{1 + \left(-\frac{3}{4}\right)} = -\sqrt{7}$$

38. Verify $\cos^2 \frac{\alpha}{2} = \frac{1 + \sec \alpha}{2 \sec \alpha}$.

Work on the left side.

$$\cos^2 \frac{\alpha}{2} = \left(\pm \sqrt{\frac{1 + \cos \alpha}{2}} \right)^2$$

$$= \frac{1 + \cos \alpha}{2}$$

$$= \frac{\dfrac{1}{\cos \alpha} + \dfrac{\cos \alpha}{\cos \alpha}}{\dfrac{2}{\cos \alpha}}$$

$$= \frac{\sec \alpha + 1}{2 \sec \alpha}$$

$$= \frac{1 + \sec \alpha}{2 \sec \alpha}$$

39. If $a = \arcsin b$, then $\sin a = b$.

$\sin (a - \pi)$
$= \sin a \cos \pi - \cos a \sin \pi$
$= \sin a(-1) - \cos a (0)$
$= -\sin a$
$= -b$

$\sin (\pi - a)$
$= \sin \pi \cos a - \cos \pi \sin a$
$= (0) \cos a - (-1) \sin a$
$= \sin a$
$= b$

$\sin (a + \pi)$
$= \sin a \cos \pi + \cos a \sin \pi$
$= \sin a (-1) + \cos a (0)$
$= -\sin a$
$= -b$

The only other solution to $\sin x = b$ is $\pi - a$.

40. $\sec \left(\text{arccot } \dfrac{3}{5} \right)$

Let $\theta = \text{arccot } \dfrac{3}{5}$. Then $\cot \theta = \dfrac{3}{5}$

and θ is in quadrant I. Find $\sec \theta$.

$$\tan \theta = \frac{1}{\dfrac{3}{5}} = \frac{5}{3}$$

$$\sec^2 \theta = \tan^2 \theta + 1$$

$$= \left(\frac{5}{3} \right)^2 + 1$$

$$= \frac{25}{9} + \frac{9}{9}$$

$$\sec^2 \theta = \frac{34}{9}$$

$$\sec \theta = \pm \frac{\sqrt{34}}{3}$$

Since θ is in quadrant I, $\sec \theta = \sqrt{34}/3$. Thus,

$$\sec \left(\text{arccot } \frac{3}{5} \right) = \frac{\sqrt{34}}{3} .$$

41. $\sec (\cot^{-1} u)$

Let $\theta = \cot^{-1} u$. Then $\cot \theta = u = \dfrac{u}{1}$. Refer to the right triangle below.

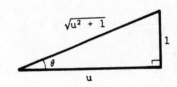

$$\sec \theta = \frac{\sqrt{u^2 + 1}}{u}$$

Therefore, $\sec (\cot^{-1} u) = \dfrac{\sqrt{u^2 + 1}}{u}$.

42.
$$\sin^2 \theta + \sin \theta = \cos^2 \theta$$
$$\sin^2 \theta + \sin \theta = 1 - \sin^2 \theta$$
$$2 \sin^2 \theta + \sin \theta - 1 = 0$$
$$(2 \sin \theta - 1)(\sin \theta + 1) = 0$$

$\sin \theta = \dfrac{1}{2}$ or $\sin \theta = -1$

$\theta = 30°, 150°$ $\theta = 270°$

$\theta = 30°, 150°,$ or $270°$

43.

$$\sin 2x \cos x + \cos 2x \sin x = 1$$
$$\sin (2x + x) = 1$$
$$\sin 3x = 1$$
$$3x = \frac{\pi}{2}, \frac{5\pi}{2}, \text{ or } \frac{9\pi}{2}$$
$$x = \frac{\pi}{6}, \frac{5\pi}{6}, \text{ or } \frac{3\pi}{2}$$

44.

$$\sin x = \frac{1}{2}\sqrt{3} \tan x$$
$$\sin x = \frac{\sqrt{3} \sin x}{2 \cos x}$$
$$\sin x - \frac{\sqrt{3} \sin x}{2 \cos x} = 0$$
$$\sin x \left(1 - \frac{\sqrt{3}}{2 \cos x}\right) = 0$$
$$\sin x = 0 \quad \text{or} \quad 1 - \frac{\sqrt{3}}{2 \cos x} = 0$$

If $\sin x = 0$, then $x = 0$ or π.

If $1 - \frac{\sqrt{3}}{2 \cos x} = 0$, then

$1 = \frac{\sqrt{3}}{2 \cos x}$ and $\cos x = \frac{\sqrt{3}}{2}$. Thus,

$x = \frac{\pi}{6}$ or $\frac{11\pi}{6}$.

Therefore,

$$x = 0, \frac{\pi}{6}, \pi, \text{ or } \frac{11\pi}{6}.$$

45. $\cos 2x = 2 \sin x \cos x$

$\cos 2x = \sin 2x$

$$1 = \frac{\sin 2x}{\cos 2x}$$

$\tan 2x = 1$

$$2x = \frac{\pi}{4}, \frac{5\pi}{4}, \frac{9\pi}{4}, \text{ or } \frac{13\pi}{4}$$
$$x = \frac{\pi}{8}, \frac{5\pi}{8}, \frac{9\pi}{8}, \text{ or } \frac{13\pi}{8}$$

46.

$$\tan 4\theta - \sec 4\theta = 1$$
$$\tan 4\theta = 1 + \sec 4\theta$$
$$(\tan 4\theta)^2 = (1 + \sec 4\theta)^2$$
$$\tan^2 4\theta = 1 + 2 \sec 4\theta + \sec^2 4\theta$$
$$\sec^2 4\theta - 1 = 1 + 2 \sec 4\theta + \sec^2 4\theta$$
$$-2 = 2 \sec 4\theta$$
$$\sec 4\theta = -1$$
$$4\theta = 180°, 540°, 900°,$$
$$\text{or } 1260°$$
$$\theta = 45°, 135°, 225°,$$
$$\text{or } 315°$$

Since both sides of the equation were squared, each solution must be checked. All solutions check.

47.

$$y = 4 \sin (3x + 5)$$
$$\sin (3x + 5) = \frac{y}{4}$$
$$3x + 5 = \arcsin \left(\frac{y}{4}\right)$$
$$3x = \arcsin \left(\frac{y}{4}\right) - 5$$
$$x = \frac{1}{3}\left[\arcsin \left(\frac{y}{4}\right) - 5\right]$$

48.

$$\arccos x = \arctan \frac{5}{12}$$
$$\cos (\arccos x) = \cos \left(\arctan \frac{5}{12}\right)$$
$$x = \cos \left(\arctan \frac{5}{12}\right)$$

Let $\theta = \arctan \frac{5}{12}$. Then $\tan \theta = \frac{5}{12}$.
Refer to the right triangle.

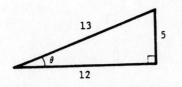

$x = \cos \theta$

$x = \dfrac{12}{13}$

CHAPTER 7 APPLICATIONS OF TRIGONOMETRY AND VECTORS

Section 7.1

1. $\dfrac{6}{\sin 30°} = \dfrac{x}{\sin 45°}$

$\dfrac{6\sin 45°}{\sin 30°} = x$

$\dfrac{6\left(\frac{\sqrt{2}}{2}\right)}{\frac{1}{2}} = x$

$6\sqrt{2} = x$

$x = 6\sqrt{2}$

3.   The measure of angle C is

$180° - (60° + 75°) = 45°.$

$\dfrac{a}{\sin A} = \dfrac{c}{\sin C}$

$\dfrac{a}{\sin 60°} = \dfrac{\sqrt{2}}{\sin 45°}$

$a = \dfrac{\sqrt{2}\sin 60°}{\sin 45°}$

$a = \dfrac{\sqrt{2}\left(\frac{\sqrt{3}}{2}\right)}{\frac{\sqrt{2}}{2}}$

$a = \sqrt{3}$

5.   A = 37°, B = 48°, c = 18 m

$C = 180° - A - B$

$C = 180° - 37° - 48°$

$C = 95°$

$\dfrac{b}{\sin B} = \dfrac{c}{\sin C}$

$b = \dfrac{18\sin 48°}{\sin 95°}$

$b = 13$ m

$\dfrac{a}{\sin A} = \dfrac{c}{\sin C}$

$a = \dfrac{18\sin 37°}{\sin 95°}$

$a = 11$ m

7.   A = 27.2°, C = 115.5°, c = 76.0 ft

$B = 180° - A - C$

$B = 180° - 27.2° - 115.5°$

$B = 37.3°$

$\dfrac{a}{\sin A} = \dfrac{c}{\sin C}$

$a = \dfrac{76.0\sin 27.2°}{\sin 115.5°}$

$a = 38.5$ ft

$\dfrac{b}{\sin B} = \dfrac{c}{\sin C}$

$b = \dfrac{76.0\sin 37.3°}{\sin 115.5°}$

$b = 51.0$ ft

9.   A = 68.41°, B = 54.23°, a = 12.75 ft

$C = 180° - A - B$

$C = 180° - 68.41° - 54.23°$

$C = 57.36°$

$\dfrac{a}{\sin A} = \dfrac{b}{\sin B}$

$b = \dfrac{12.75\sin 54.23°}{\sin 68.41°}$

$b = 11.13$ ft

$\dfrac{a}{\sin A} = \dfrac{c}{\sin C}$

$c = \dfrac{12.75\sin 57.36°}{\sin 68.41°}$

$c = 11.55$ ft

11.   A = 87.2°, b = 75.9 yd, C = 74.3°

$B = 180° - A - C$

$B = 180° - 87.2° - 74.3°$

$B = 18.5°$

$\dfrac{a}{\sin A} = \dfrac{b}{\sin B}$

$a = \dfrac{75.9\sin 87.2°}{\sin 18.5°}$

$a = 239$ yd

$$\frac{b}{\sin B} = \frac{c}{\sin C}$$

$$c = \frac{75.9 \sin 74.3°}{\sin 18.5°}$$

$$c = 230 \text{ yd}$$

13. B = 20° 50′, AC = 132 ft,
     C = 103° 10′

A = 180° − B − C

A = 180° − 20° 50′ − 103° 10′

A = 56° 00′

$$\frac{AC}{\sin B} = \frac{AB}{\sin C}$$

$$AB = \frac{132 \sin 103° 10′}{\sin 20° 50′}$$

$$AB = 361 \text{ ft}$$

$$\frac{BC}{\sin A} = \frac{AC}{\sin B}$$

$$BC = \frac{132 \sin 56° 00′}{\sin 20° 50′}$$

$$BC = 308 \text{ ft}$$

15. A = 39.70°, C = 30.35°, b = 39.74 m

B = 180° − A − C

B = 180° − 39.70° − 30.35°

B = 110.0° (rounded)

$$\frac{a}{\sin A} = \frac{b}{\sin B}$$

$$a = \frac{39.74 \sin 39.70°}{\sin 109.95°}$$

$$a = 27.01 \text{ m}$$

$$\frac{b}{\sin B} = \frac{c}{\sin C}$$

$$c = \frac{39.74 \sin 30.35°}{\sin 109.95°}$$

$$c = 21.36 \text{ m}$$

17. B = 42.88°, C = 102.40°, b = 3974 ft

A = 180° − B − C

A = 180° − 42.88° − 102.40°

A = 34.72°

$$\frac{a}{\sin A} = \frac{b}{\sin B}$$

$$a = \frac{3974 \sin 34.72°}{\sin 42.88°}$$

$$a = 3326 \text{ ft}$$

$$\frac{b}{\sin B} = \frac{c}{\sin C}$$

$$c = \frac{3974 \sin 102.40°}{\sin 42.88°}$$

$$c = 5704 \text{ ft}$$

19. A = 39° 54′, a = 268.7 m,
     B = 42° 32′

C = 180° − A − B

C = 180° − 39° 54′ − 42° 32′

C = 97° 34′

$$\frac{a}{\sin A} = \frac{b}{\sin B}$$

$$b = \frac{268.7 \sin 42° 32′}{\sin 39° 54′}$$

$$b = 283.2 \text{ m}$$

$$\frac{a}{\sin A} = \frac{c}{\sin C}$$

$$c = \frac{268.7 \sin 97° 34′}{\sin 39° 54′}$$

$$c = 415.2 \text{ m}$$

21. The law of sines cannot be used to solve a triangle if we are given the lengths of three sides of a triangle because the law of sines states

$$\frac{a}{\sin A} = \frac{b}{\sin B} = \frac{c}{\sin C}.$$

If we know a, b, and c, there is no
way to solve for A, B, or C because
we would have one equation with two
unknowns or two equations with three
unknowns.

25. Yes, the law of sines can be written
as

$$\frac{a}{b} = \frac{\sin A}{\sin B}.$$

Start with $\frac{a}{\sin A} = \frac{b}{\sin B}$.

Multiply both sides by sin A and
divide both sides by b to get

$$\frac{a}{b} = \frac{\sin A}{\sin B}.$$

27.

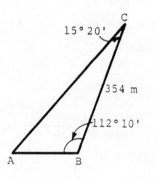

A = 180° − B − C
A = 180° − 112° 10′ − 15° 20′
A = 52° 30′

$$\frac{BC}{\sin A} = \frac{AB}{\sin C}$$

$$AB = \frac{354 \sin 15° 20′}{\sin 52° 30′}$$

AB = 118 m

29. Let C = the transmitter.

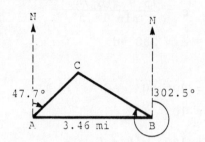

Since side AB is on an east−west
line, the angle between it and any
north−south line is 90°.

A = 90° − 47.7° = 42.3°
B = 302.5° − 270° = 32.5°
C = 180° − A − B
  = 180° − 42.3° − 32.5°
C = 105.2°

$$\frac{AC}{\sin 32.5°} = \frac{3.46}{\sin 105.2°}$$

$$AC = \frac{3.46 \sin 32.5°}{\sin 105.2°}$$

AC = 1.93 mi

31. $\frac{x}{\sin 54.8°} = \frac{12.0}{\sin 70.4°}$

$$x = \frac{12.0 \sin 54.8°}{\sin 70.4°}$$

x = 10.4 in

33. Label α in the triangle as shown.

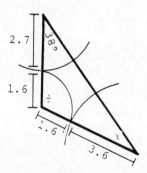

$$\frac{1.6 + 3.6}{\sin 38°} = \frac{1.6 + 2.7}{\sin \alpha}$$

$$\sin \alpha = \frac{(1.6 + 2.7)\sin 38°}{1.6 + 3.6}$$

$$\alpha = 31°$$

$$\theta = 180° - 38° - \alpha$$

$$= 180° - 38° - 31°$$

$$\theta = 111°$$

35. Let x = the distance to the lighthouse at bearing N 37° E

  y = the distance to the lighthouse at bearing N 25° E.

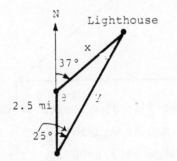

$$\theta = 180° - 37° = 143°$$
$$\alpha = 180° - \theta - 25°$$
$$= 180° - 143° - 25° = 12°$$

$$\frac{2.5}{\sin \alpha} = \frac{x}{\sin 25°}$$

$$x = \frac{2.5 \sin 25°}{\sin 12°}$$

$$= 5.1 \text{ mi}$$

$$\frac{2.5}{\sin \alpha} = \frac{y}{\sin \theta}$$

$$y = \frac{2.5 \sin 143°}{\sin 12°}$$

$$= 7.2 \text{ mi}$$

37. Using K = $\frac{1}{2}$bh,

$$K = \frac{1}{2}(1)(\sqrt{3})$$

$$= \frac{\sqrt{3}}{2}.$$

Using K = $\frac{1}{2}$ab sin C,

$$K = \frac{1}{2}(\sqrt{3})(1)\sin 90°$$

$$= \frac{1}{2}(\sqrt{3})(1)(1)$$

$$= \frac{\sqrt{3}}{2}.$$

39. Using K = $\frac{1}{2}$bh,

$$K = \frac{1}{2}(1)(\sqrt{2})$$

$$= \frac{\sqrt{2}}{2}.$$

Using K = $\frac{1}{2}$ab sin C,

$$K = \frac{1}{2}(2)(1)\sin 45°$$

$$= \frac{1}{2}(2)(1)\left(\frac{\sqrt{2}}{2}\right)$$

$$= \frac{\sqrt{2}}{2}.$$

41. A = 42.5°, b = 13.6 m, c = 10.1 m

Angle A is included between sides
b and c.

Area = $\frac{1}{2}$ bc sin A

$$= \frac{1}{2}(13.6)(10.1)\sin 42.5°$$

$$= 46.4 \text{ m}^2$$

43. B = 124.5°, a = 30.4 cm, c = 28.4 cm

Angle B is included between sides a and c.

$$\text{Area} = \frac{1}{2}ac \sin B$$

$$= \frac{1}{2}(30.4)(28.4) \sin 124.5°$$

$$= 356 \text{ cm}^2$$

45. A = 56.80°, b = 32.67 in, c = 52.89 in

Angle A is included between sides b and c.

$$\text{Area} = \frac{1}{2}bc \sin A$$

$$= \frac{1}{2}(32.67)(52.89) \sin 56.80°$$

$$= 722.9 \text{ in}^2$$

47. $\text{Area} = \frac{1}{2}ab \sin C$

$$= \frac{1}{2}(16.1)(15.2) \sin 125°$$

$$= 100 \text{ m}^2$$

49. The function y = sin x is increasing from y = 0 to y = 1 on the interval $0 \le x \le \pi/2$.

51. $\dfrac{a}{\sin A} = \dfrac{b}{\sin B}$

Solve for b.

$$\frac{a \sin B}{\sin A} = b$$

$$b = \frac{a \sin B}{\sin A}$$

52. From Exercise 51,

$$b = \frac{a \sin B}{\sin A}.$$

If $\dfrac{\sin B}{\sin A} < 1$, then

$$b = a \cdot \frac{\sin B}{\sin A} < a \cdot 1 \text{ or } a.$$

So, b < a.

55.

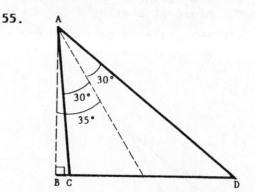

We must find the length of CD. Angle BAC is equal to 35° − 30° = 5°. Side AB = 5000 ft. Since triangle ABC is a right triangle,

$$\cos 5° = \frac{5000}{AC}$$

$$AC = \frac{5000}{\cos 5°}$$

$$\approx 5019 \text{ ft.}$$

The angular coverage of the lens is 60°, so angle CAD = 60°. From geometry, angle ACB = 85°, angle ACD = 95°, and angle ADC = 25°. We now know three angles and one side in triangle ACD and can use the law of sines to solve for the length of CD.

$$\frac{CD}{\sin 60°} = \frac{AC}{\sin 25°}$$

$$CD = \frac{5019 \sin 60°}{\sin 25°}$$

$$CD \approx 10{,}285 \text{ ft}$$

The photograph would cover a horizontal distance of approximately
10,285 ft or $\frac{10,285}{5280} \approx 1.95$ mi.

The coverage is less with a longer
focal length. Cameras that have
shorter focal lengths have a greater
ground coverage in aerial photo-
graphs.

57.

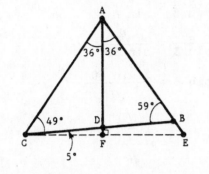

In the figure, angle CAB is 72°.
Triangle ACE is isosceles so angles
ACE and AEC are both equal to 54°.
It follows that angle ACB is equal
to 54° − 5° = 49°. Angles CAD and
DAB both equal 36° since the photograph is taken with no tilt and AD
is the bisector of angle CAB with
length 3500 ft. Since the measures
of the angles in triangle ABC total
180°, angle ABC equals

180° − (72° + 49°) = 59°.

Using the law of sines,

$$CD = \frac{3500 \sin 36°}{\sin 49°} \approx 2725.9 \text{ ft}$$

and

$$DB = \frac{3500 \sin 36°}{\sin 59°} \approx 2400.1 \text{ ft.}$$

The ground distance in the resulting
photograph will equal the length of
segment CB.

CB = CD + DB
= 2725.9 + 2400.1
CB = 5126 ft

Section 7.2

1. $\frac{\sqrt{2}}{\sin 30°} = \frac{2}{\sin \theta°}$

$\sin \theta° = \frac{2 \sin 30°}{\sqrt{2}}$

$= \frac{2(\frac{1}{2})}{\sqrt{2}}$

$= \frac{\sqrt{2}}{2}$

So, θ = 45 or 135.

3. The vertical distance from the point
(3, 4) to the x-axis is 4.

    (a) If h is more than 4, two tri-
angles can be drawn. But h must be
less than 5 for both triangles to be
on the positive x-axis. So,
4 < h < 5.

    (b) If h = 4, then exactly one tri-
angle is possible. If h > 5, then
only one triangle is possible on the
positive x-axis.

    (c) If h < 4, then no triangle is
possible, since the side of length h
would not reach the x-axis.

5. a = 50, b = 26, A = 95°

$$\frac{a}{\sin A} = \frac{b}{\sin B}$$

$$\frac{50}{\sin 95°} = \frac{26}{\sin B}$$

$$\sin B = \frac{26 \sin 95°}{50}$$

$$\sin B \approx .518$$

$$B \approx 31.2°$$

Another possible value for B is

$$180° - 31.2° = 148.8°.$$

This measure, combined with the angle of 95°, is too large, however. Therefore, only one triangle is possible.

7. a = 31, b = 26, B = 48°

$$\frac{a}{\sin A} = \frac{b}{\sin B}$$

$$\frac{31}{\sin A} = \frac{26}{\sin 48°}$$

$$\sin A = \frac{31 \sin 48°}{26}$$

$$\sin A \approx .886$$

$$A \approx 62.4°$$

Another possible value for A is

$$180° - 62.4° = 117.6°.$$

Therefore, two triangles are possible.

9. a = 50, b = 61, A = 58°

$$\frac{a}{\sin A} = \frac{b}{\sin B}$$

$$\frac{50}{\sin 58°} = \frac{61}{\sin B}$$

$$\sin B = \frac{61 \sin 58°}{50}$$

$$\sin B \approx 1.03$$

sin B > 1 is impossible. Therefore, no triangle is possible for the given parts.

11. $$\frac{b}{\sin B} = \frac{a}{\sin A}$$

$$\frac{2}{\sin B} = \frac{\sqrt{6}}{\sin 60°}$$

$$\sin B = \frac{2 \sin 60°}{\sqrt{6}}$$

$$= \frac{2\left(\frac{\sqrt{3}}{2}\right)}{\sqrt{6}}$$

$$= \frac{\sqrt{2}}{2}$$

$$B = 45°$$

Another possible value for B is

$$180° - 45° = 135°,$$

but this is too large since A = 60°. Therefore, B = 45°.

In Exercises 13-29, the number of possible triangles is found based on the given conditions. Remember that, for example, if angle B and sides a and b are given such that a > b > a sin B, then two triangles are possible.

13. A = 29.7°, b = 41.5 ft, a = 27.2 ft

b > a > b sin A tells us that there are two possible triangles.

$$\frac{\sin B}{b} = \frac{\sin A}{a}$$

$$\sin B = \frac{41.5 \sin 29.7°}{27.2}$$

$$\sin B = .75593878$$

$$B_1 = 49.1°$$

$$C_1 = 180° - A - B_1$$

$$= 180° - 29.7° - 49.1°$$

$$= 101.2°$$

$B_2 = 180° - B_1 = 130.9°$

$C_2 = 19.4°$

15. $C = 41° \; 20'$, $b = 25.9$ m, $c = 38.4$ m

$c > b$ tells us that there is only one triangle.

$$\frac{\sin B}{b} = \frac{\sin C}{c}$$

$$\sin B = \frac{25.9 \sin 41° \; 20'}{38.4}$$

$\sin B = .44545209$

$\quad B = 26° \; 30'$

Note that $B \neq 153° \; 30'$ because $153° \; 30' + C > 180°$.

$\quad A = 180° - 26° \; 30' - 41° \; 20'$

$\quad A = 112° \; 10'$

17. $B = 74.3°$, $a = 859$ m, $b = 783$ m

$$\frac{\sin A}{a} = \frac{\sin B}{b}$$

$$\sin A = \frac{859 \sin 74.3°}{783}$$

$\sin A = 1.0561331$

$\sin A > 1$ is impossible.

Note that $b < a \sin B$.

No such triangle exists.

19. $A = 142.13°$, $b = 5.432$ ft, $a = 7.297$ ft

$a > b$ tells us that there is only one triangle.

$$\frac{\sin B}{b} = \frac{\sin A}{a}$$

$$\sin B = \frac{5.432 \sin 142.13°}{7.297}$$

$\sin B = .45697581$

$\quad B = 27.19°$

B must be acute since A is obtuse.

$\quad C = 180° - 142.13° - 27.19°$

$\quad C = 10.68°$

21. $A = 42.5°$, $a = 15.6$ ft, $b = 8.14$ ft

$a > b$ tells us that there is only one triangle.

$$\frac{\sin B}{b} = \frac{\sin A}{a}$$

$$\sin B = \frac{b \sin A}{a}$$

$$= \frac{8.14 \sin 42.5°}{15.6}$$

$\sin B = .35251951$

$\quad B = 20.6°$

$B \neq 159.4°$ since $159.4° + A > 180°$.

$\quad C = 180° - 42.5° - 20.6°$

$\quad = 116.9°$

$$\frac{c}{\sin C} = \frac{a}{\sin A}$$

$$c = \frac{a \sin C}{\sin A}$$

$$= \frac{15.6 \sin 116.9°}{\sin 42.5°}$$

$\quad c = 20.6$ ft

23. $B = 72.2°$, $b = 78.3$ m, $c = 145$ m

$$\frac{\sin C}{c} = \frac{\sin B}{b}$$

$$\sin C = \frac{c \sin B}{b}$$

$$= \frac{145 \sin 72.2°}{78.3}$$

$\sin C = 1.7632025$

$\sin C > 1$ is impossible. Note that $c \sin B > b$. No such triangle exists.

25. $A = 38° \, 40'$, $a = 9.72$ km,
     $b = 11.8$ km

$b > a > b \sin A$ tells us that there are two possible triangles.

$$\frac{\sin B}{b} = \frac{\sin A}{a}$$

$$\sin B = \frac{b \sin A}{a}$$

$$= \frac{11.8 \sin 38° \, 40'}{9.72}$$

$$\sin B = .75848811$$

$B_1 = 49° \, 20'$

$C_1 = 180° - 38° \, 40' - 49° \, 20'$

$C_1 = 92° \, 00'$

$$\frac{c_1}{\sin C_1} = \frac{a}{\sin A}$$

$$c_1 = \frac{a \sin C_1}{\sin A}$$

$$= \frac{9.72 \sin 92° \, 00'}{\sin 38° \, 40'}$$

$$c_1 = 15.5 \text{ km}$$

$B_2 = 130° \, 40'$

$C_2 = 180° - 38° \, 40' - 130° \, 40'$

$C_2 = 10° \, 40'$

$$\frac{c_2}{\sin C_2} = \frac{a}{\sin A}$$

$$c_2 = \frac{a \sin C_2}{\sin A}$$

$$= \frac{9.72 \sin 10° \, 40'}{\sin 38° \, 40'}$$

$$c_2 = 2.88 \text{ km}$$

27. $A = 96.80°$, $b = 3.589$ ft,
     $a = 5.818$ ft

$a > b$ tells us that there is only one possible triangle.

$$\frac{\sin B}{b} = \frac{\sin A}{a}$$

$$\sin B = \frac{b \sin A}{a}$$

$$= \frac{3.589 \sin 96.80°}{5.818}$$

$$\sin B = .61253922$$

$$B = 37.77°$$

B must be acute since A is obtuse.

$C = 180° - 96.80° - 37.77°$

$C = 45.43°$

$$\frac{c}{\sin C} = \frac{a}{\sin A}$$

$$c = \frac{a \sin C}{\sin A}$$

$$= \frac{5.818 \sin 45.43°}{\sin 96.80°}$$

$$c = 4.174 \text{ ft}$$

29. $B = 39.68°$, $a = 29.81$ m, $b = 23.76$ m

$a > b > a \sin B$ tells us that there are two possible triangles.

$$\frac{\sin A}{a} = \frac{\sin B}{b}$$

$$\sin A = \frac{a \sin B}{b}$$

$$= \frac{29.81 \sin 39.68°}{23.76}$$

$$\sin A = .80108002$$

$A_1 = 53.23°$

$C_1 = 180° - 53.23° - 39.68°$

$C_1 = 87.09°$

$$\frac{b}{\sin B} = \frac{c_1}{\sin C_1}$$

$$c_1 = \frac{b \sin C_1}{\sin B}$$

$$= \frac{23.76 \sin 87.09°}{\sin 39.68°}$$

$$c_1 = 37.16 \text{ m}$$

$A_2 = 126.77°$

$C_2 = 180° - 126.77° - 39.68°$

$C_2 = 13.55°$

$$\frac{b}{\sin B} = \frac{c_2}{\sin C_2}$$

$$c_2 = \frac{b \sin C_2}{\sin B}$$

$$= \frac{23.76 \sin 13.55°}{\sin 39.68°}$$

$$c_2 = 8.719 \text{ m}$$

31. $a = \sqrt{5}$, $c = 2\sqrt{5}$, $A = 30°$

$$\frac{a}{\sin A} = \frac{c}{\sin C}$$

$$\sin C = \frac{2\sqrt{5} \sin 30°}{\sqrt{5}}$$

$$\sin C = 1$$

$$C = 90°$$

This is a right triangle.

33. If $A = 103° \ 20'$, $a = 14.6$ ft, $b = 20.4$ ft, since $b > a$, B should be larger than A. But $A = 103° \ 20'$, and $A + B$ would be greater than 180°. Since the sum of the angles of a triangle is 180°, the given dimensions indicate that no triangle ABC exists.

35. Let $A = 38° \ 50'$, $a = 21.9$, $b = 78.3$.

$b \sin A = 78.3 \sin 38° \ 50'$

$= 49.1$

Thus, $21.9 < 49.1$.

That is, $a < b \sin A$.

The piece of property cannot exist with the given data.

37. Prove that $\frac{a + b}{b} = \frac{\sin A + \sin B}{\sin B}$.

Start with the law of sines.

$$\frac{a}{\sin A} = \frac{b}{\sin B}$$

$$\left(\frac{\sin A}{b}\right)\left(\frac{a}{\sin A}\right) = \left(\frac{\sin A}{b}\right)\left(\frac{b}{\sin B}\right)$$

$$\frac{a}{b} = \frac{\sin A}{\sin B}$$

$$\frac{a}{b} + 1 = \frac{\sin A}{\sin B} + 1$$

$$\frac{a}{b} + \frac{b}{b} = \frac{\sin A}{\sin B} + \frac{\sin B}{\sin B}$$

$$\frac{a + b}{b} = \frac{\sin A + \sin B}{\sin B}$$

Section 7.3

3. $a^2 = 3^2 + 8^2 - 2(3)(8) \cos 60°$

$= 9 + 64 - 48\left(\frac{1}{2}\right)$

$= 73 - 24$

$= 49$

$a = 7$

5. $\cos \theta = \frac{b^2 + c^2 - a^2}{2bc}$

$= \frac{1^2 + (\sqrt{3})^2 - 1^2}{2(1)(\sqrt{3})}$

$= \frac{1 + 3 - 1}{2\sqrt{3}}$

$= \frac{3}{2\sqrt{3}}$

$= \frac{\sqrt{3}}{2}$

$\theta = 30°$

7. C = 28.3°, b = 5.71 in, a = 4.21 in

$$c^2 = a^2 + b^2 - 2ab \cos C$$
$$c = \sqrt{(4.21)^2 + (5.71)^2 - 2(4.21)(5.71) \cos 28.3°}$$
$$c = 2.83 \text{ in}$$

Keep all digits of $\sqrt{c^2}$ in the calculator for use in the next calculation. If 2.8 is used, the answer will vary slightly due to round-off error.

Find angle A next, since it is the smaller angle and must be acute.

$$\sin A = \frac{a \sin C}{c} = \frac{4.21 \sin 28.3°}{\sqrt{c^2}}$$

$$\sin A = .70581857$$
$$A = 44.9°$$

$$B = 180° - 44.9° - 28.3°$$
$$B = 106.8°$$

9. C = 45.6°, b = 8.94 m, a = 7.23 m

$$c^2 = a^2 + b^2 - 2ab \cos C$$
$$c = \sqrt{(7.23)^2 + (8.94)^2 - 2(7.23)(8.94) \cos 45.6°}$$
$$c = 6.46 \text{ m}$$

Find angle A next, since it is the smaller angle and must be acute.

$$\sin A = \frac{a \sin C}{c} = \frac{7.23 \sin 45.6°}{\sqrt{c^2}}$$

$$\sin A = .79946437$$
$$A = 53.1°$$

$$B = 180° - 53.1° - 45.6°$$
$$B = 81.3°$$

11. A = 80° 40', b = 143 cm, c = 89.6 cm

$$a^2 = b^2 + c^2 - 2bc \cos A$$
$$a = \sqrt{143^2 + (89.6)^2 - 2(143)(89.6) \cos 80° \ 40'}$$
$$a = 156 \text{ cm}$$

Find angle C next, since it is the smaller angle and must be acute.

$$\sin C = \frac{c \sin A}{a} = \frac{89.6 \sin 80° \ 40'}{\sqrt{a^2}}$$

$$\sin C = .56692713$$
$$C = 34° \ 30'$$

$$B = 180° - 80° \ 40' - 34° \ 30'$$

$$B = 64° \ 50'$$

13. $B = 74.80°$, $a = 8.919$ in, $c = 6.427$ in

$$b^2 = a^2 + c^2 - 2ac \cos B$$

$$b = \sqrt{(8.919)^2 + (6.427)^2 - 2(8.919)(6.427) \cos 74.80°}$$

$$b = 9.529 \text{ in}$$

Find angle C next, since it is the smaller angle and must be acute.

$$\sin C = \frac{c \sin B}{b} = \frac{6.427 \sin 74.80°}{\sqrt{b^2}}$$

$$\sin C = .65089219$$

$$C = 40.61°$$

$$A = 180° - 74.80° - 40.61°$$

$$A = 64.59°$$

15. $A = 112.8°$, $b = 6.28$ m, $c = 12.2$ m

$$a^2 = b^2 + c^2 - 2bc \cos A$$

$$a = \sqrt{(6.28)^2 + (12.2)^2 - 2(6.28)(12.2) \cos 112.8°}$$

$$a = 15.7 \text{ m}$$

Angle A is obtuse, so both B and C are acute. Find either angle next.

$$\sin B = \frac{b \sin A}{a} = \frac{6.28 \sin 112.8°}{\sqrt{a^2}}$$

$$\sin B = .36787456$$

$$B = 21.6°$$

$$C = 180° - 112.8° - 21.6°$$

$$C = 45.6°$$

17. $a = 3.0$ ft, $b = 5.0$ ft, $c = 6.0$ ft

Angle C is the largest, so find it first.

$$c^2 = a^2 + b^2 - 2ab \cos C$$

$$\cos C = \frac{3.0^2 + 5.0^2 - 6.0^2}{2(3.0)(5.0)}$$

$$\cos C = -.06666667$$

$$C = 94°$$

$\sin B = \dfrac{b \sin C}{c}$

$\sin B = .83147942$

$\quad B = 56°$

$\quad A = 180° - 56° - 94°$

$\quad A = 30°$

19. $a = 9.3$ cm, $b = 5.7$ cm, $c = 8.2$ cm

Angle A is the largest, so find it first.

$\quad a^2 = b^2 + c^2 - 2bc \cos A$

$\cos A = \dfrac{5.7^2 + 8.2^2 - 9.3^2}{2(5.7)(8.2)}$

$\cos A = .14163457$

$\quad A = 82°$

$\sin B = \dfrac{b \sin A}{a}$

$\sin B = .60672455$

$\quad B = 37°$

$\quad C = 180° - 82° - 37°$

$\quad C = 61°$

21. $a = 42.9$ m, $b = 37.6$ m, $c = 62.7$ m

Angle C is the largest, so find it first.

$\quad c^2 = a^2 + b^2 - 2ab \cos C$

$\cos C = \dfrac{42.9^2 + 37.6^2 - 62.7^2}{2(42.9)(37.6)}$

$\cos C = -.20988940$

$\quad C = 102.1°$

$\sin B = \dfrac{b \sin C}{c}$

$\sin B = .58632321$

$\quad B = 35.9°$

$\quad A = 180° - 35.9° - 102.1°$

$\quad A = 42.0°$

23. $AB = 1240$ ft, $AC = 876$ ft, $BC = 918$ ft

$AB = c$, $AC = b$, $BC = a$

Angle C is the largest, so find it first.

$\quad c^2 = a^2 + b^2 - 2ab \cos C$

$\cos C = \dfrac{a^2 + b^2 - c^2}{2ab}$

$\cos C = .04507765$

$\quad C = 87.4°$

$\sin B = \dfrac{b \sin C}{c}$

$\sin B = .70573350$

$\quad B = 44.9°$

$\quad A = 180° - 44.9° - 87.4°$

$\quad A = 47.7°$

25. Let a be the length of the segment from (0, 0) to (6, 8). Use the distance formula.

$a = \sqrt{(6 - 0)^2 + (8 - 0)^2}$

$\quad = \sqrt{6^2 + 8^2}$

$\quad = \sqrt{36 + 64} = \sqrt{100} = 10$

Let b be the length of the segment from (0, 0) to (4, 3).

$b = \sqrt{(4 - 0)^2 + (3 - 0)^2}$

$\quad = \sqrt{4^2 + 3^2}$

$\quad = \sqrt{16 + 9} = \sqrt{25} = 5$

Let c be the length of the segment from (4, 3) to (6, 8).

$c = \sqrt{(6 - 4)^2 + (8 - 3)^2}$

$\quad = \sqrt{2^2 + 5^2}$

$\quad = \sqrt{4 + 25} = \sqrt{29}$

$$\cos \theta = \frac{a^2 + b^2 - c^2}{2ab}$$

$$\cos \theta = \frac{10^2 + 5^2 - (\sqrt{29})^2}{2(10)(5)}$$

$$\cos \theta = \frac{100 + 25 - 29}{100}$$

$$\cos \theta = .96$$

$$\theta \approx 16.26°$$

27. Using $K = \frac{1}{2}bh$,

$$K = \frac{1}{2}(16)(3\sqrt{3})$$

$$= 24\sqrt{3}$$

$$\approx 41.57.$$

To use Heron's Formula, first find the semiperimeter.

$$s = \frac{1}{2}(a + b + c)$$

$$= \frac{1}{2}(6 + 14 + 16)$$

$$= \frac{1}{2}(36)$$

$$= 18$$

Now find the area of the triangle.

$$K = \sqrt{s(s - a)(s - b)(s - c)}$$

$$= \sqrt{18(18 - 6)(18 - 14)(18 - 16)}$$

$$= \sqrt{18(12)(4)(2)}$$

$$= \sqrt{1728}$$

$$\approx 41.57$$

Both formulas give the same area.

29. a = 12 m, b = 16 m, c = 25 m

$$s = \frac{1}{2}(12 + 16 + 25)$$

$$s = 26.5$$

$$\text{area} = \sqrt{s(s - a)(s - b)(s - c)}$$

$$= \sqrt{(26.5)(14.5)(10.5)(1.5)}$$

$$= 78 \text{ m}^2$$

31. a = 154 cm, b = 179 cm, c = 183 cm

$$s = \frac{1}{2}(154 + 179 + 183)$$

$$s = 258$$

$$\text{area} = \sqrt{s(s - a)(s - b)(s - c)}$$

$$\text{area} = \sqrt{(258)(104)(79)(75)}$$

$$= 12,600 \text{ cm}^2$$

33. a = 76.3 ft, b = 109 ft, c = 98.8 ft

$$s = \frac{1}{2}(76.3 + 109 + 98.8)$$

$$s = 142.05$$

$$\text{area} = \sqrt{s(s - a)(s - b)(s - c)}$$

$$= \sqrt{(142.05)(65.75)(33.05)(43.25)}$$

$$= 3650 \text{ ft}^2$$

35. Find the area of the region.

$$s = \frac{1}{2}(75 + 68 + 85)$$

$$s = 114$$

$$\text{area} = \sqrt{(114)(39)(46)(29)}$$

$$= 2435.3571 \text{ m}^2$$

Number of cans needed

$$= (\text{area in m}^2)/(\text{m}^2 \text{ per can})$$

$$\frac{2435.3571}{75} = 32.471428 \text{ cans}$$

She will need to open 33 cans.

37. Find the area of the Bermuda triangle.

$$s = \frac{1}{2}(850 + 925 + 1300)$$

$$s = 1537.5$$

$$\text{area} = \sqrt{1537.5(687.5)(612.5)(237.5)}$$

$$= 392,000$$

The area of the Bermuda Triangle is 392,000 mi^2.

39. Find AB, or c, in the following triangle.

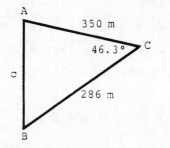

$$c^2 = a^2 + b^2 - 2ab \cos C$$

$$c = \sqrt{286^2 + 350^2 - 2(286)(350) \cos 46.3°}$$

$$c = 257$$

$$AB = 257 \text{ m}$$

41. Find AC, or b, in the following triangle.

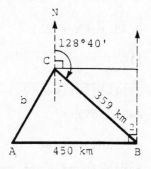

Angle 1 = 180° - 128° 40′ = 51° 20′

Angle 1 = Angle 2

Angle B = 90° - Angle 2 = 38° 40′

$$b^2 = a^2 + c^2 - 2ac \cos B$$

$$b = \sqrt{359^2 + 450^2 - 2(359)(450) \cos 38° 40′}$$

$$b = 281 \text{ km}$$

43. Find x.

$$x^2 = 25^2 + 25^2 - 2(25)(25) \cos 52°$$

$$= 480$$

$$x = 22 \text{ ft}$$

45. $AB^2 = 10^2 + 10^2 - 2(10)(10) \cos 128°$

$AB^2 = 323$

$AB = 18 \text{ ft}$

47. Let x = the distance from the tracking station to the satellite at 12:03 P.M.

Notice that the distance from the center of the earth to the satellite is

6400 + 1600 = 8000 km.

Let θ = the angle made by the movement of the satellite from noon to 12:03 P.M.

$$\frac{2\pi \text{ radians}}{2 \text{ hours}} = \frac{\theta \text{ radians}}{3 \text{ minutes}}$$

$$\frac{2\pi}{2(60)} = \frac{\theta}{3}$$

$$\theta = \frac{6\pi}{120} = \frac{\pi}{20} \text{ radian}$$

$$x^2 = 6400^2 + 8000^2 - 2(6400)(8000) \cos \frac{\pi}{20}$$

$$x = 2000$$

The distance between the satellite and the tracking station is 2000 km.

49. Let θ = the angle opposite the diagonal.

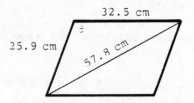

$$57.8^2 = 25.9^2 + 32.5^2$$

$$- 2(25.9)(32.5) \cos \theta$$

$$\theta = 163.5°$$

51. $AB = 22.47928$ mi, $AC = 28.14276$ mi,

$A = 58.56989°$

$BC^2 = AC^2 + AB^2 - 2(AC)(AB) \cos A$

Use a calculator and substitute.

$BC^2 = 637.5539346$

$BC = 25.24983$ mi

53. Applying the law of cosines when
$a = 3$, $b = 4$, and $c = 10$ gives

$$c^2 = a^2 + b^2 - 2ab \cos C$$
$$10^2 = 3^2 + 4^2 - 2(3)(4) \cos C$$
$$75 = -24 \cos C$$
$$-3.125 = \cos C.$$

Since $-1 \le \cos C \le 1$, a triangle
cannot have sides 3, 4, and 10.

55. Verify

$$1 + \cos A = \frac{(b + c + a)(b + c - a)}{2bc}.$$

$$1 + \cos A = \frac{2bc}{2bc} + \frac{b^2 + c^2 - a^2}{2bc}$$

$$= \frac{b^2 + 2bc + c^2 - a^2}{2bc}$$

$$= \frac{(b + c)^2 - a^2}{2bc}$$

$$= \frac{(b + c + a)(b + c - a)}{2bc}$$

57. Verify

$$\cos \frac{A}{2} = \sqrt{\frac{s(s - a)}{bc}}.$$

$$\cos \frac{A}{2} = \sqrt{\frac{1 + \cos A}{2}}$$

Use the result of Exercise 55.

$$= \sqrt{\frac{(b + c + a)}{2} \cdot \frac{(b + c - a)}{2} \cdot \frac{1}{bc}}$$

Now use the fact that

$$2s = a + b + c$$
and
$$2s - 2a = b + c - a.$$

$$= \sqrt{\left(\frac{2s}{2}\right)\left(\frac{2s - 2a}{2}\right)\frac{1}{bc}}$$

$$\cos \frac{A}{2} = \sqrt{\frac{s(s - a)}{bc}}$$

59. Verify

$$\frac{1}{2}bc \sin A$$

$$= \sqrt{\frac{1}{2}bc(1 + \cos A) \cdot \frac{1}{2}bc(1 - \cos A)}.$$

$$\frac{1}{2}bc \sin A$$

$$= \frac{1}{2}bc\sqrt{\sin^2 A}$$

$$= \frac{1}{2}bc\sqrt{1 - \cos^2 A}$$

$$= \sqrt{\frac{1}{2} \cdot \frac{1}{2} \cdot b^2 \cdot c^2} \cdot \sqrt{1 - \cos^2 A}$$

$$= \sqrt{\frac{1}{2} \cdot \frac{1}{2} \cdot b^2 \cdot c^2(1 + \cos A)(1 - \cos A)}$$

$$= \sqrt{\frac{1}{2}bc(1 + \cos A) \cdot \frac{1}{2}bc(1 - \cos A)}$$

61. Prove that if a and b are equal
sides of an isosceles triangle,
then $c^2 = 2a^2(1 - \cos C)$.

By the law of cosines,

$$c^2 = a^2 + b^2 - 2ab \cos C.$$

Since $a = b$,

$$c^2 = a^2 + a^2 - 2a^2 \cos C$$
$$c^2 = 2a^2 - 2a^2 \cos C$$
$$c^2 = 2a^2(1 - \cos C).$$

63. Let a = 2, b = $2\sqrt{3}$, A = 30°,

B = 60°.

Verify $\dfrac{\tan \frac{1}{2}(A - B)}{\tan \frac{1}{2}(A + B)} = \dfrac{a - b}{a + b}$.

$$\frac{\tan \frac{1}{2}(A - B)}{\tan \frac{1}{2}(A + B)} = \frac{\tan \frac{1}{2}(30° - 60°)}{\tan \frac{1}{2}(30° + 60°)}$$

$$= \frac{\tan(-15°)}{\tan 45°}$$

$$= -.26794919$$

$$\frac{a - b}{a + b} = \frac{2 - 2\sqrt{3}}{2 + 2\sqrt{3}} = \frac{-1.46}{5.46} = -.26794919$$

(*Note*: −.26794919 = −2 + $\sqrt{3}$.)

65. If 90° < A < 180°,

−1 < cos A < 0.

So cos A is negative if A is an

obtuse angle.

66. $a^2 = b^2 + c^2 - 2bc \cos A$

If cos A is negative, then

$b^2 + c^2 - 2bc \cos A > b^2 + c^2$.

So, $a^2 > b^2 + c^2$ for an obtuse angle

A.

67. Since $b^2 + c^2 > b^2$ and $b^2 + c^2 > c^2$,

then $a^2 > b^2$ and $a^2 > c^2$.

Thus, a > b and a > c, since a, b,

and c are all positive numbers.

68. For this obtuse triangle, side a

opposite obtuse angle A is smaller

than side c, which is impossible.

The longest side should be a, not c.

Section 7.4

3. Equal vectors have the same

magnitude and direction:

m = **p**, **n** = **r**.

5. One vector is a positive scalar

multiple of another if the two

vectors point in the same direction;

they may have different magnitudes.

m = 1**p**, **m** = 2**t**, **n** = 1**r**, **p** = 2**t**

(or **p** = 1**m**, **t** = $\frac{1}{2}$**m**, **r** = 1**n**, **t** = $\frac{1}{2}$**p**)

For Exercises 7–23, see the answer

sketches at the back of the textbook.

25. **a** + (**b** + **c**) = (**a** + **b**) + **c**

Yes, vector addition is associative.

In Exercises 27–31, sketches may vary

slightly. See the sketches at the

back of the textbook.

In Exercises 33–39, **x** is the horizontal

component and **y** is the vertical component.

33. α = 38°, |**v**| = 12

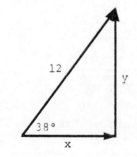

$$\cos 38° = \frac{|\mathbf{x}|}{12}$$

$$|\mathbf{x}| = 12 \cos 38°$$

$$= 9.5$$

$$\sin 38° = \frac{|\mathbf{y}|}{12}$$

$$|\mathbf{y}| = 12 \sin 38°$$

$$= 7.4$$

35. $\alpha = 50°$, $|\mathbf{v}| = 26$

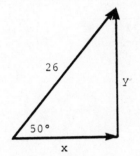

$$\cos 50° = \frac{|\mathbf{x}|}{26}$$

$$|\mathbf{x}| = 26 \cos 50°$$

$$= 17$$

$$\sin 50° = \frac{|\mathbf{y}|}{26}$$

$$|\mathbf{y}| = 26 \sin 50°$$

$$= 20$$

37. $\alpha = 27° \, 30'$, $|\mathbf{v}| = 15.4$

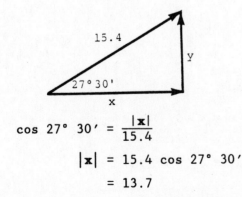

$$\cos 27° \, 30' = \frac{|\mathbf{x}|}{15.4}$$

$$|\mathbf{x}| = 15.4 \cos 27° \, 30'$$

$$= 13.7$$

$$\sin 27° \, 30' = \frac{|\mathbf{y}|}{15.4}$$

$$|\mathbf{y}| = 15.4 \sin 27° \, 30'$$

$$= 7.11$$

39. $\alpha = 146.3°$, $|\mathbf{v}| = 238$

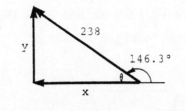

$$\theta = 180° - 146.3° = 33.7°$$

$$|\mathbf{x}| = 238 \cos 33.7° = 198$$

$$|\mathbf{y}| = 238 \sin 33.7° = 132$$

41. If $|\mathbf{a} + \mathbf{b}| = |\mathbf{a}| + |\mathbf{b}|$, then $\mathbf{a}$ and $\mathbf{b}$ have the same direction. Otherwise, $|\mathbf{a}| + |\mathbf{b}| > |\mathbf{a} + \mathbf{b}|$.

43. If the angle between $\mathbf{a}$ and $\mathbf{b}$ is acute, then $|\mathbf{a} + \mathbf{b}| > |\mathbf{a}|$, but $|\mathbf{a} - \mathbf{b}| < |\mathbf{a}|$. So $|\mathbf{a} + \mathbf{b}| > |\mathbf{a} - \mathbf{b}|$ if the angle between $\mathbf{a}$ and $\mathbf{b}$ is less than $90°$.

45. Forces of 250 newtons and 450 newtons, forming an angle of $85°$

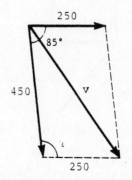

$\alpha = 180° - 85° = 95°$

$|\mathbf{v}|^2 = (250)^2 + (450)^2$

$\quad - 2(250)(450) \cos 95°$

$\quad = 284{,}610$

$|\mathbf{v}| = 530$ newtons

47. Forces of 17.9 lb and 25.8 lb, forming an angle of 105.5°

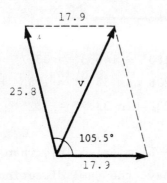

$\alpha = 180° - 105.5°$

$\quad = 74.5°$

$|\mathbf{v}|^2 = (25.8)^2 + (17.9)^2$

$\quad - 2(25.8)(17.9) \cos 74.5°$

$\quad = 739.218$

$|\mathbf{v}| = 27.2$ lb

49. Forces of 116 lb and 139 lb, forming an angle of 140° 50′

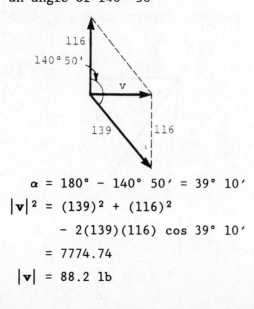

$\alpha = 180° - 140° 50′ = 39° 10′$

$|\mathbf{v}|^2 = (139)^2 + (116)^2$

$\quad - 2(139)(116) \cos 39° 10′$

$\quad = 7774.74$

$|\mathbf{v}| = 88.2$ lb

51. $\mathbf{u} = <5, 12>$

$|\mathbf{u}| = \sqrt{5^2 + 12^2}$

$\quad = \sqrt{25 + 144}$

$\quad = \sqrt{169}$

$|\mathbf{u}| = 13$

$\cos \theta = \dfrac{a}{|\mathbf{u}|} = \dfrac{5}{13}$

$\theta = 67.4°$

53. $\mathbf{u} = <-3, 4>$

$|\mathbf{u}| = \sqrt{(-3)^2 + 4^2}$

$\quad = \sqrt{9 + 16}$

$\quad = \sqrt{25}$

$|\mathbf{u}| = 5$

$\cos \theta = \dfrac{a}{|\mathbf{u}|} = -\dfrac{3}{5}$

$\theta = 126.9°$

55. $|\mathbf{u}| = 6, \theta = 30°$

$a = |\mathbf{u}| \cos \theta$

$\quad = 6 \cos 30°$

$\quad = 6 \left(\dfrac{\sqrt{3}}{2} \right)$

$a = 3\sqrt{3}$

$b = |\mathbf{u}| \sin \theta$

$\quad = 6 \sin 30°$

$\quad = 6 \left(\dfrac{1}{2} \right)$

$b = 3$

Thus,

$\mathbf{u} = <3\sqrt{3}, 3>.$

57. $|\mathbf{u}| = 4, \theta = 120°$

$a = |\mathbf{u}| \cos \theta$

$\quad = 4 \cos 120°$

$\quad = 4 \left(-\dfrac{1}{2} \right)$

$a = -2$

$b = |\mathbf{u}| \sin \theta$

$= 4 \sin 120°$

$= 4\left(\frac{\sqrt{3}}{2}\right)$

$b = 2\sqrt{3}$

Thus,

$\mathbf{u} = <-2, \ 2\sqrt{3}>.$

For Exercises 59–63, $\mathbf{u} = <a_1, \ b_1>$, $\mathbf{v} = <a_2, \ b_2>$, $\mathbf{w} = <a_3, \ b_3>$, and $\mathbf{0} = <0, \ 0>$.

59. Verify $\mathbf{u} + \mathbf{v} = \mathbf{v} + \mathbf{u}$.

$\mathbf{u} + \mathbf{v} = <a_1, \ b_1> + <a_2, \ b_2>$

$\quad = <a_1 + a_2, \ b_1 + b_2>$

$\quad = <a_2 + a_1, \ b_2 + b_1>$

$\quad = <a_2, \ b_2> + <a_1, \ b_1>$

$\quad = \mathbf{v} + \mathbf{u}$

Therefore, $\mathbf{u} + \mathbf{v} = \mathbf{v} + \mathbf{u}$.

61. Verify $-1(\mathbf{u}) = -\mathbf{u}$.

$-1(\mathbf{u}) = -1<a_1, \ b_1>$

$\quad = <-a_1, \ -b_1>$

$\quad = -\mathbf{u}$

Therefore, $-1(\mathbf{u}) = -\mathbf{u}$.

63. Verify $\mathbf{u} + \mathbf{0} = \mathbf{u}$.

$\mathbf{u} + \mathbf{0} = <a_1, \ b_1> + <0, \ 0>$

$\quad = <a_1 + 0, \ b_1 + 0>$

$\quad = <a_1, \ b_1>$

$\quad = \mathbf{u}$

Therefore, $\mathbf{u} + \mathbf{0} = \mathbf{u}$.

65. $\mathbf{u} = 6\mathbf{i} - 2\mathbf{j}$, $\mathbf{v} = -3\mathbf{i} + 2\mathbf{j}$

$\mathbf{u} \cdot \mathbf{v} = (6)(-3) + (-2)(2)$

$\quad = (-18) + (-4)$

$\quad = -22$

67. $\mathbf{u} = <-6, \ 8>$, $\mathbf{v} = <3, \ -4>$

$\mathbf{u} = -6\mathbf{i} + 8\mathbf{j}$, $\mathbf{v} = 3\mathbf{i} - 4\mathbf{j}$

$\mathbf{u} \cdot \mathbf{v} = (-6)(3) + 8(-4)$

$\quad = -18 + (-32)$

$\quad = -50$

69. Verify $\mathbf{u} \cdot \mathbf{v} = |\mathbf{u}||\mathbf{v}| \cos \alpha$.

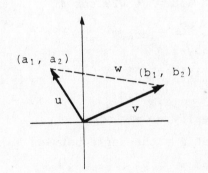

By the law of cosines,

$$|\mathbf{w}|^2 = |\mathbf{u}|^2 + |\mathbf{v}|^2 - 2|\mathbf{u}||\mathbf{v}| \cos \alpha.$$

Let $\mathbf{w} = <b_1 - a_1, \ b_2 - a_2>$.
Therefore,

$\mathbf{w} \cdot \mathbf{w} = \mathbf{u} \cdot \mathbf{u} + \mathbf{v} \cdot \mathbf{v} - 2|\mathbf{u}||\mathbf{v}| \cos \alpha.$

$(b_1 - a_1)^2 + (b_2 - a_2)^2$

$\quad = a_1{}^2 + a_2{}^2 + b_1{}^2 + b_2{}^2$

$\qquad - 2|\mathbf{u}||\mathbf{v}| \cos \alpha$

$b_1{}^2 - 2a_1b_1 + a_1{}^2 + b_2{}^2 - 2b_2a_2 + a_2{}^2$

$\quad = a_1{}^2 + a_2{}^2 + b_1{}^2 + b_2{}^2$

$\qquad - 2|\mathbf{u}||\mathbf{v}| \cos \alpha$

$-2a_1b_1 - 2a_2b_2 = -2|\mathbf{u}||\mathbf{v}| \cos \alpha$

$a_1b_1 + a_2b_2 = |\mathbf{u}||\mathbf{v}| \cos \alpha$

$\mathbf{u} \cdot \mathbf{v} = |\mathbf{u}||\mathbf{v}| \cos \alpha.$

Therefore, $\mathbf{u} \cdot \mathbf{v} = |\mathbf{u}||\mathbf{v}| \cos \alpha$.

71. $u = \langle 1, 8 \rangle = i + 8j$

$v = \langle 2, -5 \rangle = 2i - 5j$

$|u| = \sqrt{1^2 + 8^2} = \sqrt{1 + 64} = \sqrt{65}$

$|v| = \sqrt{2^2 + (-5)^2} = \sqrt{4 + 25} = \sqrt{29}$

$u \cdot v = (1)(2) + 8(-5)$

$= 2 - 40 = -38$

$u \cdot v = |u||v| \cos \theta$

$-38 = \sqrt{65}\sqrt{29} \cos \theta$

$\dfrac{-38}{\sqrt{65}\sqrt{29}} = \cos \theta$

$151° = \theta$

73. Verify $u \cdot v = v \cdot u$.

Let θ = the angle between vectors u and v.

$u \cdot v = |u||v| \cos \theta = |v||u| \cos \theta$

$v \cdot u = |v||u| \cos \theta$

Therefore, $u \cdot v = v \cdot u$.

75. Verify $u \cdot (v + w) = u \cdot v + u \cdot w$

Let $u = ai + bj$, $v = ci + dj$, and $w = ei + fj$.

$u \cdot (v + w) = u \cdot [(c + e)i + (d + f)j]$

$= a(c + e) + b(d + f)$

$= ac + ae + bd + bf$

$u \cdot v + u \cdot w = (ac + bd) + (ae + bf)$

$= ac + ae + bd + bf$

Therefore, $u \cdot (v + w) = u \cdot v + u \cdot w$.

77. Given u and v are not 0, $u \cdot v = 0$. Verify that u and v are perpendicular.

$u \cdot v = |u||v| \cos \theta$,

$0 = |u||v| \cos \theta$

Since u and v are not 0,

$0 = \cos \theta$

$90° = \theta$.

Since θ is the angle between u and v, and $\theta = 90°$, u and v are perpendicular.

Section 7.5

1. Let α = the angle between the forces.

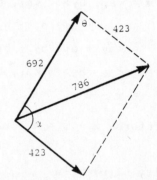

To find α, use the law of cosines to find θ.

$786^2 = 692^2 + 423^2$

$- 2(692)(423) \cos \theta$

$\cos \theta = \dfrac{786^2 - 692^2 - 423^2}{-2(692)(423)}$

$\theta = 86.1°$

$\alpha = 180° - 86.1° = 93.9°$

3. Let θ = the angle that the hill makes with the horizontal.

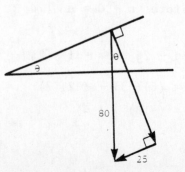

The 80-lb downward force has a 25-lb component parallel to the hill. The two right triangles are similar and have congruent angles.

$$\sin \theta = \frac{25}{80}$$

$$\theta = 18°$$

5. Find the force needed to pull a 60-ton monolith along the causeway.

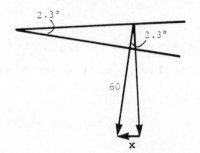

The force needed to pull 60 tons is equal to the magnitude of **x**, the component parallel to the causeway.

$$\sin 2.3° = \frac{|\mathbf{x}|}{60}$$

$$|\mathbf{x}| = 60 \sin 2.3°$$

$$= 2.4$$

The force needed is 2.4 tons.

7. Find the direction and magnitude of the equilibrant.

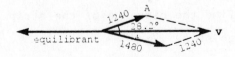

$A = 180° - 28.2° = 151.8°$
Use the law of cosines to find the magnitude of the resultant **v**.

$$|\mathbf{v}|^2 = 1240^2 + 1480^2$$
$$- 2(1240)(1480) \cos 151.8°$$
$$|\mathbf{v}| = 2640 \text{ lb}$$

Use the law of sines to find α.

$$\frac{\sin \alpha}{1240} = \frac{\sin 151.8°}{2640}$$

$$\sin \alpha = \frac{1240 \sin 151.8°}{2640}$$

$$\alpha = 12.82°$$

$$\theta = 180° - 12.82° = 167.2°$$

The equilibrant has a magnitude of 2640 lb and makes an angle of 167.2° with the 1480-lb force. The weight is 2640 lb at an angle of 167.2° with the 1480-lb force.

9. Find the weight of the crate and the tension on the horizontal rope.

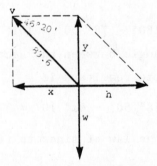

v has horizontal component **x** and vertical component **y**. The resultant **v** + **h** also has vertical component **y**. The resultant balances the weight of the crate, so its vertical component is the equilibrant of the crate's weight.

$$|\mathbf{w}| = |\mathbf{y}| = 89.6 \sin 46° 20'$$
$$|\mathbf{w}| = 64.8 \text{ lb}$$

Since the crate is not moving side-to-side, **h**, the horizontal tension on the rope, is the opposite of **x**.

$|\mathbf{h}| = |\mathbf{x}| = 89.6 \cos 46° \, 20'$

$|\mathbf{h}| = 61.9 \text{ lb}$

The weight of the crate is 64.8 lb; the tension is 61.9 lb.

11. Find the magnitude of the second force and of the resultant. Use the parallelogram rule; **v** is the resultant and **x** is the second force.

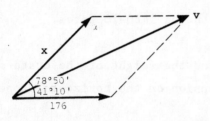

$\alpha = 180° - 78° \, 50' = 101° \, 10'$

The angle between the second force and the resultant is

$\beta = 78° \, 50' - 41° \, 10' = 37° \, 40'.$

Use the law of sines to find $|\mathbf{v}|$.

$$\frac{|\mathbf{v}|}{\sin \alpha} = \frac{176}{\sin \beta}$$

$$|\mathbf{v}| = \frac{176 \sin 101° \, 10'}{\sin 37° \, 40'}$$

$$|\mathbf{v}| = 283 \text{ lb}$$

Use the law of sines to find $|\mathbf{x}|$.

$$\frac{|\mathbf{x}|}{\sin 41° \, 10'} = \frac{176}{\sin 37° \, 40'}$$

$$|\mathbf{x}| = \frac{176 \sin 41° \, 10'}{\sin 37° \, 40'}$$

$$|\mathbf{x}| = 190 \text{ lb}$$

13. Let **v** = the groundspeed vector of the plane.
Find the actual bearing of the plane.

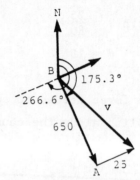

Angle A = 266.6° − 175.3° = 91.3°

Use the law of cosines to find $|\mathbf{v}|$.

$$|\mathbf{v}|^2 = 25^2 + 650^2$$
$$- 2(25)(650) \cos 91.3°$$
$$= 423,862$$
$$|\mathbf{v}| = 651$$

Use the law of sines to find the drift angle, B, the angle that **v** makes with the airspeed vector of the plane.

$$\frac{\sin B}{25} = \frac{\sin A}{651}$$
$$\sin B = .038392573$$
$$B = 2.2°$$

The bearing is 175.3° − 2.2°
= 173.1°.

15. Find the distance of the strip from point A.

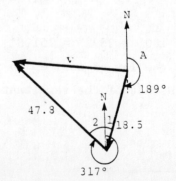

Angle 1 = 189° - 180° = 9°

Angle 2 = 360° - 317° = 43°

Angle 1 + Angle 2 = 9° + 43°
$$= 52°$$

Use the law of cosines to find $|\mathbf{v}|$.

$|\mathbf{v}|^2 = (47.8)^2 + (18.5)^2$
$$- 2(47.8)(18.5) \cos 52°$$

$|\mathbf{v}|^2 = 1538.2311$

$|\mathbf{v}| = 39.2$ km

17. Let $\mathbf{v}$ = the groundspeed vector.
Find the bearing and groundspeed
of the plane.

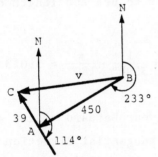

Angle A = 233° - 114° = 119°

Use the law of cosines to find $|\mathbf{v}|$.

$|\mathbf{v}|^2 = 39^2 + 450^2$
$$- 2(39)(450) \cos 119°$$

$|\mathbf{v}|^2 = 221,038$

$|\mathbf{v}| = 470$

The groundspeed is 470 mph.

Use the law of sines to find
angle B, the drift angle.

$$\frac{\sin B}{39} = \frac{\sin 119°}{470}$$

$$\sin B = .0726$$

$$B = 4°$$

The bearing is 4° + 233° = 237°.

19. Let $\mathbf{v}$ = the airspeed vector.
The groundspeed must be

$$\frac{400}{2.5} = 160 \text{ mph.}$$

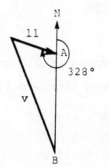

Angle A = 328° - 180° = 148°

Use the law of cosines to find $|\mathbf{v}|$.

$|\mathbf{v}|^2 = 11^2 + 160^2$
$$- 2(11)(160) \cos 148°$$

$|\mathbf{v}|^2 = 28,706$

$|\mathbf{v}| = 170$

The airspeed must be 170 mph.

Use the law of sines to find B,
the drift angle.

$$\frac{\sin B}{11} = \frac{\sin 148°}{170}$$

$$\sin B = \frac{11 \sin 148°}{170}$$

$$B = 2.0°$$

The bearing must be

$$360° - 2.0° = 358°.$$

21. Find the groundspeed and resulting
bearing.

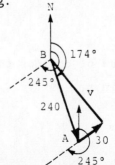

Angle A = 245° – 174° = 71°

Use the law of cosines to find $|\mathbf{v}|$.

$$|\mathbf{v}|^2 = 30^2 + 240^2$$
$$\quad - 2(30)(240) \cos 71°$$
$$|\mathbf{v}|^2 \approx 53,812$$
$$|\mathbf{v}| \approx 230$$

The groundspeed is 230 km per hr.

Use the law of sines to find angle B.

$$\frac{\sin B}{30} = \frac{\sin 71°}{230}$$
$$\sin B = .1233$$
$$B = 7°$$

The resulting bearing is

$$174° - 7° = 167°.$$

23. At what time did the pilot turn?
The plane will fly 2.6 hr before
it runs out of fuel. In 2.6 hr,
the carrier will travel (2.6)(30)
= 83.2 mi and the plane will travel
a total of (2.6)(520) = 1352 mi.
Suppose it travels x mi on its
initial bearing; then it travels
1352 – x mi after having turned.

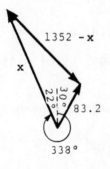

Use the law of cosines to get an
equation in x.

$$(1352 - x)^2$$
$$= x^2 + (83.2)^2 - 2(x)(83.2) \cos 52°$$
$$1,827,904 - 2704x + x^2$$
$$\quad = x^2 + 6922.24 - 102.45x$$
$$1,820,982 = 2601.55x$$
$$700 = x$$

To travel 700 mi at 520 mph requires

$\frac{700}{520} = 1.35$ hr, or 1 hr and 21 min.

The pilot turned at 1 hr 21 min

after 2 P.M., or at 3:21 P.M.

25. **(a)** First, change 10.34" to radians
in order to use the length of arc
formula.

$10.34" \cdot \dfrac{1}{3600"} \cdot \dfrac{\pi}{180°} \approx 5.013 \times 10^{-5}$

radian.

In one year Barnard's Star will move
in the tangential direction a distance of

$$s = r\theta = (35 \times 10^{12})(5.013 \times 10^{-5})$$
$$= 1,754,550,000 \text{ mi.}$$

In one second Barnard's Star moves
approximately

$$\frac{1,754,550,000}{60 \cdot 60 \cdot 24 \cdot 365} \approx 56 \text{ mi}$$

tangentially. Thus, $\mathbf{v}_t = 56$ mi/sec.

(b) The magnitude of $\mathbf{v}$ is given by

$$|\mathbf{v}|^2 = |\mathbf{v}_r|^2 + |\mathbf{v}_t|^2$$
$$|\mathbf{v}| = \sqrt{67^2 + 56^2}$$
$$\approx 87 \text{ mi/sec.}$$

Chapter 7 Review Exercises

1. Find b, given C = 74.2°, c = 96.3 m, B = 39.5°.

Use the law of sines to find b.

$$\frac{b}{\sin B} = \frac{c}{\sin C}$$

$$b = \frac{c \sin B}{\sin C}$$

$$= \frac{96.3 \sin 39.5°}{\sin 74.2°}$$

$$= 63.7 \text{ m}$$

3. Find B, given C = 51.3°, c = 68.3 m, b = 58.2 m.

Use the law of sines to find B.

$$\frac{\sin B}{b} = \frac{\sin C}{c}$$

$$\sin B = \frac{b \sin C}{c}$$

$$= \frac{58.2 \sin 51.3°}{68.3}$$

$$\sin B = .66533254$$

$$B = 41.7°$$

Angle B cannot be obtuse, since b < c, B < C, and C is acute.

5. Given B = 39° 50′, b = 268 m, a = 340 m; find A.

a > b > a sin B tells us that two triangles are possible and A can have two values.

Use the law of sines to find A.

$$\frac{\sin A}{a} = \frac{\sin B}{b}$$

$$\sin A = \frac{a \sin B}{b}$$

$$= \frac{340 \sin 39° 50′}{268}$$

$$\sin A = .81264638$$

$$A = 54° 20′ \text{ or } A = 125° 40′$$

7. If we are given a, A, and C in a triangle ABC, B also is known since B = 180° − A − C. We now have the case where we know two angles and the included side. Since we do not have two sides and the angle opposite one of the sides, the possibility of the ambiguous case does not exist.

9. a = 10, B = 30°

(a) The value of b that forms a right triangle would yield exactly one value for A. That is,

b = 10 sin 30° = 5.

Also any value of b greater than or equal to 10 would yield a unique value for A.

(b) Any value of b between 5 and 10 would yield two possible values for A.

(c) If b is less than 5, then no value for A is possible.

11. Given a = 86.14 in, b = 253.2 in, c = 241.9 in, find A.

Use the law of cosines to find A.

$$a^2 = b^2 + c^2 - 2bc \cos A$$

$$\cos A = \frac{b^2 + c^2 - a^2}{2bc}$$

$$= \frac{253.2^2 + 241.9^2 - 86.14^2}{2(253.2)(241.9)}$$

$$\cos A = .94046923$$

$$A = 19.87° \text{ or } 19° 52′$$

13. Given A = 51° 20′, c = 68.3 m, b = 58.2 m; find a.

Use the law of cosines to find a.

a² = b² + c² - 2bc cos A

$a = \sqrt{(58.2)^2 + (68.3)^2 - 2(58.2)(68.3)\cos 51°\ 20'}$

a = 55.5 m

15. Given A = 46° 10′, b = 184 cm, c = 192 cm; find a.

Use the law of cosines to find a.

a² = b² + c² - 2bc cos A

$a = \sqrt{(184)^2 + (192)^2 - 2(184)(192)\cos 46°\ 10'}$

a = 148 cm

17. Solve the triangle, given A = 25.2°, a = 6.92 yd, b = 4.82 yd.

a > b tells us that only one triangle is possible.

Use the law of sines to find b.

$$\frac{\sin B}{b} = \frac{\sin A}{a}$$

$$\sin B = \frac{b \sin A}{a}$$

$$= \frac{4.82 \sin 25.2°}{6.92}$$

$$B = 17.3°$$

(Angle B cannot be obtuse, since b < a, B < A, and A is acute.)

$$C = 180° - A - B = 137.5°$$

Use the law of sines to find c.

$$\frac{c}{\sin C} = \frac{a}{\sin A}$$

$$c = \frac{a \sin C}{\sin A}$$

$$= \frac{6.92 \sin 137.5°}{\sin 25.2°}$$

$$c = 11.0 \text{ yd}$$

19. Solve the triangle, given $a = 27.6$ cm, $b = 19.8$ cm, $C = 42° 30'$.

Using the law of cosines, we have

$c^2 = a^2 + b^2 - 2ab \cos C$

$c^2 = 27.6^2 + 19.8^2$
$\qquad - 2(27.6)(19.8) \cos 42° 30'$

$c = 18.7$ cm.   (rounded)

Note: Keep all values of $\sqrt{c^2}$ in your calculator for use in the next calculation. If 18.7 is used, the answer will vary slightly due to round-off error.

Now use the law of sines to find B so there is no ambiguity.

$$\frac{c}{\sin C} = \frac{b}{\sin B}$$

$$\frac{18.65436576}{\sin 42° 30'} = \frac{19.8}{\sin B}$$

$$\sin B = \frac{19.8 \sin 42° 30'}{18.65436576}$$

$$B = 45° 50'.$$

Since the sum of the angles of a triangle is 180°,

$A = 180° - B - C$
$\quad = 180° - 45° 50' - 42° 30'$
$A = 91° 40'.$

21. Given $b = 840.6$ m, $c = 715.9$ m, $A = 149.3°$, find the area.

Angle A is included between sides b and c.

$\text{area} = \frac{1}{2} bc \sin A$

$\quad = \frac{1}{2}(840.6)(715.9) \sin 149.3°$

$\text{area} = 153,600$ m²

23. Given $a = .913$ km, $b = .816$ km, $c = .582$ km, find the area.

$s = \frac{1}{2}(a + b + c)$

$\quad = \frac{1}{2}(.913 + .816 + .582)$

$\quad = 1.1555$

area
$= \sqrt{(1.1555)(.2425)(.3395)(.5735)}$
area
$= .234$ km²

25. Substitute $a = 7$, $b = 7\sqrt{3}$, $c = 14$, $A = 30°$, $B = 60°$, and $C = 90°$ into

$$\frac{a + b}{c} = \frac{\cos \frac{1}{2}(A - B)}{\sin \frac{1}{2}C}$$

in order to verify the formula.

$\frac{a + b}{c} = \frac{7 + 7\sqrt{3}}{14}$

$\quad = \frac{7(1 + \sqrt{3})}{14} = \frac{1 + \sqrt{3}}{2}$

$\frac{\cos \frac{1}{2}(A - B)}{\sin \frac{1}{2}C} = \frac{\cos \frac{1}{2}(30° - 60°)}{\sin \frac{1}{2}(90°)}$

$\quad = \frac{\cos \frac{1}{2}(-30°)}{\sin 45°}$

$\quad = \frac{\cos 15°}{\sin 45°}$

$\quad = \frac{\frac{\sqrt{2} + \sqrt{6}}{4}}{\frac{\sqrt{2}}{2}} = \frac{1 + \sqrt{3}}{2}$

(Use half-angle formulas from Section 5.6 to find cos 15°.)

Each expression is equal to $\frac{1 + \sqrt{3}}{2}$.

27. a = 7, b = 7, c = 6

$$s = \frac{1}{2}(7 + 7 + 6) = 10$$

$$\text{area} = \sqrt{10(3)(3)(4)}$$

$$= \sqrt{360}$$

$$= 18.97 \text{ m}^2$$

18.97 m² ÷ 7.5 m²/can ≈ 2.5 cans

He must buy 3 cans.

29. Let AC = the height of the tree.

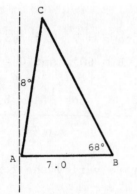

Angle A = 90° − 8.0° = 82°

Angle C = 180° − B − C = 30°

Use the law of sines to find

AC = b.

$$\frac{b}{\sin B} = \frac{c}{\sin C}$$

$$b = \frac{c \sin B}{\sin C}$$

$$= \frac{7.0 \sin 68°}{\sin 30°}$$

$$= 13$$

The tree is 13 m tall.

31. Sketch a triangle showing the situation as follows.

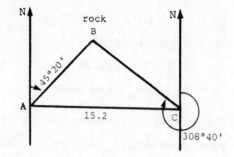

Angle A = 90° − 45° 20′ = 44° 40′

Angle C = 308° 40′ − 270° = 38° 40′

Angle B = 180° − A − C = 96° 40′

Use the law of sines to find BC = a.

$$\frac{a}{\sin A} = \frac{b}{\sin B}$$

$$a = \frac{b \sin A}{\sin B}$$

$$= \frac{15.2 \sin 44° 40′}{\sin 96° 40′}$$

$$a = 10.8$$

The distance between them is 10.8 mi.

33. Let x = the distance between the boats.

In 3 hr, the first boat travels 3(36.2) = 108.6 km and the second travels 3(45.6) = 136.8 km.

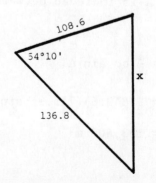

Use the law of cosines to find x.

$$x^2 = (108.6)^2 + (136.8)^2$$
$$\quad - 2(108.6)(136.8) \cos 54° \, 10'$$
$$\quad = 13,113.359$$
$$x = 115 \text{ km}$$

35. Let c = the length of the tunnel.

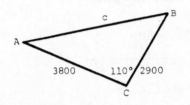

Use the law of cosines to find c.

$$c^2 = 3800^2 + 2900^2$$
$$\quad - 2(3800)(2900) \cos 110°$$
$$c^2 = 30,388,124$$
$$c = 5500$$

The tunnel is 5500 m long.

37. Let x = the distance between the ends of the two equal sides.

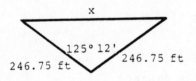

$$x^2 = (246.75)^2 + (246.75)^2$$
$$\quad - 2(246.75)(246.75) \cos 125° \, 12'$$
$$x = 438.14$$

The distance between the ends of the two equal sides is 438.14 ft.

For Exercises 39 and 41, see the answer sketches at the back of the textbook.

43. The statement: A diagonal of a parallelogram must bisect two angles of the parallelogram. Draw parallelogram ABCD so that diagonal AC forms θ_1 and θ_2 from angle A and α_1 and α_2 from angle C. Side AB $\neq$ side BC and side AD $\neq$ side CD.

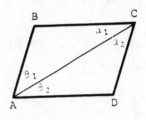

$\theta_1 = \alpha_2$ and $\theta_2 = \alpha_1$ since they are alternate interior angles.
Assume that $\theta_1 = \theta_2$ and $\alpha_1 = \alpha_2$.
Then, $\theta_1 = \alpha_1$ and $\theta_2 = \alpha_2$ and by substitution, triangles ABC and ACD are both isosceles with sides AB = BC and AD = CD. But this statement is a contradiction, so the original statement is false.

45. Find the horizontal and vertical components.

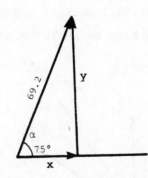

Horizontal:

$$\cos 75° = \frac{|\mathbf{x}|}{69.2}$$

$$|\mathbf{x}| = 69.2 \cos 75° = 17.9$$

Vertical:

$$\sin 75° = \frac{|\mathbf{y}|}{69.2}$$

$$|\mathbf{y}| = 69.2 \sin 75° = 66.8$$

In Exercises 47 and 49, find the magnitude of the resultant force.

47. Forces of 15 lb and 23 lb, forming an angle of 87°

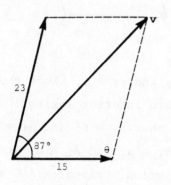

$$\theta = 180° - 87° = 93°$$

Use the law of cosines to find $|\mathbf{v}|$.

$$|\mathbf{v}|^2 = 15^2 + 23^2$$
$$- 2(15)(23) \cos 93°$$
$$|\mathbf{v}|^2 = 790$$
$$|\mathbf{v}| = 28 \text{ lb}$$

49. Forces of 85.2 newtons and 69.4 newtons, forming an angle of 58° 20'

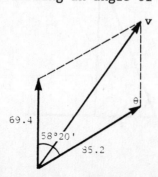

$$\theta = 180° - 58° 20'$$
$$= 121° 40'$$

Use the law of cosines to find $|\mathbf{v}|$.

$$|\mathbf{v}|^2 = (85.2)^2 + (69.4)^2$$
$$- 2(85.2)(69.4) \cos 121° 40'$$
$$|\mathbf{v}|^2 = 18,284$$
$$|\mathbf{v}| = 135 \text{ newtons}$$

51. $\mathbf{u} = <21, -20>$

$$|\mathbf{u}| = \sqrt{21^2 + (-20^2)}$$
$$= \sqrt{841} = 29$$

$$\cos \theta = \frac{a}{|\mathbf{u}|} = \frac{21}{29}$$
$$\theta \approx 43.6°$$

But $\mathbf{u}$ is in the 4th quadrant, so

$$\theta = 360° - 43.6°$$
$$= 316.4°.$$

53. $|\mathbf{u}| = 6, \theta = 60°$

$$a = |\mathbf{u}| \cos \theta$$
$$= 6 \cos 60°$$
$$= 6\left(\frac{1}{2}\right) = 3$$

$$b = |\mathbf{u}| \sin \theta$$
$$= 6 \sin 60°$$
$$= 6\left(\frac{\sqrt{3}}{2}\right) = 3\sqrt{3}$$

Thus,

$$\mathbf{u} = <3, 3\sqrt{3}>.$$

55. Let $\mathbf{x}$ = the resultant vector.

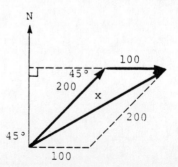

$\alpha = 180° - 45° = 135°$

$|\mathbf{x}|^2 = 200^2 + 100^2$
$\qquad - 2(200)(100) \cos 135°$

$|\mathbf{x}| = 280$

Using the law of cosines again, we get

$$\cos \theta = \frac{200^2 - 100^2 - 280^2}{-2(100)(280)}$$

$$\theta = 30.4°$$

The force is 280 newtons at an angle of 30.4° with the first rope.

57. Let θ = the angle that the hill makes with the horizontal.

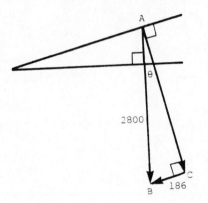

The downward force of 2800 lb has component **AC** perpendicular to the hill and component **CB** parallel to the hill. **AD** represents the force of 186 lb that keeps the car from rolling down the hill. Since vectors **AD** and **BC** are equal, $|\mathbf{BC}| = 186$. Angle B = angle EAB because they are alternate interior angles so the two right triangles are similar. Hence, angle θ = angle BAC.

$$\sin BAC = \frac{186}{2800}$$

$$BAC = 3° \ 50' = \theta$$

The angle that the hill makes with the horizontal is 3° 50′.

59. Let $\mathbf{v}$ = the resultant vector.

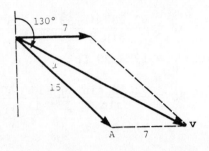

Angle A = 180° − (130° − 90°)

$\qquad = 140°$

Use the law of cosines to find the magnitude of the resultant $\mathbf{v}$.

$|\mathbf{v}|^2 = 15^2 + 7^2$
$\qquad - 2(15)(7) \cos 140°$

$|\mathbf{v}| = 21$

The resulting speed is 21 kph.
Use the law of sines to find α.

$$\frac{\sin \alpha}{7} = \frac{\sin 140°}{21}$$

$$\sin \alpha = \frac{7 \sin 140°}{21}$$

$$\alpha = 12°$$

The bearing is 130° − 12° = 118°.

61.

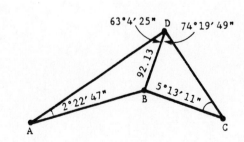

In each of the triangles ABP and PBC
we know two angles and one side.
Solve each triangle using the law of
sines.

$$AB = \frac{92.13 \sin 63° \ 4' \ 25''}{\sin 2° \ 22' \ 47''}$$

$$\approx 1978.28 \text{ ft}$$

$$BC = \frac{92.13 \sin 74° \ 19' \ 49''}{\sin 5° \ 13' \ 11''}$$

$$\approx 975.05 \text{ ft}$$

CHAPTER 7 TEST

Find the indicated part of each triangle ABC if the triangle exists.

[7.3] **1.** A = 150.4°, b = 31.5 cm, c = 22.3 cm; find a.

[7.1] **2.** A = 21° 20′, B = 62° 10′, b = 91.2 m; find a.

[7.2] **3.** A = 35.6°, a = 58.2 m, b = 104.8 m; find all possible
          values of B.

[7.2] **4.** B = 29° 15′, b = 16.3 cm, c = 20.4 cm; find all possible
          values of C.

Find the area of each triangle ABC.

[7.3] **5.** a = 45.5 yd, b = 54.2 yd, c = 63.7 yd

[7.1] **6.** B = 50.7°, a = 39.2 cm, c = 17.3 cm

[7.2] **7.** What happens if you use the law of sines to try to find
          angle B of triangle ABC if you are given a, b, and angle
          A, such that a < b sin A?

[7.4] Sketch the indicated vectors.

8. a − b   **9. 2m + n**

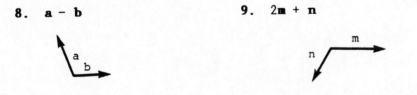

[7.4] Find the vertical component of each of the following vectors.
       In each case, α is the angle that the vector makes with the
       horizontal.

10. α = 27°, magnitude 55   **11.** α = 98° 30′, magnitude 38.6

Solve each of the following.

[7.1]   12. Two ships on a north—south line are 11 mi apart. The
             northernmost ship sights a lighthouse at a bearing of
             105°. Meanwhile, the other ship sights the lighthouse at
             a bearing of 45°. How far is the northernmost ship from
             the lighthouse?

[7.3]   13. A painter needs to cover a triangular region that is 21 m by
             32 m by 15 m. A can of paint covers 50 m² of area. How many
             cans (to the next higher number of cans) will he need?

[7.5]   14. Find the force required to pull a 65—lb weight up a ramp in-
             clined at 15° to the horizontal.

[7.5]   15. A force of 32 newtons makes an angle of 55° with a force of
             48 newtons. Find the magnitude of the equilibrant.

[7.5]   16. A plane flying 550 mph on a bearing of 75° encounters a 25—mph
             wind blowing from due south. Find the reulting bearing of the
             plane.

CHAPTER 3 TEST ANSWERS

1. 52.1 cm

2. 37.5 m

3. The triangle does not exist.

4. 37° 40′ or 142° 20′

5. 1210 yd²

6. 262 cm²

7. sin B > 1 results, which means that no such triangle exists.

8.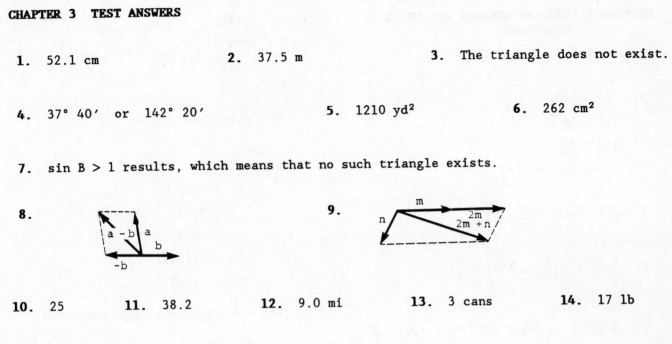

9.

10. 25

11. 38.2

12. 9.0 mi

13. 3 cans

14. 17 lb

15. 71 newtons

16. 73°

CHAPTER 8   COMPLEX NUMBERS AND POLAR EQUATIONS

Section 8.1

3. $x^2 = -25$

$\sqrt{x^2} = \sqrt{-25}$

$|x| = 5i$

$x = 5i$ or $-5i$

5. $\sqrt{-4} = \sqrt{4} \cdot \sqrt{-1} = 2i$

7. $\sqrt{-\frac{25}{9}} = i\sqrt{\frac{25}{9}} = \frac{\sqrt{25}}{\sqrt{9}}i = \frac{5}{3}i$

9. $\sqrt{-150} = i\sqrt{150} = i\sqrt{25 \cdot 6} = i\sqrt{25} \cdot \sqrt{6}$
$= 5i\sqrt{6}$

11. $\sqrt{-80} = i\sqrt{80} = i\sqrt{16 \cdot 5}$
$= 4i\sqrt{5}$

13. $\sqrt{-3} \cdot \sqrt{-3} = i\sqrt{3} \cdot i\sqrt{3}$
$= -3$

15. $\sqrt{-5} \cdot \sqrt{-6} = i\sqrt{5} \cdot i\sqrt{6}$
$= -\sqrt{30}$

17. $\frac{\sqrt{-12}}{\sqrt{-8}} = \frac{i\sqrt{12}}{i\sqrt{8}}$
$= \sqrt{\frac{3}{2}} = \frac{\sqrt{6}}{2}$

19. $\frac{\sqrt{-24}}{\sqrt{72}} = \frac{i\sqrt{24}}{\sqrt{72}}$
$= i\sqrt{\frac{1}{3}} = i\frac{1}{\sqrt{3}} \cdot \frac{\sqrt{3}}{\sqrt{3}}$
$= \frac{\sqrt{3}}{3}i$

21. $x^2 = -16$

$x = \pm\sqrt{-16}$

$x = \pm i\sqrt{16}$

$x = \pm 4i$

The solutions are $4i$ and $-4i$.

23. $z^2 + 12 = 0$

$z^2 = -12$

$z = \pm\sqrt{-12}$

$z = \pm i\sqrt{12}$

$z = \pm i(2\sqrt{3})$

$z = \pm 2i\sqrt{3}$

The solutions are $2i\sqrt{3}$ and $-2i\sqrt{3}$.

25. $3x^2 + 4x + 2 = 0$

Use the quadratic formula with
$a = 3$, $b = 4$, and $c = 2$.

$x = \frac{-4 \pm \sqrt{4^2 - 4 \cdot 3 \cdot 2}}{2 \cdot 3}$

$x = \frac{-4 \pm \sqrt{16 - 24}}{6}$

$x = \frac{-4 \pm \sqrt{-8}}{6}$

$x = \frac{-4 \pm i\sqrt{8}}{6}$

$x = \frac{2(-2 \pm i\sqrt{2})}{6}$

$x = \frac{-2 \pm i\sqrt{2}}{3}$

The solutions are $-\frac{2}{3} + \frac{\sqrt{2}}{3}i$ and

$-\frac{2}{3} - \frac{\sqrt{2}}{3}i$.

27. $m^2 - 6m + 14 = 0$

Use the quadratic formula with
$a = 1$, $b = -6$, and $c = 14$.

$$m = \frac{-(-6) \pm \sqrt{(-6)^2 - 4 \cdot 1 \cdot 14}}{2 \cdot 1}$$

$$m = \frac{6 \pm \sqrt{-20}}{2}$$

$$m = \frac{6 \pm i\sqrt{20}}{2}$$

$$m = \frac{2(3 \pm i\sqrt{5})}{2}$$

$$m = 3 \pm i\sqrt{5}$$

The solutions are $3 + i\sqrt{5}$ and
$3 - i\sqrt{5}$.

29. $4z^2 = 4z - 7$

$4z^2 - 4z + 7 = 0$

Use the quadratic formula with
$a = 4$, $b = -4$, and $c = 7$.

$$z = \frac{-(-4) \pm \sqrt{(-4)^2 - 4 \cdot 4 \cdot 7}}{2 \cdot 4}$$

$$z = \frac{4 \pm \sqrt{-96}}{8}$$

$$z = \frac{4 \pm i\sqrt{96}}{8}$$

$$z = \frac{4(1 \pm i\sqrt{6})}{8}$$

$$z = \frac{1 \pm i\sqrt{6}}{2}$$

The solutions are $\frac{1}{2} + \frac{\sqrt{6}}{2}i$ and

$\frac{1}{2} - \frac{\sqrt{6}}{2}i$.

31. $m^2 + 1 = -m$

$m^2 + m + 1 = 0$

Use the quadratic formula with
$a = 1$, $b = 1$, $c = 1$.

$$m = \frac{-1 \pm \sqrt{1^2 - 4 \cdot 1 \cdot 1}}{2 \cdot 1}$$

$$m = \frac{-1 \pm \sqrt{-3}}{2}$$

$$m = \frac{-1 \pm i\sqrt{3}}{2}$$

The solutions are $-\frac{1}{2} + \frac{\sqrt{3}}{2}i$ and

$-\frac{1}{2} - \frac{\sqrt{3}}{2}i$.

33. A quadratic equation of the form
$ax^2 + bx + c = 0$ with a, b, and c
real numbers and with $a \neq 0$ will
have nonreal complex solutions if
the discriminant, $b^2 - 4ac$, is
negative.

35. The property $\sqrt{a} \cdot \sqrt{b} = \sqrt{ab}$ is only
true if $a \geq 0$ and $b \geq 0$. The
definition $\sqrt{-a} = i\sqrt{a}$ if $a > 0$ must
be used before using other rules for
radicals.

$$\sqrt{-5} \cdot \sqrt{-5} = i\sqrt{5} \cdot i\sqrt{5}$$

Then, multiply using the rules for
radicals.

$$= i^2\sqrt{25}$$
$$= (-1)5$$
$$= -5$$

37. $(2 - 5i) + (3 + 2i)$

$= (2 + 3) + (-5 + 2)i$

$= 5 - 3i$

39. $(-2 + 3i) - (3 + i)$

$= (-2 - 3) + (3 - 1)i$

$= -5 + 2i$

41. $(1 - i) - (5 - 2i)$

$= (1 - 5) + [-1 - (-2)]i$

$= -4 + i$

43. $(2 + i)(3 - 2i)$

$= 6 - 4i + 3i - 2i^2$

$= 6 - i - 2(-1)$

$= 8 - i$

45. $(2 + 4i)(-1 + 3i)$

$= -2 + 6i - 4i + 12i^2$

$= -2 + 2i + 12(-1)$

$= -14 + 2i$

47. $(2 - i)(2 + i) = 4 + 2i - 2i - i^2$

$= 4 - (-1)$

$= 5$ or $5 + 0i$

Note that the product of conjugates is the difference of squares.

$(2 - i)(2 + i) = 2^2 - i^2$

$= 4 - (-1)$

$= 5$

49. $\dfrac{5}{i} = \dfrac{5}{0 + i} \cdot \dfrac{0 - i}{0 - i}$

$= \dfrac{5}{i} \cdot \dfrac{-i}{-i}$

$= \dfrac{-5i}{-i^2}$

$= \dfrac{-5i}{-(-1)}$

$= \dfrac{-5i}{1}$

$= -5i$

or $0 - 5i$

51. $\dfrac{4 - 3i}{4 + 3i} = \dfrac{4 - 3i}{4 + 3i} \cdot \dfrac{4 - 3i}{4 - 3i}$

$= \dfrac{16 - 24i + 9i^2}{16 - 9i^2}$

$= \dfrac{16 - 24i + 9(-1)}{16 - 9(-1)}$

$= \dfrac{7 - 24i}{25}$

$= \dfrac{7}{25} - \dfrac{24}{25}i$

53. $\dfrac{4 + i}{6 + 2i} = \dfrac{4 + i}{6 + 2i} \cdot \dfrac{6 - 2i}{6 - 2i}$

$= \dfrac{24 - 2i - 2i^2}{36 - 4i^2}$

$= \dfrac{24 - 2i - 2(-1)}{36 - 4(-1)}$

$= \dfrac{26 - 2i}{40}$

$= \dfrac{13 - i}{20}$

$= \dfrac{13}{20} - \dfrac{1}{20}i$

55.   To simplify i^n, divide n by 4 and obtain a quotient q with a remainder r, which must be 0, 1, 2, or 3. Thus,

$$n = 4q + r.$$

Therefore,

$$i^n = i^{4q+r}$$

$$= i^{4q} \cdot i^r.$$

Since $i^{4q} = 1$,

$$i^n = 1 \cdot i^r$$

$$= i^r.$$

Thus, if $r = 0$, $i^n = i^0 = 1$;

if $r = 1$, $i^n = i^1 = i$;

if $r = 2$, $i^n = i^2 = -1$; and

if $r = 3$, $i^n = i^3 = -i$.

57. $i^{12} = (i^4)^3$

$= 1^3$

$= 1$

59. $i^{18} = i^{16} \cdot i^2$

$= (i^4)^4 \cdot (-1)$

$= 1^4 \cdot (-1)$

$= 1 \cdot (-1)$

$= -1$

61. $i^{-3} = i^{-4} \cdot i^1$

$= (i^4)^{-1} \cdot i$

$= 1^{-1} \cdot i$

$= 1 \cdot i$

$= i$

63. $i^{-10} = i^{-12} \cdot i^2$

$= (i^4)^{-3} \cdot (-1)$

$= 1^{-3} \cdot (-1)$

$= 1 \cdot (-1)$

$= -1$

65. $1 + i + i^2 + i^3$

$= 1 + i + (-1) + (-i)$

$= 0$

67. $2x + yi = 4 - 3i$

$2x = 4$

$x = 2$

$y = -3$

69. $7 - 2yi = 14x - 30i$

$7 = 14x$

$\frac{1}{2} = x$

$-2y = -30$

$y = 15$

71. $x + yi = (2 + 3i)(4 - 2i)$

$= 8 - 4i + 12i - 6i^2$

$= 8 + 8i - 6(-1)$

$= 14 + 8i$

$x = 14$, $y = 8$

75. The conjugate of $a + bi$ is $a - bi$.

$(a + bi) + (a - bi)$

$= (a + a) + [b + (-b)]i$

$= 2a + 0i$

$= 2a$

$2a$ is a real number since it has no imaginary part.

77. $\left(\frac{\sqrt{2}}{2} + \frac{\sqrt{2}}{2}i\right)^2$

$= \left(\frac{\sqrt{2}}{2}\right)^2 + 2\frac{\sqrt{2}}{2} \cdot \frac{\sqrt{2}}{2}i + \left(\frac{\sqrt{2}}{2}i\right)^2$

$= \frac{2}{4} + \frac{4}{4}i + \frac{2}{4}i^2$

$= \frac{1}{2} + i + \frac{1}{2}(-1)$

$= \frac{1}{2} + i - \frac{1}{2} = i$

Since $\left(\frac{\sqrt{2}}{2} + \frac{\sqrt{2}}{2}i\right)^2 = i$, $\frac{\sqrt{2}}{2} + \frac{\sqrt{2}}{2}i$ is a square root of i.

79. **(a)** From the figure in the text, the light bulbs have resistances of 50 and 60 ohms and the motors have reactances of 15 and 17 ohms, respectively. Thus, the total impedance is given by

$$Z = 50 + 60 + 15i + 17i$$
$$= 110 + 32i.$$

(b) From part (a), a = 110 and b = 32. Then

$$\tan \theta = \frac{b}{a}$$

$$\tan \theta = \frac{32}{110}$$

$$\theta \approx 16.22°.$$

81.   I = 10 + 6i, Z = 8 + 5i

E = IZ

E = (10 + 6i)(8 + 5i)

$= 80 + 50i + 48i + 30i^2$

$= 80 + 98i + 30(-1)$

$= 50 + 98i$

83.   E = 35 + 55i, Z = 6 + 4i

E = IZ

35 + 55i = I(6 + 4i)

$$I = \frac{35 + 55i}{6 + 4i}$$

$$= \frac{35 + 55i}{6 + 4i} \cdot \frac{6 - 4i}{6 - 4i}$$

$$= \frac{210 - 140i + 330i - 220i^2}{36 - 16i^2}$$

$$= \frac{430 + 190i}{52}$$

$$= \frac{215}{26} + \frac{95}{26}i$$

Section 8.2

1. The modulus of a complex number represents the *magnitude (or length)* of the vector representing it in the complex plane.

For Exercises 3–11, see the answer graphs in the back of the textbook.

13. In order for a complex number a + bi also to be a real number, its imaginary part b must be zero.

15. 4 − 3i, −1 + 2i

(4 − 3i) + (−1 + 2i) = 3 − i

17. 5 − 6i, −2 + 3i

(5 − 6i) + (−2 + 3i) = 3 − 3i

19. −3, 3i

(−3 + 0i) + (0 + 3i) = −3 + 3i

21. 2 + 6i, −2i

(2 + 6i) + (0 − 2i) = 2 + 4i

23. 7 + 6i, 3i

(7 + 6i) + (0 + 3i) = 7 + 9i

25. 2(cos 45° + i sin 45°)

$$= 2\left(\frac{\sqrt{2}}{2} + \frac{i\sqrt{2}}{2}\right) = \sqrt{2} + i\sqrt{2}$$

27. 10(cos 90° + i sin 90°)

= 10(0 + i)

= 0 + 10i = 10i

29. $4(\cos 240° + i \sin 240°)$

$$= 4\left(-\frac{1}{2} - i\frac{\sqrt{3}}{2}\right)$$

$$= -2 - 2i\sqrt{3}$$

31. cis 30°

$$= (\cos 30° + i \sin 30°)$$

$$= \frac{\sqrt{3}}{2} + \frac{1}{2}i$$

33. 5 cis 300°

$$= 5(\cos 300° + i \sin 300°)$$

$$= 5\left[\frac{1}{2} + i\left(-\frac{\sqrt{3}}{2}\right)\right]$$

$$= \frac{5}{2} - \frac{5\sqrt{3}}{2}i$$

35. $\sqrt{2}$ cis 180°

$$= \sqrt{2}(\cos 180° + i \sin 180°)$$

$$= \sqrt{2}(-1 + i \cdot 0)$$

$$= -\sqrt{2} + 0i$$

$$= -\sqrt{2}$$

37. $3 - 3i = x + yi$

Find r and θ.

$x = 3, \quad y = -3$

$r = \sqrt{x^2 + y^2}$

$\quad = \sqrt{3^2 + (-3)^2}$

$\quad = \sqrt{18}$

$\quad = 3\sqrt{2}$

$\tan \theta = \frac{y}{x} = \frac{-3}{3} = -1$

$3 - 3i$ is in quadrant IV, so

$\quad \theta = 315°.$

$3 - 3i = 3\sqrt{2}(\cos 315° + i \sin 315°)$

39. $-3 - 3i\sqrt{3}$

$x = -3, \quad y = -3\sqrt{3}$

$r = \sqrt{(-3)^2 + (-3\sqrt{3})^2} = 6$

$\tan \theta = \frac{-3\sqrt{3}}{-3} = \sqrt{3}$

$-3 - 3i\sqrt{3}$ is in quadrant III, so

$\quad \theta = 240°.$

$-3 - 3i\sqrt{3} = 6(\cos 240° + i \sin 240°)$

41. $\sqrt{3} - i$

$x = \sqrt{3}, \quad y = -1$

$r = \sqrt{(\sqrt{3})^2 + (-1)^2} = 2$

$\tan \theta = \frac{-1}{\sqrt{3}} = -\frac{\sqrt{3}}{3}$

$\sqrt{3} - i$ is in quadrant IV, so

$\quad \theta = 330°.$

$\sqrt{3} - i = 2(\cos 330° + i \sin 330°)$

43. $-5 - 5i$

$x = -5, \quad y = -5$

$r = \sqrt{(-5)^2 + (-5)^2}$

$\quad = \sqrt{50} = 5\sqrt{2}$

$\tan \theta = \frac{-5}{-5} = 1$

$-5 - 5i$ is in quadrant III, so

$\quad \theta = 225°.$

$-5 - 5i = 5\sqrt{2}(\cos 225° + i \sin 225°)$

45. $2 + 2i$

$x = 2, \quad y = 2$

$r = \sqrt{2^2 + 2^2} = \sqrt{8} = 2\sqrt{2}$

$\tan \theta = \frac{2}{2} = 1$

2 + 2i is in quadrant I, so

$\theta = 45°$.

$2 + 2i = 2\sqrt{2}(\cos 45° + i \sin 45°)$

47. $-4 = -4 + 0i$

$x = -4, y = 0$

$r = \sqrt{(-4)^2 + 0^2} = 4$

The graph of -4 is on the negative x-axis, so $\theta = 180°$.

$-4 = 4(\cos 180° + i \sin 180°)$

49. $-2i = 0 - 2i$

$x = 0, y = -2$

$r = \sqrt{0^2 + (-2)^2} = 2$

The graph of $-2i$ is on the negative y-axis, so $\theta = 270°$.

$-2i = 2(\cos 270° + i \sin 270°)$

51. $2 + 3i$

$x = 2, y = 3$

$r = \sqrt{2^2 + 3^2} = \sqrt{13}$

$\tan \theta = \dfrac{3}{2}$

$2 + 3i$ is in quadrant I, so

$\theta = 56.31°$.

$2 + 3i$

$= \sqrt{13}(\cos 56.31° + i \sin 56.31°)$

53. $3(\cos 250° + i \sin 250°)$

$= 3[-.34202014 + (-.93969262)i]$

$= -1.0260604 - 2.8190779i$

55. $12i = 0 + 12i$

$x = 0, y = 12$

$r = \sqrt{0^2 + 12^2} = 12$

$12i$ is on the positive y-axis, so $\theta = 90°$.

$12i = 12(\cos 90° + i \sin 90°)$

57. $3 + 5i$

$x = 3, y = 5$

$r = \sqrt{3^2 + 5^2} = \sqrt{34}$

$\tan \theta = \dfrac{5}{3}$

$3 + 5i$ is in quadrant I, so

$\theta = 59.04°$.

$3 + 5i$

$= \sqrt{34}(\cos 59.04° + i \sin 59.04°)$

59. Since the modulus represents the magnitude of the vector in the complex plane, $r = 1$ would represent a circle of radius one centered at the origin.

61. Since the real part of $z = x + yi$ is 1, the graph of $1 + yi$ would be the vertical line $x = 1$.

63.

Let $z = a + bi = r(\cos \theta + i \sin \theta)$.
The conjugate of $z = a - bi$. By looking at the graph of the vectors, the modulus is still r, but the angle is $-\theta$.
The conjugate would be

$$r(\cos (-\theta) + i \sin (-\theta)).$$

Since $\cos (-\theta) = \cos \theta$ and $\sin (-\theta) = -\sin \theta$,

$$r(\cos \theta + i(-\sin \theta))$$
$$= r(\cos \theta - i \sin \theta).$$

Since $\cos \theta = \cos (360° - \theta)$ and $\sin \theta = -\sin (360° - \theta)$,

$$r(\cos \theta - i \sin \theta)$$
$$= r(\cos (360° - \theta) - i(-\sin (360° - \theta)))$$
$$= r(\cos (360° - \theta) + i(\sin (360° - \theta))).$$

65. When $x + yi$ and $x - yi$ are drawn in the same plane, the resulting two vectors have the same length, either in the first and fourth quadrants or the second and third quadrants, each creating the same angle with itself and the closest part of the x-axis. When these two vectors are added, the result is a vector twice as long as the original, lying on the x-axis. Since this resultant lies on the x-axis, it has no imaginary part and would be a real number. (In fact, the sum would be $2x$.)

67.

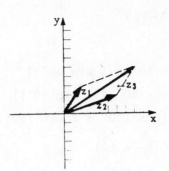

Notice that $z_1 + z_2 = z_3$. Recall the triangle inequality which states that the sum of any two sides of a triangle is greater than the third. For the modulus of the sum of two complex numbers to be equal to the sum of their moduli, we would be saying that the magnitude of two of the vectors shown in the figure would have to equal the third. The only way this can happen is if the vectors lie on the same line. Thus, the answer would have to be (C), the corresponding vectors have the same direction.

69. $z = -.2i$

$$z^2 - 1 = (-.2i)^2 - 1$$
$$= -.04 - 1$$
$$= -1.04$$

The modulus is 1.04.

$(z^2 - 1)^2 - 1 = (-1.04)^2 - 1 = .0816$
The modulus is .0816.

$[(z^2 - 1)^2 - 1]^2 - 1 = (.0816)^2 - 1$
$$= -.9933$$

The modulus is .9933.
The moduli do not exceed 2. Therefore, z is in the Julia set.

Section 8.3

3. $[3(\cos 60° + i \sin 60°)]$

    $\bullet [2(\cos 90° + i \sin 90°)]$

    $= 6(\cos 150° + i \sin 150°)$

    $= 6(-\dfrac{\sqrt{3}}{2} + i\dfrac{1}{2})$

    $= -3\sqrt{3} + 3i$

5. $[2(\cos 45° + i \sin 45°)]$

    $\bullet [2(\cos 225° + i \sin 225°)]$

    $= 4(\cos 270° + i \sin 270°)$

    $= 4(0 - i)$

    $= 0 - 4i$ or $-4i$

7. $[4(\cos 60° + i \sin 60°)]$

    $\bullet [6(\cos 330° + i \sin 330°)]$

    $= 24(\cos 390° + i \sin 390°)$

    $= 24(\cos 30° + i \sin 30°)$

    $= 24(\dfrac{\sqrt{3}}{2} + i\dfrac{1}{2})$

    $= 12\sqrt{3} + 12i$

9. $[5 \operatorname{cis} 90°][3 \operatorname{cis} 45°]$

    $= 15 \operatorname{cis} 135°$

    $= 15(\cos 135° + i \sin 135°)$

    $= 15(-\dfrac{\sqrt{2}}{2} + \dfrac{\sqrt{2}}{2}i)$

    $= -\dfrac{15\sqrt{2}}{2} + \dfrac{15\sqrt{2}}{2}i$

11. $[\sqrt{3} \operatorname{cis} 45°][\sqrt{3} \operatorname{cis} 225°]$

    $= 3 \operatorname{cis} 270°$

    $= 3(\cos 270° + i \sin 270°)$

    $= 3(0 - i)$

    $= 0 - 3i$ or $-3i$

13. $\dfrac{4(\cos 120° + i \sin 120°)}{2(\cos 150° + i \sin 150°)}$

    $= 2[\cos (-30°) + i \sin (-30°)]$

    $= 2(\dfrac{\sqrt{3}}{2} - \dfrac{1}{2}i)$

    $= \sqrt{3} - i$

15. $\dfrac{16(\cos 300° + i \sin 300°)}{8(\cos 60° + i \sin 60°)}$

    $= 2(\cos 240° + i \sin 240°)$

    $= 2(-\dfrac{1}{2} - \dfrac{\sqrt{3}}{2}i)$

    $= -1 - i\sqrt{3}$

17. $\dfrac{3 \operatorname{cis} 305°}{9 \operatorname{cis} 65°}$

    $= \dfrac{1}{3} \operatorname{cis} 240°$

    $= \dfrac{1}{3}(\cos 240° + i \sin 240°)$

    $= \dfrac{1}{3}(-\dfrac{1}{2} - \dfrac{\sqrt{3}}{2}i)$

    $= -\dfrac{1}{6} - \dfrac{\sqrt{3}}{6}i$

19. $\dfrac{8}{\sqrt{3} + i} = \dfrac{8(\sqrt{3} - i)}{(\sqrt{3} + i)(\sqrt{3} - i)}$

    $= \dfrac{8\sqrt{3} - 8i}{3 + 1} = \dfrac{8\sqrt{3} - 8i}{4}$

    $= 2\sqrt{3} - 2i$

21. $\dfrac{-i}{1 + i} = \dfrac{-i(1 - i)}{(1 + i)(1 - i)}$

    $= \dfrac{-i - 1}{1 + 1} = \dfrac{-1 - i}{2}$

    $= -\dfrac{1}{2} - \dfrac{1}{2}i$

23. $\dfrac{2\sqrt{6} - 2i\sqrt{2}}{\sqrt{2} - i\sqrt{6}}$

$= \dfrac{(2\sqrt{6} - 2i\sqrt{2})(\sqrt{2} + i\sqrt{6})}{(\sqrt{2} - i\sqrt{6})(\sqrt{2} + i\sqrt{6})}$

$= \dfrac{4\sqrt{3} + 12i - 4i + 4\sqrt{3}}{2 + 6}$

$= \dfrac{8\sqrt{3} + 8i}{8}$

$= \sqrt{3} + i$

25. $[2.5(\cos 35° + i \sin 35°)]$

$\cdot [3.0(\cos 50° + i \sin 50°)]$

$= 7.5(\cos 85° + i \sin 85°)$

$= .65366807 + 7.47146021i$

27. $(12 \text{ cis } 18.5°)(3 \text{ cis } 12.5°)$

$= 36 \text{ cis } 31°$

$= 36(\cos 31° + i \sin 31°)$

$= 30.858023 + 18.5413711i$

29. $\dfrac{45(\cos 127° + i \sin 127°)}{22.5(\cos 43° + i \sin 43°)}$

$= 2(\cos 84° + i \sin 84°)$

$= .20905693 + 1.9890438i$

31. $\left[2 \text{ cis } \dfrac{5\pi}{9}\right]^2$

$= \left[2 \text{ cis } \dfrac{5\pi}{9}\right]\left[2 \text{ cis } \dfrac{5\pi}{9}\right]$

$= 4 \text{ cis } \dfrac{10\pi}{9}$

$= 4\left(\cos \dfrac{10\pi}{9} + i \sin \dfrac{10\pi}{9}\right)$

$= -3.7587705 - 1.3680806i$

In Exercises 33-39,

$w = -1 + i$ and $z = -1 - i.$

33. $w \cdot z = (-1 + i)(-1 - i)$

$= (-1)(-1) + (-1)(-i)$

$+ (i)(-1) + (i)(-i)$

$= 1 + i - i - i^2$

$= 1 - (-1) = 2$

34. $w = -1 + i$

$r = \sqrt{(-1)^2 + 1^2} = \sqrt{2}$

$\tan \theta = \dfrac{1}{-1} = -1$

$\theta = 135°$

$w = \sqrt{2} \text{ cis } 135°$

$z = -1 - i$

$r = \sqrt{(-1)^2 + (-1)^2} = \sqrt{2}$

$\tan \theta = \dfrac{-1}{-1} = 1$

$\theta = 225°$

$z = \sqrt{2} \text{ cis } 225°$

35. $w \cdot z = (\sqrt{2} \text{ cis } 135°)(\sqrt{2} \text{ cis } 225°)$

$= (\sqrt{2})(\sqrt{2}) \text{ cis } (135° + 225°)$

$= 2 \text{ cis } 360°$

$= 2 \text{ cis } 0°$

36. $2 \text{ cis } 0° = 2(\cos 0° + i \sin 0°)$

$= 2(1 + i(0))$

$= 2(1)$

$= 2$

The result here is the same as the result in Exercise 33.

37. $\dfrac{w}{z} = \dfrac{-1 + i}{-1 - i}$

$= \dfrac{-1 + i}{-1 - i} \cdot \dfrac{-1 + i}{-1 + i}$

$= \dfrac{1 - i - i + i^2}{1 - i^2}$

$= \dfrac{-2i}{2}$

$= -i$

38. $\dfrac{w}{z} = \dfrac{\sqrt{2}\ \text{cis } 135°}{\sqrt{2}\ \text{cis } 225°}$

$= \dfrac{\sqrt{2}}{\sqrt{2}}\ \text{cis } (135° - 225°)$

$= \text{cis } (-90°)$

39. $\text{cis } (-90°)$

$= \cos (-90°) + i \sin (-90°)$

$= 0 + i(-1)$

$= -i$

The result here is the same as the result in Exercise 37.

41. $[2(\cos 45° + i \sin 45°)] \cdot$
$[5(\cos 90° + i \sin 90°)]$ and
$[2(\cos (-315°) + i \sin (-315°))] \cdot$
$[5(\cos (-270°) + i \sin (-270°))]$
are the same because 45° and -315° are coterminal angles and 90° and -270° are coterminal angles. Thus, the factors of the two products are the same numbers.

43. $E = 8(\cos 20° + i \sin 20°)$, $R = 6$,
$X_L = 3$

$I = \dfrac{E}{Z}$, $Z = R + X_L i$

Write $Z = 6 + 3i$ in trigonometric form.

$x = 6$, $y = 3$, so $r = \sqrt{6^2 + 3^2} = \sqrt{45}$.

$\tan \theta = \dfrac{3}{6} = \dfrac{1}{2}$, so $\theta = 26.6°$

$Z = \sqrt{45}\ \text{cis } 26.6°$

$I = \dfrac{8\ \text{cis } 20°}{\sqrt{45}\ \text{cis } 26.6°}$

$= 1.19\ \text{cis}(-6.6°)$

$= 1.2 - .14i$

45. (a) Since $Z_1 = 50 + 25i$ and $Z_2 = 60 + 20i$, it follows that

$\dfrac{1}{Z_1} = \dfrac{1}{50 + 25i} \cdot \dfrac{50 - 25i}{50 - 25i} = \dfrac{50 - 25i}{3125}$

$= \dfrac{2}{125} - \dfrac{1}{125}i$;

$\dfrac{1}{Z_2} = \dfrac{1}{60 + 20i} \cdot \dfrac{60 - 20i}{60 - 20i} = \dfrac{60 - 20i}{4000}$

$= \dfrac{3}{200} - \dfrac{1}{200}i$.

$\dfrac{1}{Z_1} + \dfrac{1}{Z_2}$

$= \left[\dfrac{2}{125} - \dfrac{1}{125}i\right] + \left[\dfrac{3}{200} - \dfrac{1}{200}i\right]$

$= \dfrac{31}{1000} - \dfrac{13}{1000}i$

$Z = \dfrac{1}{\dfrac{1}{Z_1} + \dfrac{1}{Z_2}}$

$= \dfrac{1}{\dfrac{31}{1000} - \dfrac{13}{1000}i}$

$= \dfrac{1000}{31 - 13i} \cdot \dfrac{31 + 13i}{31 + 13i}$

$= \dfrac{31,000 + 13,000i}{1130}$

$= \dfrac{3100}{113} + \dfrac{1300}{113}i$

$\approx 27.43 + 11.5i$

(b) $\tan \theta = \dfrac{1300/113}{3100/113}$

$\theta = \tan^{-1} \dfrac{1300}{3100} \approx 22.75°$

Section 8.4

1. $[3(\cos 30° + i \sin 30°)]^3$
    $= 3^3[\cos (3)(30°) + i \sin (3)(30°)]$
    $= 27(\cos 90° + i \sin 90°)$
    $= 0 + 27i$ or $27i$

3. $(\cos 45° + i \sin 45°)^8$
    $= [\cos (45°)(8) + i \sin (8)(45°)]$
    $= 1 + 0i$ or 1

5. $[3 \text{ cis } 100°]^3$
    $= 3^3 \text{ cis}(3 \cdot 100°)$
    $= 27 \text{ cis } 300°$
    $= 27(\cos 300° + i \sin 300°)$
    $= 27(\frac{1}{2} - \frac{\sqrt{3}}{2}i)$
    $= \frac{27}{2} - \frac{27\sqrt{3}}{2}i$

7. $(\sqrt{3} + i)^5$
    $x = \sqrt{3}, y = 1$
    $r = \sqrt{(\sqrt{3})^2 + 1^2} = 2$
    $\tan \theta = \frac{1}{\sqrt{3}}$ so $\theta = 30°$.
    $(\sqrt{3} + i)^5$
    $= [2(\cos 30° + i \sin 30°)]^5$
    $= 2^5(\cos 150° + i \sin 150°)$
    $= 32(-\frac{\sqrt{3}}{2} + i\frac{1}{2})$
    $= -16\sqrt{3} + 16i$

9. $(2 - 2i\sqrt{3})^4$
    $r = \sqrt{2^2 + (-2\sqrt{3})^2} = 4$
    $\tan \theta = \frac{-2\sqrt{3}}{2} = -\sqrt{3}$ so $\theta = 300°$.

$(2 - 2i\sqrt{3})^4$
$= [4(\cos 300° + i \sin 300°)]^4$
$= 4^4(\cos 1200° + i \sin 1200°)$
$= 256(-\frac{1}{2} + \frac{i\sqrt{3}}{2})$
$= -128 + 128i\sqrt{3}$

11. $(-2 - 2i)^5$
    $r = \sqrt{(-2)^2 + (-2)^2}$
    $= \sqrt{8} = 2\sqrt{2}$
    $\tan \theta = \frac{-2}{-2} = 1$ so $\theta = 225°$.

$(-2 - 2i)^5$
$= (2\sqrt{2})^5(\cos 225° + i \sin 225°)^5$
$= 32\sqrt{32}(\cos 1125° + i \sin 1125°)$
$= 128\sqrt{2}(\cos 45° + i \sin 45°)$
$= 128\sqrt{2}(\frac{\sqrt{2}}{2} + \frac{\sqrt{2}}{2}i)$
$= 128 + 128i$

For Exercises 13–29, see the answer graphs in the back of the textbook.

13. $(\cos 0° + i \sin 0°)$
    $= 1(\cos 0° + i \sin 0°)$
    $r = 1, \theta = 0°, n = 3$
    $r^{1/3} = 1^{1/3} = 1$
    $\alpha = \frac{\theta + 360° \cdot k}{n}$
    $= \frac{0° + 360° \cdot k}{3}$
    If $k = 0$, $\alpha = \frac{0° + 360° \cdot 0}{3} = 0°$.
    If $k = 1$, $\alpha = \frac{0° + 360° \cdot 1}{3} = 120°$.
    If $k = 2$, $\alpha = \frac{0° + 360° \cdot 2}{3} = 240°$.

So the cube roots are

$(\cos 0° + i \sin 0°)$,

$(\cos 120° + i \sin 120°)$,

$(\cos 240° + i \sin 240°)$.

So the cube roots are

$4(\cos 60° + i \sin 60°)$,

$4(\cos 180° + i \sin 180°)$,

$4(\cos 300° + i \sin 300°)$.

15. 8 cis 60°

$r = 8$, $\theta = 60°$, $n = 3$

$r^{1/3} = 8^{1/3} = 2$

$\alpha = \dfrac{60° + 360 \cdot k}{3}$; $k = 0, 1, 2$

$\alpha = 20°, 140°, 260°$

So the cube roots are

2 cis 20°,

2 cis 140°,

2 cis 260°.

17. $-8i = 8(\cos 270° + i \sin 270°)$

$r = 8$, $\theta = 270°$, $n = 3$

$r^{1/3} = 8^{1/3} = 2$

$\alpha = \dfrac{270° + 360° \cdot k}{3}$; $k = 0, 1, 2$

$\alpha = 90°, 210°, 330°$

So the cube roots are

$2(\cos 90° + i \sin 90°)$,

$2(\cos 210° + i \sin 210°)$,

$2(\cos 330° + i \sin 330°)$.

19. $-64 = 64(\cos 180° + i \sin 180°)$

$r = 64$, $\theta = 180°$, $n = 3$

$r^{1/3} = 64^{1/3} = 4$

$\alpha = \dfrac{180° + 360° \cdot k}{3}$; $k = 0, 1, 2$

$\alpha = 60°, 180°, 300°$

21. $1 + i\sqrt{3}$

$r = \sqrt{1^2 + (\sqrt{3})^2} = 2$

$\tan \theta = \dfrac{\sqrt{3}}{1}$ so $\theta = 60°$.

$1 + i\sqrt{3} = 2(\cos 60° + i \sin 60°)$

$r^{1/3} = 2^{1/3} = \sqrt[3]{2}$

$\alpha = \dfrac{60° + 360° \cdot k}{3}$; $k = 0, 1, 2$

$\alpha = 20°, 140°, 260°$

The cube roots are

$\sqrt[3]{2}(\cos 20° + i \sin 20°)$,

$\sqrt[3]{2}(\cos 140° + i \sin 140°)$,

$\sqrt[3]{2}(\cos 260° + i \sin 260°)$.

23. $-2\sqrt{3} + 2i$

$r = \sqrt{(-2\sqrt{3})^2 + 2^2} = 4$

$\tan \theta = \dfrac{2}{-2\sqrt{3}}$ so $\theta = 150°$.

$-2\sqrt{3} + 2i = 4(\cos 150° + i \sin 150°)$

$r^{1/3} = 4^{1/3} = \sqrt[3]{4}$

$\alpha = \dfrac{150° + 360° \cdot k}{3}$; $k = 0, 1, 2$

$\alpha = 50°, 170°, 290°$

The cube roots are

$\sqrt[3]{4}(\cos 50° + i \sin 50°)$,

$\sqrt[3]{4}(\cos 170° + i \sin 170°)$,

$\sqrt[3]{4}(\cos 290° + i \sin 290°)$.

25. Find all the second (or square) roots of 1.

$1 = 1(\cos 0° + i \sin 0°)$

$r = 1$, $\theta = 0°$, $n = 2$

$r^{1/2} = 1^{1/2} = 1$

$\alpha = \dfrac{0° + 360° \cdot k}{2}$; $k = 0, 1$

$\alpha = 0°, 180°$

The second roots of 1 are

$(\cos 0° + i \sin 0°)$,
$(\cos 180° + i \sin 180°)$.

27. Find all the sixth roots of 1.

$1 = 1(\cos 0° + i \sin 0°)$

$r = 1$, $\theta = 0°$, $n = 6$

$r^{1/6} = 1^{1/6} = 1$

$\alpha = \dfrac{0° + 360° \cdot k}{6}$;

$k = 0, 1, 2, 3, 4, 5$

$\alpha = 0°, 60°, 120°, 180°, 240°, 300°$

The sixth roots of 1 are

$(\cos 0° + i \sin 0°)$,
$(\cos 60° + i \sin 60°)$,
$(\cos 120° + i \sin 120°)$,
$(\cos 180° + i \sin 180°)$,
$(\cos 240° + i \sin 240°)$,
$(\cos 300° + i \sin 300°)$.

29. Find all the second (square) roots of i.

$i = \cos 90° + i \sin 90°$

$r = 1$, $\theta = 90°$, $n = 2$

$r^{1/2} = 1^{1/2} = 1$

$\alpha = \dfrac{90° + 360° \cdot k}{2}$; $k = 0, 1$

$\alpha = 45°, 225°$

The second roots of i are

$(\cos 45° + i \sin 45°)$,
$(\cos 225° + i \sin 225°)$.

31. If a is any real number, then a will have one real nth root if n is odd for the following reasons.
If a is positive, $\sqrt[n]{a}$ for n odd will be a positive real number.
If a is zero, $\sqrt[n]{a} = 0$ for n odd.
If a is negative, $\sqrt[n]{a}$ for n odd will be a negative real number.
In any case, a will have a real nth root if n is odd.

33. The statement, "Every real number must have two real square roots," is false. Consider, for example, the real number −4. Its two square roots are 2i and −2i which are not real.

39. $x^3 - 1 = 0$

$\quad x^3 = 1$

$\quad\quad = 1 + 0i$

$\quad\quad = 1(\cos 0° + i \sin 0°)$

The modulus of the solutions is

$1^{1/3} = 1$.

The arguments of the solutions are

$\alpha = \dfrac{0° + 360° \cdot k}{3}$; $k = 0, 1, 2$.

If $k = 0$, $\dfrac{0° + 360° \cdot 0}{3} = 0°$.

If $k = 1$, $\dfrac{0° + 360° \cdot 1}{3} = 120°$.

If $k = 2$, $\dfrac{0° + 360° \cdot 2}{3} = 240°$.

x = (cos 0° + i sin 0°),

(cos 120° + i sin 120°),

(cos 240° + i sin 240°)

41. $x^3 + i = 0$

$x^3 = -i$

$= 0 - i$

$= 1(\cos 270° + i \sin 270°)$

$1^{1/3} = 1$

$\alpha = \dfrac{270° + 360° \cdot k}{3}$; k = 0, 1, 2

$\alpha = 90°, 210°, 330°$

x = (cos 90° + i sin 90°),

(cos 210° + i sin 210°),

(cos 330° + i sin 330°)

43. $x^3 - 8 = 0$

$x^3 = 8$

$= 8 + 0i$

$= 8(\cos 0° + i \sin 0°)$

$8^{1/3} = 2$

$\alpha = \dfrac{0° + 360° \cdot k}{3}$; k = 0, 1, 2

$\alpha = 0°, 120°, 240°$

x = 2(cos 0° + i sin 0°),

2(cos 120° + i sin 120°),

2(cos 240° + i sin 240°)

45. $x^4 + 1 = 0$

$x^4 = -1$

$= -1 + 0i$

$= 1(\cos 180° + i \sin 180°)$

$1^{1/4} = 1$

$\alpha = \dfrac{180° + 360° \cdot k}{4}$; k = 0, 1, 2, 3

$\alpha = 45°, 135°, 225°, 315°$

x = (cos 45° + i sin 45°),

(cos 135° + i sin 135°),

(cos 225° + i sin 225°),

(cos 315° + i sin 315°)

47. $x^4 - i = 0$

$x^4 = i$

$= 1 + i$

$= 1(\cos 90° + i \sin 90°)$

$1^{1/4} = 1$

$\alpha = \dfrac{90° + 360° \cdot k}{4}$; k = 0, 1, 2, 3

$\alpha = 22.5°, 112.5°, 202.5°, 292.5°$

x = (cos 22.5° + i sin 22.5°),

(cos 112.5° + i sin 112.5°),

(cos 202.5° + i sin 202.5°),

(cos 292.5° + i sin 292.5°)

49. $x^3 - (4 + 4i\sqrt{3}) = 0$

$x^3 = 4 + 4i\sqrt{3}$

$r = \sqrt{(4)^2 + (4\sqrt{3})^2} = 8$

$\tan \theta = \dfrac{4\sqrt{3}}{4}$; $\theta = 60°$

$x^3 = 8(\cos 60° + i \sin 60°)$

$8^{1/3} = 2$

$\alpha = \dfrac{60° + 360° \cdot k}{3}$; k = 0, 1, 2

$\alpha = 20°, 140°, 260°$

x = 2(cos 20° + i sin 20°),

2(cos 140° + i sin 140°),

2(cos 260° + i sin 260°)

51. $x^3 + 4 - 5i = 0$

$x^3 = -4 + 5i$

$r = \sqrt{(-4)^2 + 5^2} = \sqrt{41}$

$\tan \theta = -\dfrac{5}{4}$, θ is in quadrant II.

$\theta = 128.65981°$

$x^3 = \sqrt{41}(\cos 128.65981°$
$+ i \sin 128.65981°)$

$(\sqrt{41})^{1/3} = 1.8569376$

$\alpha = \dfrac{\theta + 360° \cdot k}{3}$; $k = 0, 1, 2$

If $k = 0$, $\alpha = \dfrac{128.65981° + 0 \cdot 360°}{3}$

$= 42.886603°$.

If $k = 1$, $\alpha = \dfrac{128.65981° + 1 \cdot 360°}{3}$

$= 162.8866°$.

If $k = 2$, $\alpha = \dfrac{128.65981° + 2 \cdot 360°}{3}$

$= 282.8866°$.

$x = 1.8569376$

$\cdot (\cos 42.886603° + i \sin 42.886603°)$,

1.8569376

$\cdot (\cos 162.8866° + i \sin 162.8866°)$,

1.8569376

$\cdot (\cos 282.8866° + i \sin 282.8866°)$

In standard form,

$x = 1.3606 + 1.2637i$,

$-1.7747 + .5464i$,

$.4141 - 1.8102i$.

53. $x^3 - 1 = 0$

$(x - 1)(x^2 + x + 1) = 0$

$x - 1 = 0$

$x = 1$

or

$x^2 + x + 1 = 0$

Use the quadratic formula with
$a = 1$, $b = 1$, and $c = 1$.

$x = \dfrac{-1 \pm \sqrt{1^2 - 4 \cdot 1 \cdot 1}}{2 \cdot 1}$

$x = \dfrac{-1 \pm \sqrt{-3}}{2}$

$x = \dfrac{-1 \pm i\sqrt{3}}{2}$

$x = -\dfrac{1}{2} \pm \dfrac{\sqrt{3}}{2}i$

$x = 1, -\dfrac{1}{2} + \dfrac{\sqrt{3}}{2}i, -\dfrac{1}{2} - \dfrac{\sqrt{3}}{2}i$

We see that the solutions are the
same as Exercise 39.

For Exercise 39, the solutions are

$(\cos 0° + i \sin 0°)$

$(\cos 120° + i \sin 120°)$

$(\cos 240° + i \sin 240°)$.

In standard form, these are

$1, -\dfrac{1}{2} + \dfrac{\sqrt{3}}{2}i, -\dfrac{1}{2} - \dfrac{\sqrt{3}}{2}i$.

55. $2 + 2\sqrt{3}i$ is one cube root.

$r = \sqrt{2^2 + (2\sqrt{3})^2} = \sqrt{4 + 12} = 4$

$\tan \theta = \dfrac{2\sqrt{3}}{2} = \sqrt{3}$; $\theta = 60°$

This root is

$4 \text{ cis } 60° = 4 \text{ cis } (60° \cdot 1)$.

Since the graphs of the other roots
must be equally spaced around a
circle and the graphs of these roots
are all on a circle that has center
at the origin and radius 4, the
other roots are

$4 \text{ cis } (60° + 120°) = 4 \text{ cis } 180°$

and

$4 \text{ cis } (60° + 2 \cdot 120°) = 4 \text{ cis } 300°$.

4 cis 180° = 4(cos 180° + i sin 180°)

$\qquad$ = 4(-1 + i · 0) = -4

4 cis 300° = 4(cos 300° + i sin 300°)

$\qquad = 4\left(\dfrac{1}{2} + i\left(-\dfrac{\sqrt{3}}{2}\right)\right)$

$\qquad = 2 - 2i\sqrt{3}$

The roots are $2 - 2i\sqrt{3}$, -4, and $2 + 2i\sqrt{3}$.

56. $(\cos\theta + i\sin\theta)^2$

$\qquad = 1^2(\cos 2\theta + i\sin 2\theta)$

$\qquad = \cos 2\theta + i\sin 2\theta$

57. $(\cos\theta + i\sin\theta)^2$

$\qquad = \cos^2\theta + 2i\sin\theta\cos\theta$

$\qquad\quad + i^2\sin^2\theta$

$\qquad = \cos^2\theta + 2i\sin\theta\cos\theta - \sin^2\theta$

$\qquad = (\cos^2\theta - \sin^2\theta)$

$\qquad\quad + i(2\sin\theta\cos\theta)$

$\qquad = \cos 2\theta + i\sin 2\theta$

58. Two complex numbers a + bi and c + di are equal only if a = c and b = d. Therefore,

$\qquad \cos^2\theta - \sin^2\theta = \cos 2\theta$

and

$\qquad 2\sin\theta\cos\theta = \sin 2\theta.$

59. By tracing the pentagon, the following roots are found:

$\qquad$ 1,

$\qquad$ .30901699 + .95105652i,

$\qquad$ -.809017 + .587785251,

$\qquad$ -.809017 - .58778531,

$\qquad$ .30901699 - .95105565i.

61. **(a)** If z = 0 + 0i, we must make the following calculations:

$\qquad 0^2 + 0 = 0;\ 0^2 + 0 = 0.$

The calculations repeat as 0, 0, 0, ..., and will never exceed a modulus of 2. The point (0, 0) is part of the Mandelbrot set. The pixel at the origin should be turned on.

(b) If z = 1 - 1i,

$\qquad (1 - i)^2 + (1 - i) = 1 - 3i.$

The modulus of 1 - 3i is $\sqrt{10} > 2$. Therefore, 1 - 1i is not part of the Mandelbrot set, and the pixel at (1, -1) should be left off.

(c) If z = -.5i,

$(-.5i)^2 - .5i = -.25 - .5i;$

$(-.25 - .5i)^2 + (-.25 - .5i)$

$\quad = -.4375 - .25i;$

$(-.4375 - .25i)^2 + (-.4375 - .25i)$

$\quad = -.308593 - .03125i;$

$(-.308593 - .03125i)^2$

$\ + (-.308593 - .03125i)$

$\quad = -.214339 - .0119629i;$

$(-.214339 - .0119629i)^2$

$\ + (-.214339 - .0119629i)$

$\quad = -.16854 - .006834466i.$

This sequence appears to be approaching the origin, and no number has a modulus greater than 2. Thus, -.5i is part of the Mandelbrot set, and the pixel at (0, -.5) should be turned on.

Section 8.5

For Exercises 3-25, see the answer graphs in the back of the textbook (except for Exercise 13). Answers may vary for Exercises 3-11.

3. Two other pairs of polar coordinates for (1, 45°) are (1, 405°) and (-1, 225°).

5. Two other pairs of polar coordinates for (-2, 135°) are (-2, 495°) and (2, 315°).

7. Two other pairs of polar coordinates for (5, -60°) are (5, 300°) and (-5, 120°).

9. Two other pairs of polar coordinates for (-3, -210°) are (-3, 150°) and (3, -30°).

11. Two other pairs of polar coordinates for (3, 300°) are (3, 660°) and (-3, 120°).

13. If point P lies on an axis in the Cartesian plane, then θ must be a quadrantal angle if (r, θ) represents the point P in polar coordinates.

15. $r = 2 + 2 \cos \theta$ (cardioid)

θ	0°	30°	60°	90°	120°	150°
$\cos \theta$	1	.9	.5	0	-.5	-.9
r	4	3.8	3	2	1	.2

θ	180°	210°	240°	270°	300°	330°
$\cos \theta$	-1	-.9	-.5	0	.5	.9
r	0	.3	1	2	3	3.7

17. $r = 3 + \cos \theta$ (limaçon)

θ	0°	30°	60°	90°	120°	150°
r	4	3.9	3.5	3	2.5	2.1

θ	180°	210°	240°	270°	300°	330°
r	2	2.1	2.5	3	3.5	3.9

19. $r = 4 \cos 2\theta$ (four-leaved rose)

θ	0°	30°	45°	60°	90°	120°	135°	150°
r	4	2	0	-2	-4	-2	0	2

θ	180°	210°	225°	240°	270°	300°	315°	330°
r	4	2	0	-2	-4	-2	0	2

21. $r^2 = 4 \cos 2\theta$ (lemniscate)
$r = \pm 2\sqrt{\cos 2\theta}$
Graph only exists for [0°, 45°], [135°, 225°], and [315°, 360°] because $\cos 2\theta$ must be positive.

θ	0°	30°	45°	135°	150°
r	±2	±1.4	0	0	±1.4

θ	180°	210°	225°	315°	330°
r	±2	±1.4	0	0	±1.4

23. $r = 4(1 - \cos \theta)$ (cardioid)

θ	0°	30°	60°	90°	120°	150°
r	0	.5	2	4	6	7.5

θ	180°	210°	240°	270°	300°	330°
r	8	7.5	6	4	2	.5

25. $r = 2 \sin \theta \tan \theta$ (cissoid)

r is undefined at $\theta = 90°$ and $\theta = 270°$.

θ	0°	30°	45°	60°	90°	120°	135°	150°	180°
r	0	.6	1.4	3	–	-3	-1.4	-.6	0

Notice that for [180°, 360°), the graph retraces the path traced for [0°, 180°).

27. Suppose (r, θ) is on a graph. $-\theta$ reflects this point with respect to the x-axis, and $-r$ reflects the resulting point with respect to the origin. The net result is that the original point is reflected with respect to the y-axis (B).

29. If (r, θ) lies on a graph, $(r, -\theta)$ would reflect that point across the x-axis. Therefore, there is symmetry about the x-axis (C).

33. $r = 4 \cos 2\theta$, $0° \le \theta < 360°$

Since the largest value of $\cos 2\theta$ is 1 (the range of the cosine function is [-1, 1]), the largest value of r is $4 \cdot 1 = 4$.

$4 \cos 2\theta = 0$

$\cos 2\theta = 0$

Since $0° \le \theta < 360°$, $0° \le 2\theta < 720°$. The cosine is 0 at 90°, 270°, 450°, 630°. Thus,

$2\theta = 90°$ or $2\theta = 270°$

$\theta = 45°$ or $\theta = 135°$

$2\theta = 450°$ or $2\theta = 630°$

$\theta = 225°$ or $\theta = 315°$.

35. In the screens, R = 2.6666667, $\theta = 30°$,

$X = 2.3094011$, $Y = 1.3333333$.

$X = R \cos \theta$

$= 2.6666667 \cos 30°$

$= 2.3094011$

$Y = R \sin \theta$

$= 2.6666667 \sin 30°$

$= 1.3333333$

37. $r = 2 + \sin \theta$, $r = 2 + \cos \theta$,

$0 \le \theta < 2\pi$

$2 + \sin \theta = 2 + \cos \theta$

$\sin \theta = \cos \theta$

$\theta = \dfrac{\pi}{4}$ or $\dfrac{5\pi}{4}$

$r = 2 + \sin \dfrac{\pi}{4} = 2 + \dfrac{\sqrt{2}}{2} = \dfrac{4 + \sqrt{2}}{2}$

$r = 2 + \sin \dfrac{5\pi}{4} = 2 - \dfrac{\sqrt{2}}{2} = \dfrac{4 - \sqrt{2}}{2}$

The points of intersection are

$\left(\dfrac{4 + \sqrt{2}}{2}, \dfrac{\pi}{4}\right)$, $\left(\dfrac{4 - \sqrt{2}}{2}, \dfrac{5\pi}{4}\right)$.

39. To plot a point (r, θ) in polar coordinates with $r < 0$, move $|r|$ units in the direction opposite the ray for θ.

For Exercises 41–49, see the answer graphs in the back of the textbook.

41. $r = 2 \sin \theta$

Multiply both sides by r.

$$r^2 = 2r \sin \theta$$

Since $r^2 = x^2 + y^2$ and $y = r \sin \theta$, $x^2 + y^2 = 2y$.

Complete the square on y.

$$x^2 + y^2 - 2y + 1 = 1$$
$$x^2 + (y - 1)^2 = 1$$

The graph is a circle with center at (0, 1) and radius 1.

43. $r = \dfrac{2}{1 - \cos \theta}$

Multiply both sides by $1 - \cos \theta$.

$r - r \cos \theta = 2$

$$\sqrt{x^2 + y^2} - x = 2$$
$$\sqrt{x^2 + y^2} = 2 + x$$
$$x^2 + y^2 = (2 + x)^2$$
$$x^2 + y^2 = 4 + 4x + x^2$$
$$y^2 = 4(1 + x)$$

The graph is a parabola with vertex at (-1, 0) and axis y = 0.

45.
$$r + 2 \cos \theta = -2 \sin \theta$$
$$r^2 = -2r \sin \theta - 2r \cos \theta$$
$$x^2 + y^2 = -2y - 2x$$
$$x^2 + y^2 + 2y + 2x = 0$$
$$(x + 1)^2 + (y + 1)^2 = 2$$

The graph is a circle with center (-1, -1) and radius $\sqrt{2}$.

47.
$$r = 2 \sec \theta$$
$$r = \dfrac{2}{\cos \theta}$$
$$r \cos \theta = 2$$
$$x = 2$$

The graph is a vertical line.

49.
$$r(\cos \theta + \sin \theta) = 2$$
$$r \cos \theta + r \sin \theta = 2$$
$$x + y = 2$$

The graph is the line with intercepts (0, 2) and (2, 0).

51.
$$x + y = 4$$
$$r \cos \theta + r \sin \theta = 4$$
$$r(\cos \theta + \sin \theta) = 4$$

53.
$$x^2 + y^2 = 16$$
$$r^2 = 16$$
$$r = 4$$

55.
$$y = 2$$
$$r \sin \theta = 2 \text{ or }$$
$$r = \dfrac{2}{\sin \theta} = 2 \csc \theta$$

57. Graph $r = \theta$, a spiral of Archimedes.

θ	-360°	-270°	-180°	-90°
θ (radians)	-6.3	-4.7	-3.1	-1.6
r	-6.3	-4.7	-3.1	-1.6

θ	0°	90°	180°	270°	360°
θ (radians)	0	1.6	3.1	4.7	6.3
r	0	1.6	3.1	4.7	6.3

See the answer graph at the back of the textbook.

59. (1, 0°); x = 1 cos 0°; y = 1 sin 0°

This is equivalent to the rectangular point (1, 0).

(2, 90°) is equivalent to the rectangular point (0, 2).

$$m = \frac{2 - 0}{0 - 1} = \frac{2}{-1} = -2$$

y = −2x + 2 is the rectangular equation of the line. Since x = r cos θ
and y = r sin θ,

$$r \sin \theta = -2(r \cos \theta) + 2$$

r sin θ + 2r cos θ = 2

r(sin θ + 2 cos θ) = 2

$$r = \frac{2}{\sin \theta + 2 \cos \theta}.$$

61. (a) Plot the following polar equations on the same polar axis:

Mercury: $r = \frac{.39(1 - .206^2)}{1 + .206 \cos \theta}$;

Venus: $r = \frac{.78(1 - .007^2)}{1 + .007 \cos \theta}$;

Earth: $r = \frac{1(1 - .017^2)}{1 + .017 \cos \theta}$;

Mars: $r = \frac{1.52(1 - .093^2)}{1 + .093 \cos \theta}$.

Use a viewing window of [−2.4, 2.4]
by [−1.6, 1.6] with Xscl = .2,
Yscl = .2. See the answer graph in
the back of the textbook.

(b) Plot the following polar equations on the same polar axis:

Earth: $r = \frac{1(1 - .017^2)}{1 + .017 \cos \theta}$;

Jupiter: $r = \frac{5.2(1 - .048^2)}{1 + .048 \cos \theta}$;

Uranus: $r = \frac{19.2(1 - .047^2)}{1 + .047 \cos \theta}$;

Pluto: $r = \frac{39.4(1 - .249^2)}{1 + .249 \cos \theta}$.

Use a viewing window of [−60, 60] by
[−40, 40] with Xscl = 10, Yscl = 10.
See the answer graph in the back of
the textbook. From the graph, Earth
is closest to the sun of these four
planets.

(c) We must determine if the orbit
of Pluto is always outside the
orbits of the other planets. Since
Neptune is closest to Pluto, plot
the orbits of Neptune and Pluto on
the same polar axes.

Neptune: $r = \frac{30.1(1 - .009^2)}{1 + .009 \cos \theta}$

Pluto: $r = \frac{39.4(1 - .249^2)}{1 + .249 \cos \theta}$

Use a viewing window of [−60, 60] by
[−40, 40] with Xscl = 10, Yscl = 10.

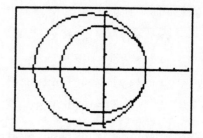

The graph shows that their orbits
are very close near the polar axis.
Use ZOOM to determine that the orbit
of Pluto does indeed pass inside the
orbit of Neptune.

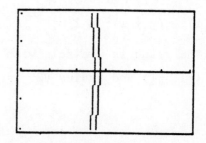

Therefore, there are times when Neptune, not Pluto, is the farthest planet from the sun. (However, Pluto's average distance from the sun is considerably greater than Neptune's average distance.)

t	$x = \cos t$	$y = -\sin t$
0	$\cos 0 = 1$	$-\sin 0 = 0$
$\dfrac{\pi}{6}$	$\cos \dfrac{\pi}{6} = \dfrac{\sqrt{3}}{2}$	$-\sin \dfrac{\pi}{6} = -\dfrac{1}{2}$
$\dfrac{\pi}{4}$	$\cos \dfrac{\pi}{4} = \dfrac{\sqrt{2}}{2}$	$-\sin \dfrac{\pi}{4} = -\dfrac{\sqrt{2}}{2}$
$\dfrac{\pi}{3}$	$\cos \dfrac{\pi}{3} = \dfrac{1}{2}$	$-\sin \dfrac{\pi}{3} = -\dfrac{\sqrt{3}}{2}$
$\dfrac{\pi}{2}$	$\cos \dfrac{\pi}{2} = 0$	$-\sin \dfrac{\pi}{2} = -1$

Section 8.6

1. $X = -2 + 6T$, $Y = 1 + 2T$, T in $[0, 1]$
    When $T = 0$,

    $X = -2 + 6(0) = -2$,
    $Y = 1 + 2(0) = 1$.

    The first point graphed is $(-2, 1)$.
    When $T = 1$,

    $X = -2 + 6(1) = 4$,
    $Y = 1 + 2(1) = 3$.

    The last point graphed is $(4, 3)$.
    When $T = .6$,

    $X = -2 + 6(.6) = 1.6$,
    $Y = 1 + 2(.6) = 2.2$.

    The point plotted when $T = .6$ is $(1.6, 2.2)$.

3. The second set of equations $x = \cos t$, $y = -\sin t$, t in $[0, 2\pi]$ trace the circle out clockwise. A table of values confirms this.

For Exercises 5–23, see the answer graphs in the back of the textbook.

5. $x = 2t$, $y = t + 1$, for t in $[-2, 3]$

t	$x = 2t$	$y = t + 1$
-2	$2(-2) = -4$	$-2 + 1 = -1$
-1	$2(-1) = -2$	$-1 + 1 = 0$
0	$2(0) = 0$	$0 + 1 = 1$
1	$2(1) = 2$	$1 + 1 = 2$
2	$2(2) = 4$	$2 + 1 = 3$
3	$2(3) = 6$	$3 + 1 = 4$

$x = 2t$, $y = t + 1$

Since $\dfrac{x}{2} = t$, $y = \dfrac{x}{2} + 1$.

Since t is in $[-2, 3]$, x is in $[2(-2), 2(3)]$ or $[-4, 6]$.

7. $x = \sqrt{t}$, $y = 3t - 4$, for t in $[0, 4]$

t	$x = \sqrt{t}$	$y = 3t - 4$
0	$\sqrt{0} = 0$	$3(0) - 4 = -4$
1	$\sqrt{1} = 1$	$3(1) - 4 = -1$
2	$\sqrt{2} = 1.4$	$3(2) - 4 = 2$
3	$\sqrt{3} = 1.7$	$3(3) - 4 = 5$
4	$\sqrt{4} = 2$	$3(4) - 4 = 8$

$$x = \sqrt{t} \quad y = 3t - 4$$

Since $x^2 = t$, $y = 3x^2 - 4$.

Since t is in $[0, 4]$, x is in $[\sqrt{0}, \sqrt{4}]$ or $[0, 2]$.

9. $x = t^3 + 1$, $y = t^3 - 1$, for t in $(-\infty, \infty)$

t	$x = t^3 + 1$	$y = t^3 - 1$
-2	$(-2)^3 + 1 = -7$	$(-2)^3 - 1 = -9$
-1	$(-1)^3 + 1 = 0$	$(-1)^3 - 1 = -2$
0	$0^3 + 1 = 1$	$0^3 - 1 = -1$
1	$1^3 + 1 = 2$	$1^3 - 1 = 0$
2	$2^3 + 1 = 9$	$2^3 - 1 = 7$
3	$3^3 + 1 = 28$	$3^3 - 1 = 26$

Since
$$x = t^3 + 1,$$
$$x - 1 = t^3.$$

Since
$$y = t^3 - 1,$$
$$y = (x - 1) - 1 = x - 2.$$

Since t is in $(-\infty, \infty)$, x is in $(-\infty, \infty)$.

11. $x = 2 \sin \theta$, $y = 2 \cos \theta$, for θ in $[0, 2\pi]$

t	$x = 2 \sin \theta$	$y = 2 \cos \theta$
0	$2 \sin 0 = 0$	$2 \cos \theta = 2$
$\frac{\pi}{6}$	$2 \sin \frac{\pi}{6} = 1$	$2 \cos \frac{\pi}{6} = \sqrt{3}$
$\frac{\pi}{4}$	$2 \sin \frac{\pi}{4} = \sqrt{2}$	$2 \cos \frac{\pi}{4} = \sqrt{2}$
$\frac{\pi}{3}$	$2 \sin \frac{\pi}{3} = \sqrt{3}$	$2 \cos \frac{\pi}{3} = 1$
$\frac{\pi}{2}$	$2 \sin \frac{\pi}{2} = 2$	$2 \cos \frac{\pi}{2} = 0$

(Table can be continued.)

Since

$x = 2 \sin \theta$ and $y = 2 \cos \theta$,

$\frac{x}{2} = \sin \theta$ and $\frac{y}{2} = \cos \theta$.

Since $\sin^2 \theta + \cos^2 \theta = 1$,

$$\left(\frac{x}{2}\right)^2 + \left(\frac{y}{2}\right)^2 = 1$$

$$\frac{x^2}{4} + \frac{y^2}{4} = 1$$

$$x^2 + y^2 = 4.$$

Since θ is in $[0, 2\pi]$, x is in $[-2, 2]$ because the graph is a circle, centered at the origin, with radius 2.

13. $x = 3 \tan \theta$, $y = 2 \sec \theta$, for θ in $(-\pi/2, \pi/2)$

t	$x = 3 \tan \theta$	$y = 2 \sec \theta$
$-\frac{\pi}{3}$	$3 \tan\left(-\frac{\pi}{3}\right) = -3\sqrt{3}$	$2 \sec\left(-\frac{\pi}{3}\right) = 4$
$-\frac{\pi}{6}$	$3 \tan\left(-\frac{\pi}{6}\right) = -\sqrt{3}$	$2 \sec\left(-\frac{\pi}{6}\right) = \frac{4\sqrt{3}}{3}$
0	$3 \tan 0 = 0$	$2 \sec 0 = 2$
$\frac{\pi}{6}$	$3 \tan \frac{\pi}{6} = \sqrt{3}$	$2 \sec \frac{\pi}{6} = \frac{4\sqrt{3}}{3}$
$\frac{\pi}{3}$	$3 \tan \frac{\pi}{3} = 3\sqrt{3}$	$2 \sec \frac{\pi}{3} = 4$

Since

$\frac{x}{3} = \tan \theta$ and $\frac{y}{2} = \sec \theta$,

and $1 + \tan^2 \theta = \sec^2 \theta$,

$$1 + \left(\frac{x}{3}\right)^2 = \left(\frac{y}{2}\right)^2$$

$$1 + \frac{x^2}{9} = \frac{y^2}{4}$$

$$y^2 = 4\left(1 + \frac{x^2}{9}\right)$$

$$y = 2\sqrt{1 + \frac{x^2}{9}}.$$

Since this graph is the top half of a hyperbola, x is in $(-\infty, \infty)$.

15. $x = \sin \theta$, $y = \csc \theta = \dfrac{1}{\sin \theta}$

Therefore,

$$y = \frac{1}{x}.$$

Since θ is in $(0, \pi)$, x is in $(\sin 0, \sin \pi)$ or $(0, 1]$.

17. Since $x = t$ and $y = \sqrt{t^2 + 2}$,

$$y = \sqrt{x^2 + 2}.$$

Since t is in $(-\infty, \infty)$ and $x = t$, x is in $(-\infty, \infty)$.

19. Since $x = 2 + \sin \theta$

and $y = 1 + \cos \theta$,

$$x - 2 = \sin \theta$$

and $y - 1 = \cos \theta$.

Since $\sin^2 \theta + \cos^2 \theta = 1$,

$$(x - 2)^2 + (y - 1)^2 = 1.$$

Since this is a circle centered at $(2, 1)$ with radius 1, and θ is in $[0, 2\pi]$, x is in $[1, 3]$.

21. Since $x = t + 2$ and $y = \dfrac{1}{t + 2}$,

$$y = \frac{1}{x}.$$

Since $t \neq -2$, $x \neq -2 + 2$, $x \neq 0$.

23. (a) $x = \sin t$, $y = \cos t$

This graph is a circle centered at the origin with radius 1.

(b) $x = t$, $y = \dfrac{\sqrt{4 - 4t^2}}{2}$

This graph is the semicircle formed by the top half of the graph of part (a).

25. $x = \theta - \sin \theta$, $y = 1 - \cos \theta$, θ in $[0, 4\pi]$

See the answer graph in the back of the textbook.

27. $x = (400 \cos 45°)t = 400 \cdot \dfrac{\sqrt{2}}{2}t$

$= 200\sqrt{2}t$

$y = (400 \sin 45°)t - 16t^2$

$= 200\sqrt{2}t - 16t^2$

(a) $200\sqrt{2}t - 16t^2 = 0$

$t(200\sqrt{2} - 16t) = 0$

$t = 0$ or $200\sqrt{2} - 16t = 0$

$200\sqrt{2} = 16t$

$t = \dfrac{200\sqrt{2}}{16}$

$= 17.7$ sec

The projectile strikes the ground in 17.7 sec.

(b) $x = 200\sqrt{2}(17.7) \approx 5000$ ft

(c) To find the maximum height, find the value of t which maximizes y.

$y = -16t^2 + 200\sqrt{2}t$

$= -16\left(t^2 - \dfrac{25\sqrt{2}}{2}t \qquad \right)$

$= -16\left(t^2 - \dfrac{25\sqrt{2}}{2}t + \dfrac{625}{8}\right) + 1250$

$y = -16\left(t - \dfrac{25\sqrt{2}}{4}\right)^2 + 1250$

The maximum height is the y-value of the vertex of this downward parabola, namely 1250 ft.

29.
$$x = (v_0 \cos \theta)t$$

$$\frac{x}{v_0 \cos \theta} = t$$

$$y = (v_0 \sin \theta)t - 16t^2$$

Substitute $\dfrac{x}{v_0 \cos \theta}$ for t.

$$y = (v_0 \sin \theta)\frac{x}{v_0 \cos \theta}$$

$$- 16\left(\frac{x}{v_0 \cos \theta}\right)^2$$

$$= x \tan \theta - \frac{16x^2}{v_0{}^2 \cos^2 \theta}$$

31. The equation of a line with slope m through (x_1, y_1) is

$$y - y_1 = m(x - x_1).$$

To find two parametric representations, let $x = t$. Then

$$y - y_1 = m(t - x_1)$$

$$y = m(t - x_1) + y_1.$$

Or, let $x = t^2$.

Then

$$y - y_1 = m(t^2 - x_1)$$

$$y = m(t^2 - x_1) + y_1.$$

33. $\dfrac{x^2}{a^2} - \dfrac{y^2}{b^2} = 1$

To find a parametric representation, let

$$x = a \sec \theta.$$

Then

$$\frac{(a \sec \theta)^2}{a^2} - \frac{y^2}{b^2} = 1$$

$$\sec^2 \theta - \frac{y^2}{b^2} = 1$$

$$\sec^2 \theta - 1 = \frac{y^2}{b^2}$$

$$\tan^2 \theta = \frac{y^2}{b^2}$$

$$b^2 \tan^2 \theta = y^2$$

$$b \tan \theta = y.$$

Or, let

$$x = t.$$

Then

$$\frac{t^2}{a^2} - \frac{y^2}{b^2} = 1$$

$$\frac{y^2}{b^2} = \frac{t^2}{a^2} - 1$$

$$y^2 = b^2\left(\frac{t^2}{a^2} - 1\right)$$

$$y^2 = \frac{b^2}{a^2}(t^2 - a^2).$$

35. $r^2 = x^2 + y^2$, $x = a\theta \cos \theta$,
$y = a\theta \sin \theta$

$$r^2 = (a\theta \cos \theta)^2 + (a\theta \sin \theta)^2$$
$$r^2 = a^2\theta^2 \cos^2 \theta + a^2\theta^2 \sin^2 \theta$$
$$r^2 = a^2\theta^2(\cos^2 \theta + \sin^2 \theta)$$
$$r^2 = a^2\theta^2$$
$$r = a\theta$$

Chapter 8   Review Exercises

1. $\sqrt{-9} = i\sqrt{9} = 3i$

3. $x^2 = -81$

   $x = \pm\sqrt{-81}$

   $x = \pm i\sqrt{81}$

   $x = \pm i(9)$

   $x = \pm 9i$

   The solutions are $9i$ and $-9i$.

5. $(1 - i) - (3 + 4i) + 2i$

   $= (1 - 3) + (-1 - 4 + 2)i$

   $= -2 - 3i$

7. $(6 - 5i) + (2 + 7i) - (3 - 2i)$

   $= (6 + 2 - 3) + (-5 + 7 + 2)i$

   $= 5 + 4i$

9. $(3 + 5i)(8 - i)$

   $= 24 - 3i + 40i - 5i^2$

   $= 24 + 37i - 5(-1)$

   $= 29 + 37i$

11. $(2 + 6i)^2 = 4 + 24i + 36i^2$

    $= 4 + 24i + 36(-1)$

    $= -32 + 24i$

13. $(1 - i)^3 = (1 - i)^2(1 - i)$

    $= (1 - 2i + i^2)(1 - i)$

    $= -2i(1 - i)$

    $= -2i + 2i^2$

    $= -2i - 2$

    or $-2 - 2i$

15. $\dfrac{6 + 2i}{3 - i} = \dfrac{6 + 2i}{3 - i} \cdot \dfrac{3 + i}{3 + i}$

    $= \dfrac{18 + 12i + 2i^2}{9 - i^2}$

    $= \dfrac{18 + 12i + 2(-1)}{9 - (-1)}$

    $= \dfrac{16 + 12i}{10}$

    $= \dfrac{8 + 6i}{5}$

    $= \dfrac{8}{5} + \dfrac{6}{5}i$

17. $\dfrac{2 + i}{1 - 5i} = \dfrac{2 + i}{1 - 5i} \cdot \dfrac{1 + 5i}{1 + 5i}$

    $= \dfrac{2 + 11i + 5i^2}{1 - 25i^2}$

    $= \dfrac{2 + 11i + 5(-1)}{1 - 25(-1)}$

    $= \dfrac{-3 + 11i}{26}$

    $= -\dfrac{3}{26} + \dfrac{11}{26}i$

19. $i^{53} = i^{52} \cdot i$

    $= (i^4)^{13}i$

    $= 1^{13}i$

    $= 1 \cdot i$

    $= i$

    or $0 + i$

21. $1 \cdot i \cdot i^2 \cdot i^3 = 1 \cdot i^6 = i^6 = i^4 \cdot i^2$

    $= (1)(-1) = -1$

23. $[5(\cos 90° + i \sin 90°)]$

    $\cdot [6(\cos 180° + i \sin 180°)]$

    $= 5 \cdot 6[\cos (90° + 180°)$

    $+ i \sin (90° + 180°)]$

    $= 30(\cos 270° + i \sin 270°)$

    $= 30(0 - i)$

    $= 0 - 30i$ or $-30i$

25. $\dfrac{2(\cos 60° + i \sin 60°)}{8(\cos 300° + i \sin 300°)}$

$\quad = \dfrac{1}{4}[\cos (-240°) + i \sin (-240°)]$

$\quad = -\dfrac{1}{8} + \dfrac{\sqrt{3}}{8}i$

27. $(\sqrt{3} + i)^3$

$\quad r = \sqrt{(\sqrt{3})^2 + 1^2} = 2$

$\quad \tan \theta = \dfrac{1}{\sqrt{3}} = \dfrac{\sqrt{3}}{3}, \ \theta = 30°$

$\quad (\sqrt{3} + i)^3 = [2(\cos 30° + i \sin 30°)]^3$

$\qquad\qquad = 2^3(\cos 90° + i \sin 90°)$

$\qquad\qquad = 0 + 8i \quad \text{or} \quad 8i$

29. $(\cos 100° + i \sin 100°)^6$

$\quad = \cos 600° + i \sin 600°$

$\quad = -\dfrac{1}{2} - \dfrac{\sqrt{3}}{2}i$

31. The vector representing a real number will lie on the x-axis in the complex plane.

For Exercises 33 and 35, see the answer graphs in the back of the textbook.

37. $7 + 3i$ and $-2 + i$

$\quad (7 + 3i) + (-2 + i) = 5 + 4i$

See the answer graph in the back of the textbook.

39. $-2 + 2i$

$\quad r = \sqrt{(-2)^2 + 2^2}$

$\quad\ = \sqrt{8} = 2\sqrt{2}$

$\quad \tan \theta = \dfrac{2}{-2} = -1$

θ is in quadrant II so $\theta = 135°$.

$-2 + 2i$

$\quad = 2\sqrt{2}(\cos 135° + i \sin 135°)$

41. $2(\cos 225° + i \sin 225°)$

$\quad = 2\left(-\dfrac{\sqrt{2}}{2} - \dfrac{i\sqrt{2}}{2}\right)$

$\quad = -\sqrt{2} - i\sqrt{2}$

43. $1 - i$

$\quad r = \sqrt{1^2 + (-1)^2} = \sqrt{2}$

$\quad \tan \theta = \dfrac{-1}{1} = -1$

θ is in quadrant IV so $\theta = 315°$.

$1 - i = \sqrt{2}(\cos 315° + i \sin 315°)$

45. $-4i$

$\quad r = 4$

$\quad \theta = 270°$

$-4i = 4(\cos 270° + i \sin 270°)$

47. Since the modulus of z is 2, the graph would be a circle, centered at the origin, with radius 2.

49. If we square $a + bi$, we get

$$a^2 + 2abi - b^2.$$

For this to be a real number, there cannot be an i-term, that is, $2abi$ must be 0. Therefore, either $a = 0$ or $b = 0$. If $b = 0$, this would mean $a + bi$ is a real complex number. Since it must be a nonreal complex number, a must be 0. Since $a = 0$, the vector (a, b) lies on the y-axis.

51. Find the fifth roots of $-2 + 2i$.

$r = \sqrt{(-2)^2 + 2^2} = \sqrt{8}$

$\tan \theta = \dfrac{2}{-2} = -1$

θ is in quadrant II so $\theta = 135°$.

$-2 + 2i = \sqrt{8}(\cos 135° + i \sin 135°)$

$(\sqrt{8})^{1/5} = (8^{1/2})^{1/5} = 8^{1/10} = \sqrt[10]{8}$

$\alpha = \dfrac{135° + 360 \cdot k}{5}$; $k = 0, 1, 2, 3, 4$

$\alpha = 27°, 99°, 171°, 243°, 315°$

The fifth roots are

$\sqrt[10]{8}(\cos 27° + i \sin 27°)$,

$\sqrt[10]{8}(\cos 99° + i \sin 99°)$,

$\sqrt[10]{8}(\cos 171° + i \sin 171°)$,

$\sqrt[10]{8}(\cos 243° + i \sin 243°)$,

$\sqrt[10]{8}(\cos 315° + i \sin 315°)$.

53. The real number -32 has one real fifth root. The one real fifth root is -2, and all other fifth roots are not real.

55. $x^3 + 125 = 0$

$x^3 = -125$

$-125 = -125 + 0i$

$-125 = 125(\cos 180° + i \sin 180°)$

$125^{1/3} = 5$

$\alpha = \dfrac{180° + 360° \cdot k}{3}$; $k = 0, 1, 2$

$\alpha = 60°, 180°, 300°$

$x = 5(\cos 60° + i \sin 60°)$,

$5(\cos 180° + i \sin 180°)$,

$5(\cos 300° + i \sin 300°)$

57. $x^2 + i = 0$

$x^2 = -i$

$= 0 - i$

$= 1(\cos 270° + i \sin 270°)$

$\sqrt[2]{1} = 1$

$\alpha = \dfrac{270° + 360° \cdot k}{2}$; $k = 0, 1$

$\alpha = 135°, 315°$

$x = (\cos 135° + i \sin 135°)$,

$(\cos 315° + i \sin 315°)$

For Exercise 59 and 61, see the answer graphs in the back of the textbook.

59. $r = -1 + \cos \theta$

θ	0°	30°	45°	60°	90°
r	0	-.7	-.3	-.5	0

θ	120°	135°	150°	180°	270°	315°
r	-1.5	-1.7	-1.9	-2	-1	-.3

61. $r = 2 \sin 4\theta$

θ	0°	7.5°	15°	22.5°	0°	37.5°	45°
r	0	1	$\sqrt{3}$	2	$\sqrt{3}$	1	0

θ	52.5°	60°	67.5°	75°	82.5°	90°
r	-1	$-\sqrt{3}$	-2	$-\sqrt{3}$	-1	0

The graph continues to form eight petals for the interval $[0°, 360°)$.

63. $r = \dfrac{3}{1 + \cos \theta}$

$r(1 + \cos \theta) = 3$

$r + r \cos \theta = 3$

$\sqrt{x^2 + y^2} + x = 3$

$$x^2 + y^2 = (3 - x)^2$$

$$x^2 + y^2 = 9 - 6x + x^2$$

$$y^2 = 9 - 6x$$

$$y^2 + 6x - 9 = 0$$

or $$y^2 = -6x + 9$$

$$y^2 = -6\left(x - \frac{3}{2}\right)$$

65. $$r = \sin \theta + \cos \theta$$

$$r^2 = r \sin \theta + r \cos \theta$$

$$x^2 + y^2 = x + y$$

$$x^2 + y^2 - x - y = 0$$

or

$$(x^2 - x \quad) + (y^2 - y \quad) = 0$$

$$\left(x^2 - x + \frac{1}{4}\right) + \left(y^2 - y + \frac{1}{4}\right) = \frac{1}{4} + \frac{1}{4}$$

$$\left(x - \frac{1}{2}\right)^2 + \left(y - \frac{1}{2}\right)^2 = \frac{1}{2}$$

67. $$y = x$$

$$r \sin \theta = r \cos \theta$$

$$\sin \theta = \cos \theta$$

or

$$\tan \theta = 1$$

69. $$x = y^2$$

$$r \cos \theta = (r \sin \theta)^2$$

$$r \cos \theta = r^2 \sin^2 \theta$$

$$r = \frac{\cos \theta}{\sin^2 \theta}$$

or $$r = \cos \theta \csc^2 \theta$$

71. The line shown is a vertical line,

x = 2. Since x = 2 and x = r cos θ,

$$r \cos \theta = 2$$

$$r = \frac{2}{\cos \theta}$$

$$r = 2 \cdot \frac{1}{\cos \theta}$$

$$r = 2 \sec \theta.$$

73. x = 4t – 3, y = t², for t in [-3, 4]

t	x = 4t - 3	y = t²
-3	4(-3) - 3 = -15	(-3)² = 9
-2	4(-2) - 3 = -11	(-2)² = 4
-1	4(-1) - 3 = -7	(-1)² = 1
0	4(0) - 3 = -3	0² = 0
1	4(1) - 3 = 1	1² = 1
2	4(2) - 3 = 5	2² = 4
3	4(3) - 3 = 9	3² = 9
4	4(4) - 3 = 13	4² = 16

See the answer graph in the back of the textbook.

75. x = 3t + 2, y = t – 1, for t in [-5, 5]

Solve for t in terms of y.

$$y + 1 = t$$

Substitute y + 1 for t in the equation for x.

$$x = 3(y + 1) + 2$$

$$x = 3y + 3 + 2$$

$$x = 3y + 5$$

$$x - 3y = 5$$

Since t is in [-5, 5], x is in [3(-5) + 2, 3(5) + 2] or [-13, 17].

77. $x = \sqrt{t - 1}$, $y = \sqrt{t}$, for t in [1, ∞)

Solve for t in terms of x.

$$x^2 = t - 1$$

$$x^2 + 1 = t$$

Substitute x² + 1 for t in the equation for y.

$$y = \sqrt{x^2 + 1}$$

Since t is in $[1, \infty)$, x is in
$[\sqrt{1 - 1}, \infty)$ or $[0, \infty)$.

79. $x = 5 \tan t$, $y = 3 \sec t$, for t in
$(-\pi/2, \pi/2)$
Then

$\frac{x}{5} = \tan t$ and $\frac{y}{3} = \sec t$.

Since $1 + \tan^2 t = \sec^2 t$,

$$1 + \left(\frac{x}{5}\right)^2 = \left(\frac{y}{3}\right)^2$$

$$1 + \frac{x^2}{25} = \frac{y^2}{9}$$

$$9\left(1 + \frac{x^2}{25}\right) = y^2$$

$$\sqrt{9\left(1 + \frac{x^2}{25}\right)} = y$$

$$3\sqrt{1 + \frac{x^2}{25}} = y.$$

Since t is in $(-\pi/2, \pi/2)$, and
$x = 5 \tan t$ is undefined at $-\pi/2$ and
$\pi/2$, x is in $(-\infty, \infty)$.

81. $x = a + r \cos t$, $y = b + r \sin t$

$\frac{x - a}{r} = \cos t$, $\frac{y - b}{r} = \sin t$

Since $\sin^2 t + \cos^2 t = 1$,

$$\left(\frac{y - b}{r}\right)^2 + \left(\frac{x - a}{r}\right)^2 = 1$$

$$\frac{(y - b)^2}{r^2} + \frac{(x - a)^2}{r^2} = 1$$

$$(y - b)^2 + (x - a)^2 = r^2$$

$$(x - a)^2 + (y - b)^2 = r^2.$$

This is the standard equation of a
circle, centered at (a, b), with
radius r.

83. $z = -.5i$

(a) The complex conjugate is
$z = .5i$.   Both have a modulus of
.5.

(b) $z_1^2 + z = (-.5i)^2 + (-.5i)$
$\qquad\qquad = -.25 - .5i$
modulus of $\sqrt{(-.25)^2 + (-.5)^2} = .559$
$z_2^2 + z = (.5i)^2 + (.5i)$
$\qquad\qquad = -.25 + .5i.$

modulus of .559

Since the complex conjugates have
the same modulus, and (a, b) is in
the set, $(a, -b)$ must be in the set.
If both (a, b) and $(a, -b)$ are in
the set, this shows symmetry about
the x-axis.

CHAPTER 8   TEST

[8.1]   Write as a multiple of i.

1. $\sqrt{-256}$

2. $\sqrt{-\dfrac{100}{27}}$

[8.1]   3. Simplify i^{47}.

[8.1]   4. Solve $4x^2 + 3x + 1 = 0$.

[8.2]   Write in standard form.

5. $4(\cos 120° + i \sin 120°)$

6. $6 \text{ cis } (-30°)$

[8.2]   Write in trigonometric form.

7. $-\dfrac{3\sqrt{2}}{2} + \dfrac{3\sqrt{2}}{2}i$

8. $-1 - i\sqrt{3}$

Perform the indicated operations. Write answers in standard form.

[8.1]   9. $(3 - 2i)(7 + 4i)$

[8.1]   10. $\dfrac{4 - 3i}{2 + i}$

[8.1]   11. $\dfrac{4 - 5i}{-i}$

[8.3]   12. $(5 \text{ cis } 15°)(8 \text{ cis } 45°)$

[8.3]   13. $\dfrac{\cos 240° + i \sin 240°}{\cos 60° + i \sin 60°}$

[8.4]   14. $(3 - 3i\sqrt{3})^3$

[8.4]   15. $[7(\cos 30° + i \sin 30°)]^2$

[8.4]   Find all solutions of each equation. Write answers in trigonometric form.

16. $x^4 - 1 = 0$

17. $x^2 + 9i = 0$

18. $x^3 = 6\sqrt{2} - 6i\sqrt{2}$

[8.4]   19. Find all third roots of $-8i$. Write answers in standard form.

[8.4] 20. How may real fifth roots does −31 have?

[8.5] 21. Write an equivalent equation in polar coordinates for
$x^2 + y^2 = 25$.

[8.5] 22. Write an equivalent equation in rectangular coordinates for
$r = \dfrac{4}{1 - \sin \theta}$, and describe its graph.

[8.5] 23. Write an equivalent equation in rectangular coordinates for
$r = -7 \sec \theta$, and describe its graph.

[8.6] Write a rectangular equation for each plane curve given by the parametric equations, and graph each curve.

24. $x = t - 3$, $y = t^2$, for t in $[-2, 1]$

25. $x = 4 \sin \theta$, $y = 3 \cos \theta$, for θ in $[0, 2\pi]$

CHAPTER 8   TEST ANSWERS

1. 16i   **2.** $\frac{10\sqrt{3}}{9}$i   **3.** −i   **4.** $-\frac{3}{8} + \frac{\sqrt{7}}{8}$i, $-\frac{3}{8} - \frac{\sqrt{7}}{8}$i

5. −2 + 2i$\sqrt{3}$ **6.** 3$\sqrt{3}$ − 3i   **7.** 3(cos 135° + i sin 135°)

8. 2(cos 240° + i sin 240°)   **9.** 29 − 2i   **10.** 1 − 2i

11. 5 + 4i   **12.** 20 + 20i$\sqrt{3}$ **13.** −1 or −1 + 0i   **14.** −216 or −216 + 0i

15. $\frac{49}{2} + \frac{49\sqrt{3}}{2}$i   **16.** (cos 0° + i sin 0°), (cos 90° + i sin 90°),

(cos 180° + i sin 180°), (cos 270° + i sin 270°)   **17.** 3(cos 135° + i sin 135°),

3(cos 315° + i sin 315°)   **18.** $\sqrt[3]{12}$(cos 105° + i sin 105°), $\sqrt[3]{12}$(cos 225° + i sin 225°),

$\sqrt[3]{12}$(cos 345° + i sin 345°)   **19.** 2i (or 0 + 2i), $-\sqrt{3}$ − i, $\sqrt{3}$ − i   **20.** 1

21. r = 5   **22.** x² = 8(2 + y), a parabola with vertex (0, −2) and axis x = 0

23. x = −7; a vertical line with intercept (−7, 0)

24. y = (x + 3)²,   **25.** $\frac{x^2}{16} + \frac{y^2}{9} = 1$,

for x in [−5, −2]   for x in [−4, 4]

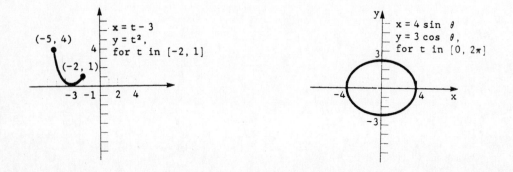

CHAPTER 9   EXPONENTIAL AND LOGARITHMIC FUNCTIONS

Section 9.1

1. If $x = 2$, then $5^x = 5^2$.

   This statement is true by additional property (b) of exponents. The two expressions are equal if and only if the exponents are equal.

3. If $x < 0$, then $x^3 > x^2$.

   This statement is false. If $x < 0$, then x^3 will always be negative and x^2 will be positive. A negative number is less than a positive number.

5. The range of $f(x) = .5^x$ is $(0, \infty)$.

   This statement is true. The range of an exponential function is $(0, \infty)$.

For Exercises 7–13, see the answer graphs in the back of the textbook.

7. (a) $f(x) = 3^{-x} - 2$

   This graph is obtained by translating the graph of $f(x) = 3^{-x}$ down 2 units.

x	-2	-1	0	1	2
y	7	1	-1	$-\frac{5}{3}$	$-\frac{17}{9}$

   (b) $f(x) = 3^{-x} + 4$

   This graph is obtained by translating the graph of $f(x) = 3^{-x}$ up 4 units.

x	-2	-1	0	1	2	3
y	13	7	5	$4\frac{1}{3}$	$4\frac{1}{9}$	$4\frac{1}{27}$

   (c) $f(x) = 3^{-x-2}$

   This equation may also be written as $f(x) = 3^{-(x+2)}$. The graph is obtained by translating the graph of $f(x) = 3^{-x}$ 2 units to the left.

x	-5	-3	-2	-1	0	1	2
y	27	3	1	$\frac{1}{3}$	$\frac{1}{9}$	$\frac{1}{27}$	$\frac{1}{81}$

   (d) $f(x) = 3^{-x+4}$

   This equation may also be written as $f(x) = 3^{-(x-4)}$. This graph is obtained by translating the graph of $f(x) = 3^{-x}$ 4 units to the right.

9. $f(x) = 3^x$

   Make a table of values.

x	-2	-1	0	1	2
y	$\frac{1}{9}$	$\frac{1}{3}$	1	3	9

   Plot these points and draw a smooth curve through them. This is an increasing function. The domain is $(-\infty, \infty)$ and the range is $(0, \infty)$. The x-axis is a horizontal asymptote.

11. $f(x) = \left(\frac{3}{2}\right)^x$

    The domain is $(-\infty, \infty)$ and the range is $(0, \infty)$. Make a table of values.

x	-2	-1	0	1	2
y	$\frac{4}{9}$	$\frac{2}{3}$	1	$\frac{3}{2}$	$\frac{9}{4}$

13. $f(x) = 2^{|x|}$

Make a table of values.

x	-3	-2	-1	0	1	2	3
y	8	4	2	1	2	4	8

Notice that for $x < 0$, $|x| = -x$, so the graph is the same as that of $f(x) = 2^{-x}$. For $x \geq 0$, $|x| = x$, so the graph is the same as that of $f(x) = 2^x$. Since $|-x| = |x|$, the graph is symmetric with respect to the y-axis.

15. $a = 2.3$

Since graph A is increasing, $a > 1$. Since graph A is the middle of the three increasing graphs, the value of a must be the middle of the three values of a greater than 1.

17. $a = .75$

Since graph C is decreasing, $0 < a < 1$. Since graph C decreases at the slowest rate of the three decreasing graphs, the value of a must be the closest to 1 of the three values of a less than 1.

19. $a = .31$

Since graph E is decreasing, $0 < a < 1$. Since graph E decreases at the fastest rate of the three decreasing graphs, the value of a must be the closest to 0 of the three values of a less than 1.

21. For $a > 1$, the value of $f(x) = a^x$ increases as x increases. For $0 < a < 1$, the value of $f(x) = a^x$ decreases as x decreases.

25. If the graph of the exponential function $f(x) = a^x$ contains the point $(3, \quad 8)$ we have

$$a^3 = 8$$
$$(a^3)^{1/3} = 8^{1/3}$$
$$a = 2.$$

Thus, the equation which satisfies the given condition is

$$f(x) = 2^x.$$

27. $f(t) = 3^{2t+3}$
$$= 3^{2t}3^3$$
$$= 27 \cdot 3^{2t}$$
$$= 27 \cdot (3^2)^t$$
$$= 27 \cdot 9^t$$

31. $125^r = 5$

Write both sides as powers of 5.
$$(5^3)^r = 5^1$$
$$5^{3r} = 5^1$$
$$3r = 1 \quad \textit{Property (b)}$$
$$r = \frac{1}{3}$$

The solution is 1/3.

33. $\left(\frac{2}{3}\right)^x = \frac{9}{4}$
$$\left(\frac{2}{3}\right)^x = \left(\frac{3}{2}\right)^2$$
$$\left(\frac{2}{3}\right)^x = \left(\frac{2}{3}\right)^{-2}$$
$$x = -2 \quad \textit{Property (b)}$$

The solution is -2.

35. $5^{2p+1} = 25$

$5^{2p+1} = 5^2$

$2p + 1 = 2$ *Property (b)*

$2p = 1$

$p = \dfrac{1}{2}$

The solution is 1/2.

37. $\dfrac{1}{81} = k^{-4}$

$81^{-1} = k^{-4}$

$[(\pm 3)^4]^{-1} = k^{-4}$

$(\pm 3)^{-4} = k^{-4}$

$[(\pm 3)^{-4}]^{-1/4} = (k^{-4})^{-1/4}$

$\pm 3 = k$

The solutions are −3, 3.

39. $z^{5/2} = 32$

Raise both sides to the 2/5 power.

$(z^{5/2})^{2/5} = (32)^{2/5}$

$z = (32)^{2/5}$

$= (32^{1/5})^2$

$= 2^2$

$= 4$

The solution is 4.

41. $32^t = 16^{1-t}$

Write both sides as powers of 2.

$(2^5)^t = (2^4)^{1-t}$

$2^{5t} = 2^{4-4t}$

$5t = 4 - 4t$ *Property (b)*

$9t = 4$

$t = \dfrac{4}{9}$

The solution is 4/9.

43. $\left(\dfrac{2}{3}\right)^{k-1} = \left(\dfrac{81}{16}\right)^{k+1}$

$\left(\dfrac{2}{3}\right)^{k-1} = \left[\left(\dfrac{3}{2}\right)^4\right]^{(k+1)}$

$\left(\dfrac{2}{3}\right)^{k-1} = \left[\left(\dfrac{2}{3}\right)^{-4}\right]^{(k+1)}$

$k - 1 = -4k - 4$

$5k = -3$

$k = -\dfrac{3}{5}$

The solution is −3/5.

For Exercises 45 and 47, see the answer graphs in the back of the textbook.

45. $f(x) = \dfrac{e^x + e^{-x}}{2}$

Graph this function in the standard viewing window.

47. $f(x) = x^2 \cdot 2^{-x}$

This function may be graphed in the standard viewing window, but a better picture of the graph is obtained by using the window [−2, 8] by [−2, 5].

49. $P = 56{,}780$, $t = \dfrac{23}{4}$, $m = 4$, $r = .053$

$A = P\left(1 + \dfrac{r}{m}\right)^{tm}$

$= 56{,}780\left(1 + \dfrac{.053}{4}\right)^{(23/4)(4)}$

$= 56{,}780(1.01325)^{23}$

$= 76{,}855.95$

The future value is $76,855.95.

51. Use the compound interest formula to find the present value P if the future value A = 45,678.93, r = .096, m = 12, and t = $\frac{11}{12}$.

$$A = P\left(1 + \frac{r}{m}\right)^{tm}$$

$$45{,}678.93 = P\left(1 + \frac{.096}{12}\right)^{(11/12)(12)}$$

$$45{,}678.93 = P(1.008)^{11}$$

$$P = \frac{45{,}678.93}{(1.008)^{11}}$$

$$P = 41{,}845.63$$

The present value is $41,845.63.

53. Use the compound interest formula to find r if P = $1200, A = $1780, m = 4, and t = 5.

$$A = P\left(1 + \frac{r}{m}\right)^{tm}$$

$$1780 = 1200\left(1 + \frac{r}{4}\right)^{5(4)}$$

$$1780 = 1200\left(1 + \frac{r}{4}\right)^{20}$$

$$1.4833 = \left(1 + \frac{r}{4}\right)^{20}$$

Now, take the 20th root of both sides. This can be done on a calculator by using an exponential key to find

$$(1.4833)^{1/20} = (1.4833)^{.05}$$
$$\approx 1.01991.$$

$$1.01991 = 1 + \frac{r}{4}$$

$$.01991 = \frac{r}{4}$$

$$.07964 = r$$

The interest rate, to the nearest tenth, is 8.0%.

55. **(a)** T = 50,000(1 + .06)^n

After 4 years, n = 4.

$$T = 50{,}000(1 + .06)^4$$
$$\approx 63{,}000$$

The total population after 4 years is about 63,000.

(b) T = 30,000(1 + .12)^n

After 3 years, n = 3.

$$T = 30{,}000(1 + .12)^3$$
$$\approx 42{,}000$$

There would be about 42,000 deer after 3 years.

(c) T = 45,000(1 + .08)^n

After 5 years, n = 5.

$$T = 45{,}000(1 + .08)^5$$
$$\approx 66{,}000$$

We can expect about 66,000 − 45,000 = 21,000 additional deer after 5 years.

57. A(t) = 500e$^{-.032t}$

(a) $A(4) = 500e^{-.032(4)}$
$$= 500e^{-.128}$$
$$\approx 500(.8799)$$
$$\approx 440$$

After 4 yr, about 440 g will remain.

(b) $A(8) = 500e^{-.032(8)}$
$$= 500e^{-.256}$$
$$\approx 500(.7741)$$
$$\approx 387$$

After 8 yr, about 387 g will remain.

(c) $A(20) = 500e^{-.032(20)}$

$= 500e^{-.64}$

$\approx 500(.5273)$

≈ 264

After 20 yr, about 264 g will remain.

(d) The domain is $[0, \infty)$ and the graph passes through $(4, 440)$, $(8, 387)$, and $(20, 264)$.
See the answer graph in the back of the textbook.

59. **(a)** T is a linear function, not an exponential function.

(b) The graph of T(R) passes through the points $(0, 0)$ and $(20, 20.6)$. The slope between these points is

$$m = \frac{20.6 - 0}{20 - 0} = 1.03.$$

Since the slope is 1.03 and the y-intercept is 0, the equation is

$$T(R) = 1.03R.$$

(c) $T(5) = 1.03(5) = 5.15$

When $R = 5$ w/m^2, the global temperature increase is 5.15°F.

61. $x = 2^x$

Use a graphing calculator to graph the line $f(x) = x$ and the exponential function $g(x) = 2^x$ on the same screen. These two graphs do not intersect, so the given equation has no solution.

63. $6^{-x} = 1 - x$

Use a graphing calculator to graph the exponential function $f(x) = 6^{-x}$ and the line $g(x) = 1 - x$ on the same screen. These two graphs intersect in two points whose coordinates may be found by using the "intersect" option in the CALC menu. The x-coordinates of these intersection points are the solutions of the given equation: $x = 0$ and $x \approx .73$.

65. Graph $f(x) = (1 + 1/x)^x$ and $y = 2.71828$ on the same screen using the window $[1, 25]$ by $[0, 3]$. As x gets large, $f(x)$ approaches the line $y = 2.71828$.
Since 2.71828 is a decimal approximation for e, this graph illustrates that as x increases, the value of $(1 + 1/x)^x$ approaches e.

Section 9.2

1. $y = \log_a x$ if and only if $x = a^y$.

3. The statement $\log_5 125 = 3$ tells us that *3* is the power of *5* that equals *125*.

5. $3^4 = 81$ is equivalent to $\log_3 81 = 4$.

7. $\left(\frac{2}{3}\right)^{-3} = \frac{27}{8}$ is equivalent to

$$\log_{2/3} \left(\frac{27}{8}\right) = -3.$$

9. $\log_6 36 = 2$ is equivalent to

$$6^2 = 36.$$

11. $\log_{\sqrt{3}} 81 = 8$ is equivalent to

$$(\sqrt{3})^8 = 81.$$

15. Let $y = \log_5 25$.

Write the equation in exponential form.
$$5^y = 25$$
$$5^y = 5^2$$
$$y = 2$$

Thus, $\log_5 25 = 2$.

17. Let $y = \log_{10} .001$.
$$10^y = .001$$
$$10^y = (10)^{-3}$$
$$y = -3$$

Thus, $\log_{10} .001 = -3$.

19. Let $y = \log_4 \left(\frac{\sqrt[3]{4}}{2}\right)$.

$$4^y = \frac{\sqrt[3]{4}}{2}$$

$$(2^2)^y = \frac{\sqrt[3]{2^2}}{2}$$

$$2^{2y} = \frac{2^{2/3}}{2}$$

$$2^{2y} = 2^{(2/3)-1}$$

$$2^{2y} = 2^{-1/3}$$

$$2y = -\frac{1}{3}$$

$$y = -\frac{1}{6}$$

Thus, $\log_4 \left(\frac{\sqrt[3]{4}}{2}\right) = -\frac{1}{6}$.

21. $2^{\log_2 9} = 9$

This is true by the theorem on inverses.

23. $x = \log_2 32$

Write the equation in exponential form.

$$2^x = 32$$
$$2^x = 2^5$$
$$x = 5$$

The solution is 5.

25. $\log_x 25 = -2$
$$x^{-2} = 25$$
$$(x^{-2})^{-1/2} = (25)^{-1/2}$$
$$x = \frac{1}{25^{1/2}}$$
$$x = \frac{1}{5}$$

The solution is 1/5.

For Exercises 29–35, see the answer graphs in the back of the textbook.

29. $f(x) = \log_{1/2} x$

(a) $f(x) = (\log_{1/2} x) - 2$

This is the graph of $f(x) = \log_{1/2} x$ translated down 2 units.

(b) $f(x) = \log_{1/2} (x - 2)$

This is the graph of $f(x) = \log_{1/2} x$ translated to the right 2 units. The graph has a vertical asymptote at $x = 2$.

(c) $f(x) = \left| \log_{1/2} (x - 2) \right|$

This is the same graph as (b) with the part below the x–axis reflected about the x–axis.

31. $f(x) = \log_3 x$

Write $y = \log_3 x$ in exponential form as $x = 3^y$ to find ordered pairs that satisfy the equation. It is easier to choose values for y and find the corresponding values of x.
Make a table of values.

x	$\frac{1}{9}$	$\frac{1}{3}$	1	3	9
y	-2	-1	0	1	2

The graph can also be found by reflecting the graph of $f(x) = 3^x$ about the line $y = x$. The graph has the y–axis as a vertical asymptote. Sketch the graph.

33. $f(x) = \log_{1/2} (1 - x)$
Make a table of values.

x	-7	-3	-1	0	$\frac{1}{2}$	$\frac{3}{4}$	$\frac{7}{8}$
1 - x	8	4	2	1	$\frac{1}{2}$	$\frac{1}{4}$	$\frac{1}{8}$
y	-3	-2	-1	0	1	2	3

The graph has a vertical asymptote at $x = 1$. Sketch the graph.

35. $f(x) = \log_3 (x - 1)$

To graph the function, translate the graph of $f(x) = \log_3 x$ (from Exercise 31) 1 unit to the right. The vertical asymptote will be $x = 1$.

37. Graph $y = \log_{10} x^2$ and $y = 2 \log_{10} x$ on separate viewing screens, using the window $[-5, 5]$ by $[-3, 3]$ for each. See the answer graphs in the back of the textbook. The graphs of $y = \log_{10} x^2$ and $y = 2 \log_{10} x$ are not the same because the domains are not the same. The domain of $y = \log_{10} x^2$ is $(-\infty, 0) \cup (0, \infty)$, while the domain of $y = 2 \log_{10} x$ is $(0, \infty)$.

39. $f(x) = \log_2 (2x)$

The graph will be similar to that of $f(x) = \log_2 x$ (graph E) but will increase more rapidly. Note that while the x–intercept for $f(x) = \log_2 x$ is 1, the x–intercept for $f(x) = \log_2 2x$ will be $1/2$, since $\log_2 2(1/2) = \log_2 1 = 0$. The correct graph is D.

41. $f(x) = \log_2 \left(\frac{x}{2}\right)$

$\qquad = \log_2 \frac{1}{2}x$

The graph will be similar to that of $f(x) = \log_2 x$ but will increase less rapidly. The x–intercept is 2 since $\log_2 (2/2) = \log_2 1 = 0$. The correct graph is C.

43. $f(x) = \log_2 (-x)$

The graph of this function is the
reflection of the graph of $f(x) = \log_2 x$ about the y-axis. Note that
the x-intercept is -1 since
$\log_2 [-(-1)] = \log_2 1 = 0$. The correct graph is A.

45. $f(x) = x^2 \log_{10} x$

Graph $f(x) = x^2 \log x$ in the window
$[-1, 5]$ by $[-5, 2]$. See the answer
graph in the back of the textbook.

47. $2^{-x} = \log_{10} x$

Graph $y_1 = 2^{-x}$ and $y_2 = \log_{10} x$ on a
graphing calculator. The x-coordinate of the intersection point will
be the solution of the original
equation. Using the "intersect"
option in the CALC menu, we find
that the solution is $x \approx 1.87$.

48. If x and y are positive numbers,

$\log_a \frac{x}{y} = \log_a x - \log_a y$.

49. Since $\log_2 \left(\frac{x}{4}\right) = \log_2 x - \log_2 4$
by the quotient rule, the graph of
$y = \log_2 \left(\frac{x}{4}\right)$ can be obtained by
shifting the graph of $y = \log_2 x$
down $\log_2 4 = 2$ units.

50. Graph $f(x) = \log_2 \left(\frac{x}{4}\right)$ and $g(x) = \log_2 x$ on the same axes. See
the answer graph in the back of the
textbook. The graph of f is 2 units
below the graph of g. This supports
the answer in Exercise 49.

51. If x = 4, $\log_2 \left(\frac{x}{4}\right) = 0$; since
$\log_2 x = 2$ and $\log_2 4 = 2$, $\log_2 x - \log_2 4 = 0$. By the quotient rule,

$$\log_2 \left(\frac{x}{4}\right) = \log_2 x - \log_2 4.$$

Both sides should equal 0. Since
$2 - 2 = 0$, they do.

53. $\log_3 \left(\frac{4p}{q}\right)$

$= \log_3 4p - \log_3 q$
         Logarithm of a quotient
$= \log_3 4 + \log_3 p - \log_3 q$
         Logarithm of a product

55. $\log_2 \left(\frac{2\sqrt{3}}{5}\right)$

$= \log_2 2\sqrt{3} - \log_2 5$
         Logarithm of a quotient
$= \log_2 2 + \log_2 \sqrt{3} - \log_2 5$
         Logarithm of a product
$= 1 + \left(\frac{1}{2}\right) \log_2 3 - \log_2 5$
         Logarithm of a power

57. $\log_6 (7m + 3q)$

Since this is a sum, none of the
logarithm properties apply, so this
expression cannot be simplified.

59. $\log_p \sqrt[3]{\dfrac{m^5 n^4}{t^2}}$

$= \dfrac{1}{3} \log_p \left(\dfrac{m^5 n^4}{t^2}\right)$

$= \dfrac{1}{3}(\log_p m^5 + \log_p n^4 - \log_p t^2)$

$= \dfrac{1}{3}(5 \log_p m + 4 \log_p n - 2 \log_p t)$

$= \dfrac{5}{3} \log_p m + \dfrac{4}{3} \log_p n - \dfrac{2}{3} \log_p t$

$= \left(\dfrac{1}{3}\right)(5 \log_p m + 4 \log_p n - 2 \log_p t)$

61. $(\log_b k - \log_b m) - \log_b a$

$= \log_b \dfrac{k}{m} - \log_b a$ *Logarithm of a quotient*

$= \log_b \left(\dfrac{k}{ma}\right)$ *Logarithm of a quotient*

63. $\dfrac{1}{2} \log_y (p^3 q^4) - \dfrac{2}{3} \log_y (p^4 q^3)$

$= \log_y (p^3 q^4)^{1/2} - \log_y (p^3 q^4)^{2/3}$

$= \log_y \dfrac{(p^3 q^4)^{1/2}}{(p^4 q^3)^{2/3}}$

$= \log_y \dfrac{p^{3/2} \cdot q^2}{p^{8/3} \cdot q^2}$

$= \log_y (p^{-7/6})$

65. $\log_b (2y + 5) - \dfrac{1}{2} \log_b (y + 3)$

$= \log_b \dfrac{2y + 5}{(y + 3)^{1/2}}$

$= \log_b \left(\dfrac{2y + 5}{\sqrt{y + 3}}\right)$

67. $\log_{10} 12 = \log_{10} (3 \cdot 2^2)$

$= \log_{10} 3 + 2 \cdot \log_{10} 2$

$= .4771 + 2(.3010)$

$= 1.0791$

69. $\log_{10} \left(\dfrac{20}{27}\right)$

$= \log_{10} 20 - \log_{10} 27$

$= \log_{10} 2 \cdot 10 - \log_{10} 3^3$

$= \log_{10} 2 + \log_{10} 10 - 3 \log_{10} 3$

$\approx .3010 + 1 - 3(.4771)$

$= -.1303$

71. Since f is a logarithmic function,

$f(27) = f(3^3)$

$= 3f(3)$

$= 3(2)$

$= 6.$

Section 9.3

1. $\log 43 = 1.6335$

   To find this value, enter 43 and press the log key.

3. $\log .014 = -1.8539$

5. $\ln 580 = 6.3630$

   To find this value, enter 580 and press the ln key.

7. $\ln .7 = -.3567$

9. The graph of $y = \log x$ has coordinates $x = 8$, $y = .90308999$. Thus,

   $\log 8 \approx .90308999.$

11. $\log_3 4$ is the logarithm to the base 3 of 4.

   ($\log_4 3$ would be the logarithm to the base 4 of 3.)

12. The exact value of $\log_3 9$ is 2 since $3^2 = 9$.

13. The exact value of $\log_3 27$ is 3 since $3^3 = 27$.

14. $\log_3 16$ must lie between 2 and 3. Because the function defined by $y = \log_3 x$ is increasing and $9 < 16 < 27$, we have

   $$\log_3 9 < \log_3 16 < \log_3 27.$$

15. By the change-of-base theorem,
   $$\log_3 16 = \frac{\log 16}{\log 3} = \frac{\ln 16}{\ln 3}$$
   $$\approx 2.523719014.$$

   This value is between 2 and 3, as predicted in Exercise 14.

16. The exact value of $\log_5 (1/5)$ is -1 since $5^{-1} = \frac{1}{5}$.

   The exact value of $\log_5 1$ is 0 since $5^0 = 1$.

17. $\log_5 .68$ must lie between -1 and 0. Since the function defined by $y = \log_5 x$ is increasing and

   $$\frac{1}{5} = .2 < .68 < 1,$$

   we must have

   $$\log_5 .2 < \log_5 .68 < \log_5 1.$$

By the change-of-base theorem,

$$\log_5 .68 = \frac{\log .68}{\log 5} = \frac{\ln .68}{\ln 5}$$
$$\approx -.239625573.$$

This value is between -1 and 0, as predicted above.

19. To find $\log_9 12$, use the change-of-base theorem with $a = 9$, $b = e$, and $x = 12$.

   $$\log_a x = \frac{\log_b x}{\log_b a}$$

   $$\log_9 12 = \frac{\ln 12}{\ln 9}$$
   $$\approx \frac{2.4849}{2.19722}$$
   $$\approx 1.13$$

21. $\log_{1/2} 3 = \dfrac{\ln 3}{\ln .5}$
   $$\approx \frac{1.0986}{-.6931}$$
   $$\approx -1.58$$

23. $\log_{200} 175 = \dfrac{\ln 175}{\ln 200}$
   $$\approx \frac{5.1648}{5.2983}$$
   $$\approx .97$$

25. $\log_{5.8} 12.7 = \dfrac{\ln 12.7}{\ln 5.8}$
   $$\approx \frac{2.5416}{1.7579}$$
   $$\approx 1.45$$

27. The table for $Y_1 = \log_3 (4 - x)$ shows "ERROR" for $X \geq 4$. This is because the function is undefined for $X \geq 4$. The domain of $y = \log_a X$ is $X > 0$, which means that for Y_1, the domain is $4 - X > 0$, or $X < 4$.

29. To graph $f(x) = \log_x 5$, use the change-of-base theorem; enter $y_1 = (\log 5)/(\log x)$. Graph this function in the window $[-1, 5]$ by $[-3, 3]$. See the answer graph in the back of the textbook.
The vertical line simulates an asymptote at $x = 1$. The base must be greater than 0 and not equal to 1.

31. Grapefruit, 6.3×10^{-4}

$$\text{pH} = -\log [H_3O^+]$$
$$= -\log (6.3 \times 10^{-4})$$
$$= -(\log 6.3 + \log 10^{-4})$$
$$= -(.7793 - 4)$$
$$= -.7993 + 4$$
$$\text{pH} = 3.2$$

The answer is rounded to the nearest tenth because it is customary to round pH values to the nearest tenth.
The pH of grapefruit is 3.2.

33. Limes, 1.6×10^{-2}

$$\text{pH} = -\log [H_3O^+]$$
$$= -\log (1.6 \times 10^{-2})$$
$$= -(\log 1.6 + \log 10^{-2})$$
$$= -(.2041 - 2)$$
$$= -(-1.7959)$$

$$\text{pH} = 1.8$$
The pH of limes is 1.8.

35. Soda pop, 2.7

$$\text{pH} = -\log [H_3O^+]$$
$$2.7 = -\log [H_3O^+]$$
$$-2.7 = \log [H_3O^+]$$
$$[H_3O^+] = 2.0 \times 10^{-3}$$

37. Beer, 4.8

$$\text{pH} = -\log [H_3O^+]$$
$$4.8 = -\log [H_3O^+]$$
$$-4.8 = \log [H_3O^+]$$
$$[H_3O^+] = 10^{-4.8}$$
$$[H_3O^+] = 1.6 \times 10^{-5}$$

39. Let r = the decibel rating of a sound.

$$r = 10 \cdot \log_{10} \frac{I}{I_0}$$

(a) $r = 10 \cdot \log_{10} \dfrac{100 \cdot I_0}{I_0}$

$$= 10 \cdot \log_{10} 100$$
$$= 10 \cdot 2$$
$$= 20$$

(b) $r = 10 \cdot \log_{10} \dfrac{1000 \cdot I_0}{I_0}$

$$= 10 \cdot \log_{10} 1000$$
$$= 10 \cdot 3$$
$$= 30$$

(c) $r = 10 \cdot \log_{10} \dfrac{100,000 \cdot I_0}{I_0}$

$$= 10 \cdot 5$$
$$= 50$$

(d) $r = 10 \cdot \log_{10} \dfrac{1,000,000 \cdot I_0}{I_0}$

$$= 10 \cdot 6$$
$$= 60$$

41. Let r = the Richter scale rating of an earthquake.

$$r = \log_{10}\left(\frac{I}{I_0}\right)$$

(a) $r = \log_{10}\frac{1000 \cdot I_0}{I_0}$

$= \log_{10} 1000$

$= 3$

(b) $r = \log_{10}\frac{1,000,000 \cdot I_0}{I_0}$

$= \log_{10} 1,000,000$

$= 6$

(c) $r = \log_{10}\frac{100,000,000 \cdot I_0}{I_0}$

$= \log_{10} 100,000,000$

$= 8$

43. (a) $8.3 = \log_{10}\frac{I}{I_0}$

$\frac{I}{I_0} = 10^{8.3}$

$I = 10^{8.3} \cdot I_0$

$I = 200,000,000 I_0$

The magnitude was about $200,000,000 I_0$.

(b) $7.1 = \log_{10}\frac{I}{I_0}$

$\frac{I}{I_0} = 10^{7.1}$

$I = 10^{7.1} \cdot I_0$

$I = 13,000,000 I_0$

The magnitude was about $13,000,000 I_0$.

(c) $\frac{200,000,000 \cdot I_0}{13,000,000 \cdot I_0} \approx 15.38$

The 1906 earthquake had a magnitude more than 15 times greater than the 1989 earthquake.

45. $S_n = a \ln\left(1 + \frac{n}{a}\right)$

$= .36 \ln\left(1 + \frac{n}{.36}\right)$

(a) $S(100) = .36 \ln\left(1 + \frac{100}{.36}\right)$

$= .36 \ln 278.77$

$= (.36)(5.6304)$

≈ 2

(b) $S(200) = .36 \ln\left(1 + \frac{200}{.36}\right)$

$= .36 \ln (556.555)$

$= .36(6.322)$

≈ 2

(c) $S(150) = .36 \ln\left(1 + \frac{150}{.36}\right)$

$= .36 \ln 417.666$

$= .36(6.0347)$

≈ 2

(d) $S(10) = .36 \ln\left(1 + \frac{10}{.36}\right)$

$= .36 \ln 28.777$

$= .36(3.3596)$

≈ 1

47. The index of diversity M for 2 species is given by

$$H = -[P_1 \log_2 P_1 + P_2 \log_2 P_2].$$

$P_1 = \frac{50}{100}$

$= .5$

$P_2 = \frac{50}{100}$

$= .5$

Substituting into the formula gives

$H = -[.5 \log_2 .5 + .5 \log_2 .5].$

Since $\log_2 .5 = \log_2 \frac{1}{2} = -1$, we have

$$H = -[.5(-1) + .5(-1)]$$
$$= -(-1)$$
$$= 1.$$

The index of diversity is 1.

49. $g(x) = e^x$

(a) By the theorem on inverses,

$$g(\ln 3) = e^{\ln 3}$$
$$= 3.$$

(b) By the theorem on inverses,

$$g[\ln (5^2)] = e^{\ln 5^2}$$
$$= 5^2 = 25.$$

(c) By the theorem on inverses,

$$g[\ln \left(\frac{1}{e}\right)] = e^{\ln(1/e)}$$
$$= \frac{1}{e}.$$

51. $f(x) = \ln x$

(a) By the theorem on inverses,

$$f(e^5) = \ln e^5$$
$$= 5.$$

(b) By the theorem on inverses,

$$f(e^{\ln 3}) = \ln e^{\ln 3}$$
$$= \ln 3.$$

(c) By the theorem on inverses,

$$f(e^{2\ln 3}) = \ln e^{2\ln 3}$$
$$= 2 \ln 3$$
$$\text{or } \ln 9.$$

53. $f(x) = -266 + 72 \ln x$

In the year 2000, $x = 100$.

$$f(100) = -266 + 72 \ln 100$$
$$\approx 66$$

Thus, the number of visitors in the year 2000 will be about 66 million. Beyond 1993, we must assume that the rate of increase continues to be logarithmic.

55. From Example 6,

$$T(R) = 1.03R$$

and $\quad R = k \ln \left(\frac{C}{C_0}\right).$

By substitution, we have

$$T(k) = 1.03k \ln \left(\frac{C}{C_0}\right).$$

Since $10 \leq k \leq 16$ and $\frac{C}{C_0} = 2$, the range for

$$T = 1.03k \ln \left(\frac{C}{C_0}\right)$$

will be between
$$T = 1.03(10) \ln 2 \approx 7.1$$
and $\quad T = 1.03(16) \ln 2 \approx 11.4.$

The predicted increased global temperature due to the greenhouse effect from a doubling of the carbon dioxide in the atmosphere is between 7°F and 11°F.

57. (a) The table of natural logarithms takes the following form. Let $x = \ln D$ and $y = \ln P$ for each planet.

Planet	ln D	ln P
Mercury	−.94	−1.43
Venus	−.33	−.48
Earth	0	0
Mars	.42	.64
Jupiter	1.65	2.48
Saturn	2.26	3.38
Uranus	2.95	4.43
Neptune	3.40	5.10

Plot this data in the window $[-2, 4]$ by $[-2, 6]$. See the answer graph in the back of the textbook.

From the plot, the data appear to be linear.

(b) Choose two points from the table showing ln D and ln P and find the equation through them. If we use $(0, 0)$, representing Earth and $(3.40, 5.10)$ representing Neptune, we obtain

$$m = \frac{5.10 - 0}{3.40 - 0} = 1.5.$$

Since the y−intercept is 0, the equation is

$$y = 1.5x$$

or $\ln P = 1.5 \ln D$.

See the answer graph in the back of the textbook.

Since the points lie approximately but not exactly on a line, a slightly different equation will be found if a different pair of points is used.

(c) For Pluto, $D = 39.5$, so

$$\ln P = 1.5 \ln D$$
$$= 1.5 \ln 39.5.$$

Then

$$P = e^{1.5 \ln 39.5}$$
$$= e^{\ln 39.5^{1.5}}$$
$$= (39.5)^{1.5}$$
$$\approx 248.3.$$

The linear equation predicts that the period of the planet Pluto is 248.3 years, which is very close to the true value of 248.5 years.

Section 9.4

5. $3^x = 6$

Take base e (natural) logarithms of both sides.

$$\ln 3^x = \ln 6$$
$$x \ln 3 = \ln 6 \quad \textit{Logarithm of a power}$$
$$x = \frac{\ln 6}{\ln 3} \quad \textit{Divide by ln 3}$$
$$\approx \frac{1.7918}{1.0986}$$
$$\approx 1.631$$

The solution is 1.631.

7. $6^{1-2k} = 8$

Take base 10 (common) logarithms of both sides. (This exercise can also be done using natural logarithms.)

$\log 6^{1-2k} = \log 8$

$(1 - 2k) \log 6 = \log 8$

$1 - 2k = \dfrac{\log 8}{\log 6}$

$2k = 1 - \dfrac{\log 8}{\log 6}$

$k = \dfrac{1}{2}\left(1 - \dfrac{\log 8}{\log 6}\right)$

$\approx \dfrac{1}{2}\left(1 - \dfrac{.9031}{.7782}\right)$

$\approx -.080$

The solution is −.080.

9. $e^{k-1} = 4$

$\ln e^{k-1} = \ln 4$

$k - 1 = \ln 4$ *Theorem on inverses*

$k = \ln 4 + 1$

$= 1.3863 + 1$

$= 2.386$

The solution is 2.386.

11. $2e^{5a+2} = 8$

$e^{5a+2} = 4$

$\ln e^{5a+2} = \ln 4$

$5a + 2 = \ln 4$ *Theorem on inverses*

$5a = \ln 4 - 2$

$a = \dfrac{1}{5}(\ln 4 - 2)$

$\approx \dfrac{1}{5}(1.3863 - 2)$

$\approx -.123$

The solution is −.123.

13. $2^x = -3$ has no solution since 2 raised to any power is positive.

15. $e^{2x} \cdot e^{5x} = e^{14}$

$e^{7x} = e^{14}$ *Product rule for exponents*

$7x = 14$ *Property 1*

$x = 2$

The solution is 2.

17. $100(1 + .02)^{3+n} = 150$

$(1.02)^{3+n} = 1.5$

$(3 + n) \log 1.02 = \log 1.5$ *Logarithm of a power*

$3 + n = \dfrac{\log 1.5}{\log 1.02}$

$3 + n \approx \dfrac{.1761}{.0086}$

$3 + n \approx 20.475$

$n \approx 17.475$

The solution is 17.475.

19. $\log (t - 1) = 1$

$t - 1 = 10^1$ *Exponential form*

$t - 1 = 10$

$t = 11$

The solution is 11.

21. $\ln (y + 2) = \ln (y - 7) + \ln 4$

$\ln (y + 2) = \ln [(y - 7) \cdot 4]$ *Logarithm of a product*

$y + 2 = 4(y - 7)$ *Property 2*

$y + 2 = 4y - 28$

$-3y = -30$

$y = 10$

The solution is 10.

23. $\ln(5 + 4y) - \ln(3 + y) = \ln 3$

$$\ln \frac{5 + 4y}{3 + y} = \ln 3$$

Logarithm of a quotient

$$\frac{5 + 4y}{3 + y} = 3$$

Property 2

$$5 + 4y = 3(3 + y)$$

$$5 + 4y = 9 + 3y$$

$$y = 4$$

The solution is 4.

25. $2\ln(x - 3) = \ln(x + 5) + \ln 4$

$$\ln(x - 3)^2 = \ln[4(x + 5)]$$

Logarithm of a power;
logarithm of a product

$$(x - 3)^2 = 4(x + 5) \quad \textit{Property 2}$$

$$x^2 - 6x + 9 = 4x + 20$$

$$x^2 - 10x - 11 = 0$$

$$(x - 11)(x + 1) = 0$$

$$x - 11 = 0 \quad \text{or} \quad x + 1 = 0$$

$$x = 11 \quad \text{or} \qquad x = -1$$

If $x = -1$, $x - 3 = -4$, so -1 is
not in the domain of $\ln(x - 3)$
and cannot be used.

The solution is 11.

27. $\log_3(a - 3) = 1 + \log_3(a + 1)$

$$\log_3(a - 3) = \log_3 3 + \log_3(a + 1)$$
$$\log_3 3 = 1$$

$$\log_3(a - 3) = \log_3[3(a + 1)]$$

Logarithm of a
product

$$a - 3 = 3(a + 1)$$

Property 2

$$a - 3 = 3a + 3$$

$$-2a = 6$$

$$a = -3$$

-3 is not in the domain of
$\log_3(a - 3)$ or $\log_3(a + 1)$
and therefore cannot be used.
This equation has no solution.

29. $\ln e^x - \ln e^3 = \ln e^5$

$$x - 3 = 5 \quad \textit{Theorem on}$$
$$\textit{inverses}$$

$$x = 8$$

The solution is 8.

31. $\log_2 \sqrt{2y^2} - 1 = \frac{1}{2}$

$$\log_2 \sqrt{2y^2} = \frac{3}{2}$$

Add 1

$$2^{3/2} = \sqrt{2y^2}$$

Change to
exponential form

$$(2^{3/2})^2 = (\sqrt{2y^2})^2$$

Square both sides

$$2^3 = 2y^2$$

$$8 = 2y^2$$

$$4 = y^2$$

$$\pm 2 = y$$

Since the solution involves squaring
both sides, both proposed solutions
must be checked in the original
equation. Both answers check.
The solutions are -2, 2.

33. $$\log z = \sqrt{\log z}$$

$$(\log z)^2 = (\sqrt{\log z})^2$$

Square both sides

$$(\log z)^2 = \log z$$

$$(\log z)^2 - \log z = 0$$

$$\log z(\log z - 1) = 0$$

Factor out log z

log z = 0   or   log z - 1 = 0

$\qquad\qquad\qquad$ log z = 1

10^0 = z $\qquad\qquad$ 10^1 = z

$\quad$ 1 = z   or $\qquad$ 10 = z

Since the work involves squaring both sides, both proposed solutions must be checked in the original equation. Both answers check.

The solutions are 1, 10.

37. $\qquad\qquad I = \frac{E}{R}(1 - e^{-Rt/2})$ for t

$$I = \frac{E}{R} - \frac{E}{R}e^{-Rt/2}$$

$$I - \frac{E}{R} = -\frac{E}{R}e^{-Rt/2}$$

$$\frac{RI - E}{R}\left(-\frac{R}{E}\right) = e^{-Rt/2}$$

$$\frac{E - RI}{E} = e^{-Rt/2}$$

$$\ln\left(\frac{E - RI}{E}\right) = \ln e^{-Rt/2}$$

$$\ln\left(\frac{E - RI}{E}\right) = -\frac{Rt}{2}$$

$$-\frac{2}{R}\ln\left(1 - \frac{RI}{E}\right) = t$$

39. $p = a + \dfrac{k}{\ln x}$ for x

$$p - a = \frac{k}{\ln x}$$

$$(p - a)\ln x = k$$

$$\ln x = k/(p - a)$$

To solve for x, change this equation from logarithmic to exponential form.

$$x = e^{k/(p-a)}$$

42. $(e^x)^2 - 4e^x + 3 = 0$

$(e^x - 1)(e^x - 3) = 0$

43. $(e^x - 1)(e^x - 3) = 0$

Set each factor to 0 and solve.

$e^x - 1 = 0$ or $e^x - 3 = 0$

$\quad e^x = 1$ $\qquad\qquad e^x = 3$

$\ln e^x = \ln 1$ $\quad$ $\ln e^x = \ln 3$

$\quad x = 0$ or $\qquad x = \ln 3$

The solutions are 0, ln 3.

44. Graph $y = e^{2x} - 4e^x + 3$ on a graphing calculator, using the window [-5, 5] by [-5, 10]. See the answer graph in the back of the textbook. The graph intersects the x-axis at 0 and $1.099 \approx \ln 3$.

45. From the graph, we see that the intervals where y > 0 are $(-\infty, 0)$ and $(\ln 3, \infty)$, so the solutions of the inequality

$$e^{2x} - 4e^x + 3 > 0$$

are in the interval $(-\infty, 0) \cup (\ln 3, \infty)$.

46. From the graph we see that y < 0 on the interval (0, ln 3), so the solutions of the inequality

$$e^{2x} - 4e^x + 3 < 0$$

are in the interval (0, ln 3).

47. $e^x + \ln x = 5$

Graph $y_1 = e^x + \ln x$
and $\quad y_2 = 5$

on the same screen.

Using the "intersect" option in the CALC menu, we find that the two graphs intersect at approximately (1.52, 5). The x-coordinate of this point is the solution of the equation.
The solution is 1.52.

49. $2e^x + 1 = 3e^{-x}$

Graph $y_1 = 2e^x + 1$

and $y_2 = 3e^{-x}$

on the same screen. The two curves intersect at the point (0, 3). The x-coordinate of this point is the solution of the equation.
The solution is 0.

51. $\log x = x^2 - 8x + 14$

Graph $y_1 = \log x$ and
    $y_2 = x^2 - 8x + 14$.

The intersection points are at x = 2.45 and x = 5.66.
The solutions are 2.45, 5.66.

53. $f(x) = e^{3x+1}$
    $y = e^{3x+1}$

To solve for x, take natural logarithms on both sides.

$$\ln y = \ln e^{3x+1}$$
$$\ln y = 3x + 1$$
$$3x = \ln y - 1$$
$$x = \frac{1}{3}(\ln y - 1) = f^{-1}(y)$$

Exchange x and y.

$$f^{-1}(x) = \frac{1}{3}(\ln (x) - 1)$$

The domain of f^{-1} is $(0, \infty)$ and the range is $(-\infty, \infty)$.

55. $\log_3 x > 3$

$\log_3 x - 3 > 0$

Graph $Y_1 = \log_3 x - 3$.

The graph is positive in the interval $(27, \infty)$, so the solution to the original inequality is $(27, \infty)$.

57. $d = 10 \log \frac{I}{I_0}$

For 89 decibels, we have

$$89 = 10 \log \frac{I}{I_0}$$
$$8.9 = \log \frac{I}{I_0}.$$

Change this equation to exponential form.

$$\frac{I}{I_0} = 10^{8.9}$$
$$I = 10^{8.9} I_0$$

For 86 decibels, we have

$$86 = 10 \log \frac{I}{I_0}$$
$$8.6 = \log \frac{I}{I_0}$$
$$\frac{I}{I_0} = 10^{8.6}$$
$$I = 10^{8.6} I_0.$$

To compare these intensities, find their ratio.

$$\frac{10^{8.9} I_0}{10^{8.6} I_0} = \frac{10^{8.9}}{10^{8.6}} \approx 2$$

From this calculation, we see that 89 decibels is about twice as loud as 86 decibels, for a 100% increase.

59. $A = P\left(1 + \frac{r}{m}\right)^{tm}$

To solve for t, substitute
A = 2063.40, P = 1786, r = .116,
and m = 12.

$$2063.40 = 1786\left(1 + \frac{.116}{12}\right)^{(t)(12)}$$

$$\frac{2063.40}{1786} = \left(1 + \frac{.116}{12}\right)^{12t}$$

$$\ln \frac{2063.40}{1786} = 12t \ln \left(1 + \frac{.116}{12}\right)$$

$$\frac{\ln \frac{2063.40}{1786}}{12 \ln \left(1 + \frac{.116}{23}\right)} = t$$

$$1.25 \approx t$$

To the nearest hundredth, t = 1.25
yr.

61. $A = P\left(1 + \frac{r}{m}\right)^{tm}$

$$20,000 = 16,000\left(1 + \frac{r}{4}\right)^{(5.25)(4)}$$

$$20,000 = 16,000\left(1 + \frac{r}{4}\right)^{21}$$

$$1.25 = \left(1 + \frac{r}{4}\right)^{21}$$

$$\sqrt[21]{1.25} = 1 + \frac{r}{4}$$

$$4(\sqrt[21]{1.25} - 1) = r$$

$$.0427 \approx r$$

The interest rate is about 4.27%.

63. $f(x) = 6.2(10)^{-12}(1.4)^x$

Find x when f(x) = 2000.

$$2000 = 6.2(10)^{-12}(1.4)^x$$

$$322.6 \times 10^{12} = (1.4)^x$$

$$\log (322.6 \times 10^{12}) = x \log 1.4$$

$$99 \approx x$$

Software exports will double their
1997 value in 1999.

65. (a) $\ln (1 - P) = -.0034 - .0053T$

Change this equation to exponential
form.

$$1 - P = e^{-.0034-.0053T}$$

$$P(T) = 1 - e^{-.0034-.0053T}$$

(b) See the answer graph in the back
of the textbook.
From the graph one can see that
initially there is a rapid reduction of carbon dioxide emissions.
However, after a while there is
little benefit in raising taxes
further.

(c) $P(T) = 1 - e^{-.0034-.0053T}$

$$P(60) = 1 - e^{-.0034-.0053(60)}$$

$$\approx .275 \text{ or } 27.5\%$$

The reduction in carbon emissions
from a tax of $60 per ton of carbon
is 27.5%.

(d) We must determine T when P =
.05.

$$P(T) = 1 - e^{-.0034-.0053T} = .5$$

$$.5 = 1 - e^{-.0034-.0053T}$$

$$.5 = e^{-.0034-.0053T}$$

$$\ln .5 = -.0034 - .0053T$$

$$T = \frac{\ln .5 + .0034}{-.0053} \approx 130.14$$

The value T = $130.14 will give a
50% reduction in carbon emissions.

Chapter 9   Review Exercises

1. $f(x) = \left(\frac{5}{4}\right)^x$ defines an *increasing* function since the base, 5/4, is larger than 1.

3. $y = \log_{.3} x$

    The point (1, 0) is on the graph of every function of the form $y = \log_a x$, so the correct choice must be either B or C.
    Since the base is a = .3 and $0 < .3 < 1$, $y = \log_{.3} x$ is a decreasing function, and so the correct choice must be B.

5. $y = \ln x$
    or
    $y = \log_e x$

    The point (1, 0) is on the graph of every function of the form $y = \log_a x$, so the correct choice must either B or C.
    Since the base is a = e and e > 1, $y = \ln x$ is an increasing function, and so the correct choice must be C.

7. $2^5 = 32$ is written in logarithmic form as
    $$\log_2 32 = 5.$$

9. $\left(\frac{1}{16}\right)^{1/4} = \frac{1}{2}$ is written in logarithmic form as
    $$\log_{1/16} \left(\frac{1}{2}\right) = \frac{1}{4}.$$

11. $10^{.4771} = 3$ is written in logarithmic form as
    $$\log_{10} 3 = .4771$$
    $$\text{or}\quad \log 3 = .4771.$$

15. The domain of f is $(-\infty, \infty)$.

17. From the graph, $f(0) = 1$.

19. Since $f(x) = a^x$, the expression for $f^{-1}(x)$ is $\log_a x$.
    $$f^{-1}(x) = \log_a x$$

21. $\log_2 \sqrt{32} = \frac{5}{2}$ is written in exponential form as $2^{5/2} = \sqrt{32}$.

23. $\ln 45 = 3.806662$ is written in exponential form as $e^{3.806662} = 45$.

27. Let $f(x) = a^x$ be the required function. Then
    $$f(-4) = \frac{1}{16}$$
    $$a^{-4} = \frac{1}{16}$$
    $$a^{-4} = 2^{-4}$$
    $$a = 2.$$

    The base is 2.

29. $\log_2 \left(\frac{\sqrt{7}}{15}\right)$

    $= \log_2 \sqrt{7} - \log_2 15$
        Logarithm of a quotient

    $= \left(\frac{1}{2}\right) \log_2 7 - \log_2 15$
        Logarithm of a power

31. $\log_7 (7k + 5r^2)$

Since this is the logarithm of a sum, this expression cannot be simplified.

33. $\log 45.6 = 1.659$

35. $\ln 470 = 6.153$

37. To find $\log_3 769$, use the change-of-base theorem.

$$\log_3 769 = \frac{\ln 769}{\ln 3}$$
$$\approx 6.049$$

39. $y = 2e^{.02t}$

(a) To find out how long it will take for the population to triple, you need to solve $2e^{.02t} = 3 \cdot 2$ for t. The correct choice is B.

(b) To find out when the population will reach 3 million, you need to solve $2e^{.02t} = 3$ for t. The correct choice is D.

(c) To find out how large the population will be in 3 years, you need to evaluate $2e^{.02(3)}$. The correct choice is C.

(d) To find out how large the population will be in 4 months, you need to evaluate $2e^{.02(1/3)}$. The correct choice is A.

41. In 220 years, there will be $\frac{220}{44} = 5$ doubling periods.
Therefore, the population will grow by a factor of 2^5 or 32.

43. $\frac{8}{27} = b^{-3}$

$$\left(\frac{2}{3}\right)^3 = \left(\frac{1}{b}\right)^3$$
$$\frac{2}{3} = \frac{1}{b}$$
$$\frac{3}{2} = b$$

The solution is 3/2.

45.
$$e^{p+1} = 10$$
$$\ln e^{p+1} = \ln 10$$
$$(p + 1) \ln e = \ln 10$$
$$p + 1 = \ln 10$$
$$p = \ln 10 - 1$$
$$= 2.3026 - 1$$
$$= 1.303$$

The solution is 1.303.

47. $\ln (6x) - \ln (x + 1) = \ln 4$

$$\ln \left(\frac{6x}{x + 1}\right) = \ln 4$$
Logarithm of a quotient

$$\frac{6x}{x + 1} = 4$$
$$4x + 4 = 6x$$
$$4 = 2x$$
$$2 = x$$

The solution is 2.

49. $\ln x + 3 \ln 2 = \ln \dfrac{2}{x}$

$\ln x + \ln 2^3 = \ln \dfrac{2}{x}$ *Logarithm of a power*

$\ln (x \cdot 2^3) = \ln \dfrac{2}{x}$ *Logarithm of a product*

$\ln 8x = \ln \dfrac{2}{x}$

Thus,

$$8x = \dfrac{2}{x}$$

$$8x^2 = 2$$

$$x^2 = \dfrac{1}{4}$$

$$x = \pm\dfrac{1}{2}.$$

$-\dfrac{1}{2}$ is not in the domain of $\ln x$ and

$\ln \dfrac{2}{x}$, so it is not a solution.

The solution is $1/2$.

51. $A = 8780, \ P = 3500, \ t = 10, \ m = 1$

$$A = P\left(1 + \dfrac{r}{m}\right)^{tm}$$

$$8780 = 3500\left(1 + \dfrac{r}{1}\right)^{10(1)}$$

$$2.50857 = (1 + r)^{10}$$

$$(2.50857)^{1/10} = 1 + r$$

$$(2.50857)^{1/10} - 1 = r$$

$$.09631 = r$$

The annual interest rate, to the nearest tenth, is 9.6%.

53. $A = P\left(1 + \dfrac{r}{m}\right)^{tm}$

First, substitute $P = 10,000$, $r = .12$, $t = 12$, and $m = 1$ into the formula.

$A = 10,000\left(1 + \dfrac{.12}{1}\right)^{12(1)}$

$= 10,000(1.12)^{12}$

$\approx 10,000(3.8960)$

$\approx 38,959.76$

After the first 12 yr, there would be about \$38,959.76 in the account. To finish off the 21-year period, substitute $P = 38,959.76$, $r = .10$, $t = 9$, and $m = 2$ into the original formula.

$A = 38,959.76\left(1 + \dfrac{.10}{2}\right)^{9(2)}$

$= 38,959.76(1.05)^{18}$

$\approx 38,959.76(2.4066)$

$\approx 93,761.31$

At the end of the 21-year period, about \$93,761.31 would be in the account.

55. $A(t) = (5 \times 10^{12})e^{-.04t}$

(a) Use $t = 1990 - 1970 = 20$.

$A(20) = (5 \times 10^{12})e^{-.04(20)}$

$= (5 \times 10^{12})e^{-.8}$

$\approx (5 \times 10^{12})(.4493)$

$= (5 \times .4493) \times 10^{12}$

$\approx 2.2 \times 10^{12}$

In 1990, about 2.2×10^{12} tons of coal were available, according to this formula.

(b) Solve $A(t) = \dfrac{1}{2}A(0)$ for t.

Note that $\dfrac{1}{2}A(0) = \dfrac{1}{2}(5 \times 10^{12})$.

$\dfrac{1}{2}(5 \times 10^{12}) = (5 \times 10^{12})e^{-.04t}$

$.5 = e^{-.04t}$

$$\ln .5 = -.04t$$

$$-.6931 \approx -.04t$$

$$17 \approx t$$

According to this formula, coal reserves were half of their 1970 levels 17 yr after 1970, in 1987.

57. Graph $Y_1 = x^2$ and $Y_2 = 2^x$ on the same screen. Since the range of both functions is $(0, \infty)$, there is no need to include Quadrants III and IV in the viewing window. A good choice for the window is $[-5, 5]$ by $[0, 20]$.
The graph shows that the two curves intersect in three points. Two of these points are $(2, 4)$ and $(4, 16)$, as given in the exercise. To find the coordinates of the third point, use the "intersect" option in the CALC menu. The calculator displays the coordinates

$$(-.7666647, .58777476).$$

59. Graph $Y_3 = 3^{x+4} - 27^{x+1}$ in the window $[0, 3]$ by $[-20, 60]$. See the answer graph in the back of the textbook.
The x-intercept is .5, which agrees with the x-value found in Exercise 58.

61. To graph

$$Y_1 = \log_2 x + \log_2 (x + 2) - 3,$$

use the change-of-base theorem with base 10 to write the function as

$$Y_1 = \frac{\log x}{\log 2} + \frac{\log (x + 2)}{\log 2} - 3.$$

63. **(a)–(d)** See the answer graphs in the back of the textbook.
Both (a) and (b) are increasing functions. The amount A should decrease as time progresses. Function (d) is equal to 100 rather than 350 when $t = 0$. Function (c) is initially 350 at $t = 0$ and then exponentially decreases to zero as time progresses. Function (c) best describes the function $A(t)$.

CHAPTER 9   TEST

[9.1]   Use the properties of exponents to simplify each of the following. Leave answers in exponential form, with positive exponents only.

1. $4^9 \cdot 4^4$

2. $\dfrac{(3^4)^2 \cdot 3^3}{3^{15}}$

Write a table of values for x and y that could be used to graph each equation.

[9.1]   3. $y = 2 + 3^x$ [9.2]   4. $y = -1 + \log_3 x$

[9.2]   Write each equation in logarithmic form.

5. $4^{-1} = \dfrac{1}{4}$

6. $25^{1/2} = 5$

[9.2]   Write each equation in exponential form.

7. $\log_3 243 = 5$

8. $\log_{16} 2 = .25$

[9.2]   Use the properties of logarithms to simplify each expression.

9. $\log_4 10 + \log_4 3$

10. $\dfrac{1}{4} \log_6 41$

11. $5 \log_a 2 - \log_a 8$

12. $\log_3 x - 4 \log_3 y$

[9.3]   Use a calculator to find each logarithm. Give answers to four decimal places.

13. $\log 13.7$

14. $\log 56{,}700$

15. $\ln 4.35$

16. $\ln .007$

Solve each problem.

[9.1] 17. A population of rodents is growing according to the formula

$$y = 150(2)^{.09x},$$

where x is the time in months and y is the number of rodents. Find the number of rodents present after 15 months.

[9.4] 18. The average number of children per family in a particular culture is decreasing over time according to the formula

$$y = \log_{1/2} x + 5,$$

where x is the number of years from the present and y is the average number of children per family. Find the number of years from now, to the nearest tenth of a year, when the average number of children per family will be 2.5.

[9.4] Solve each equation. Use logarithms as necessary. Give answers to the nearest thousandth.

19. $3^{2x-1} = 42$ 20. $\log_x 175 = 5$

[9.3] Use common logarithms or natural logarithms and a calculator to find each logarithm to four decimal places.

21. $\log_5 36$ 22. $\log_{.2} 1.7$

[9.4] Solve each equation without using a calculator.

23. $729^x = 27$ 24. $5^{3x+1} = 625^x$

[9.4] 25. What values of x could not possibly be solutions of the following equation? (Do not solve.)

$$\log (2x - 3) + \log x^2 = 0$$

CHAPTER 9   TEST ANSWERS

1. 4^{13}

2. $\dfrac{1}{3^4}$

3.

x	-2	-1	0	1	2
$y = 2 + 3^x$	$\dfrac{19}{9}$	$\dfrac{7}{3}$	3	5	11

4.

$x = 3^{y+1}$	$\dfrac{1}{9}$	$\dfrac{1}{3}$	1	3	9
y	-3	-2	-1	0	1

5. $\log_4 \dfrac{1}{4} = -1$

6. $\log_{25} 5 = \dfrac{1}{2}$

7. $3^5 = 243$

8. $16^{.25} = 2$

9. $\log_4 30$

10. $\log_6 41^{1/4}$

11. $\log_a 4$

12. $\log_3 \dfrac{x}{y^4}$

13. 1.1367

14. 4.7536

15. 1.4702

16. -4.9618

17. 382 rodents

18. 5.7 yr

19. 2.2011

20. 2.8094

21. 2.2266

22. -.3297

23. 1/2

24. 1

25. Any x less than or equal to 3/2

CUMULATIVE REVIEW EXERCISES (Chapters 1-9)

1. Find the domain of $y = \dfrac{2x + 3}{\sqrt{4 - 5x}}$.

2. Find the angle of smallest positive measure coterminal with the angle $-1234°$.

3. Find the measures of the marked angles. Assume lines m and n are parallel.

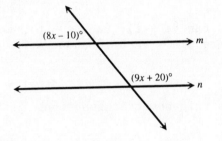

4. If the terminal side of an angle α is in quadrant IV, what is the sign of each of the trigonometric function values of α?

5. Find $\cot \beta$, if $\csc \beta = 5$ and β is in the second quadrant.

6. Find the values of the six trigonometric functions for angle A in the right triangle.

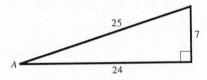

7. Without using a calculator, evaluate $\tan^2 210° + \sin^2 330° - \cos 180°$.

8. A television camera is to be mounted on a bank wall so as to have a good view of the head teller. (See the figure.) Find the angle of depression that the lens should make with the horizontal.

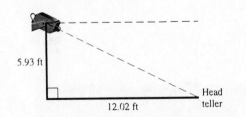

9. A phonograph record turns at the rate of 33 1/3 revolutions per min. How fast is the needle moving when the record begins playing, at which time the needle is about 6 in from the center of the record? When the needle is less than 6 in from the center, is the needle moving faster or slower?

10. Graph the function y = sin 2x over a two–period interval.

11. Graph the function $y = .5 \cos \left(4x - \frac{\pi}{2}\right)$ over a one–period interval.

12. Use the fundamental identities to simplify the expression $\frac{1}{\cot^2 \beta} + \tan \beta \cot \beta$ to a constant, a single circular function, or a power of a circular function.

13. Find cos (s + t) and cos (s – t), where cos s = –15/17 and sin t = 4/5, s in quadrant II and t in quadrant I.

14. Find sin (s + t), sin (s – t), tan (s + t), tan (s – t), the quadrant of s + t, and the quadrant of s – t, where sin s = –8/17 and cos t = –8/17, s and t in quadrant III.

15. Write $\frac{\tan 20°}{1 - \tan^2 20°}$ as a single trigonometric function.

16. Use a half–angle identity to find cos 75°.

17. Given cos β = 5/7, with β in quadrant IV, find cos β/2, sin β/2, and tan β/2.

18. Without using a calculator, determine the value of $\cos \left(\arcsin \frac{12}{13}\right)$.

19. Find all solutions to 4 sin² x tan x = tan x in the interval [0, 2π).

20. Find all solutions to 2 sin² 2θ + 5 sin 2θ = –2 in the interval [0°, 360°).

21. Solve $\arctan x = \arcsin \frac{7}{25}$.

22. A ship sailing parallel to shore sights a lighthouse at an angle of 30° from its direction of travel. (See the figure.) After traveling 2 mi farther, the angle has increased to 60°. At that time, how far is the ship from the lighthouse?

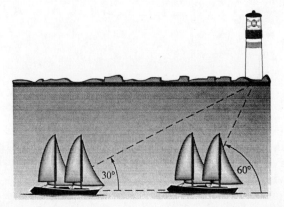

23. A surveyor standing 48.0 m from the base of a building measures the angle to the top of the building and finds it to be 37.4°. (See the figure.) The surveyor then measures the angle to the top of a clock tower on the building, finding that it is 45.6°. Find the height of the clock tower.

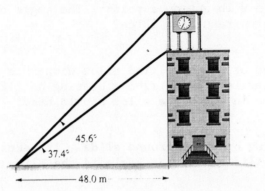

24. Determine the number of triangles possible with a = 9, b = 10, and c = 75°.

25. Find the unknown angles in the triangle ABC where A = 30.2°, b = 20.3 ft, and c = 13.5 ft.

26. To find the distance between two small towns, an Electronic Distance Measuring (EDM) instrument is placed on a hill from which both towns are visible. The distance to each town from the EDM and the angle between the two lines of sight are measured. (See the figure.) Find the distance between the towns.

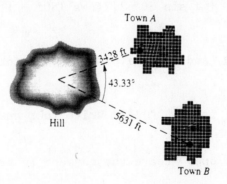

27. Three cylindrical rods are soldered together as shown in the figure. Find angle θ.

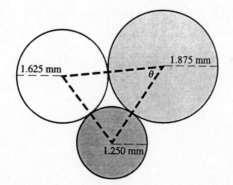

28. The vector **v** has magnitude 14 and an inclination of 25° from the horizontal. Find the magnitude of the horizontal and vertical components of **v**.

29. Forces of 12.5 lb and 5.0 lb act on a point. The angle between them is 135°. Find the magnitude of the resultant force.

30. A plane has an airspeed of 520 mph. The pilot wishes to fly on a bearing of 310.0°. A 37.0 mph wind is blowing from a bearing of 212.0°. Find the ground-speed and the direction in which the pilot should head.

31. A 20-lb child is sitting on a playground slide. It takes an 11-lb force to hold the child on the slide. At what angle is the slide inclined?

32. Simplify $\sqrt{-\dfrac{16}{25}}$.

33. Find the sum, product, and quotient of $2 + 3i$ and $3 - 2i$.

34. Write the complex number $-\dfrac{5}{2}\sqrt{3} - \dfrac{5}{2}i$ in trigonometric form $r(\cos\theta + i\sin\theta)$, with θ in the interval $[0°, 360°]$.

35. Find the product $[6(\cos 30° + i \sin 30°)][7(\cos 150° + i \sin 150°)]$.

36. Find the quotient $\dfrac{8 \text{ cis } 85°}{2 \text{ cis } 55°}$, and write it in standard form.

37. Find the power $[2(\cos 30° + i \sin 30°)]^5$. Write your answer in standard form.

38. Find the six sixth roots of one. Leave the answers in trigonometric form.

39. Graph $r = 2 \sin 3\theta$, for θ in $[0°, 360°)$.

40. Find an equivalent equation in polar coordinates for $3x - 2y = 4$.

41. Use a table of values to graph the plane curve defined by the parametric equations $x = t + 4$, $y = t^3$, for t in $[-1, 1]$. Find a rectangular equation for the curve.

42. Find a rectangular equation for the curve defined parametrically by $x = e^{2t}$, $y = e^{-t}$, for t in $(-\infty, \infty)$, and graph the curve.

43. Solve $81^x = 3^{2-x}$.

44. Find the present value of \$20,000 if interest is 4% compounded quarterly for 2 yr.

45. Find $5^{\log_5 12}$.

46. Write $\log_2 \dfrac{3\sqrt{2}}{4}$ as a sum, difference, or product of logarithms. Simplify the result if possible.

47. Evaluate $\dfrac{\log_2 25}{\log_2 5}$.

48. Use natural logarithms to find $\log_3 8$.

49. Solve $\log (2 + 3x) - \log (4 + x) = \log 2$.

50. Solve $5e^{2x-3} = 10$.

SOLUTIONS TO CUMULATIVE REVIEW EXERCISES (Chapters 1–9)

1. $y = \dfrac{2x + 3}{\sqrt{4 - 5x}}$

    Since the quantity under the radical in the denominator must be positive,

    $$4 - 5x > 0$$
    $$-5x > -4$$
    $$x < \frac{4}{5}.$$

    The domain is $(-\infty, 4/5)$.

2. $-1234° + 4(360°) = -1234° + 1440°$
    $$= 206°$$

    The smallest positive angle coterminal with $-1234°$ is $206°$.

3. The two marked angles are supplementary. Thus,

    $$(8x - 10) + (9x + 20) = 180$$
    $$17x + 10 = 180$$
    $$17x = 170$$
    $$x = 10.$$

    If $x = 10$, then

    $$8x - 10 = 8(10) - 10 = 70.$$

    If $x = 10$, then

    $$9x + 20 = 9(10) + 20 = 110.$$
    The measures of the two angles are $70°$ and $110°$.

4. If the terminal side of α is in quadrant IV, then $\cos \alpha$ and $\sec \alpha$ are positive and all of the other trigonometric functions are negative.

5. $\csc \beta = 5$, β in quadrant II

    $$1 + \cot^2 \beta = \csc^2 \beta$$
    $$1 + \cot^2 \beta = 5^2$$
    $$\cot^2 \beta = 24$$
    $$\cot \beta = \pm\sqrt{24} = \pm 2\sqrt{6}$$

    Since β is in quadrant II, $\cot \beta = -2\sqrt{6}$.

6. $\sin A = \dfrac{\text{side opposite}}{\text{hypotenuse}} = \dfrac{7}{25}$

    $\cos A = \dfrac{\text{side adjacent}}{\text{hypotenuse}} = \dfrac{24}{25}$

    $\tan A = \dfrac{\text{side opposite}}{\text{side adjacent}} = \dfrac{7}{24}$

    $\cot A = \dfrac{\text{side adjacent}}{\text{side opposite}} = \dfrac{24}{7}$

    $\sec A = \dfrac{\text{hypotenuse}}{\text{side adjacent}} = \dfrac{25}{24}$

    $\csc A = \dfrac{\text{hypotenuse}}{\text{side opposite}} = \dfrac{25}{7}$

7. $\tan^2 210° + \sin^2 330° - \cos 180°$

    $$= \left(\frac{\sqrt{3}}{3}\right)^2 + \left(-\frac{1}{2}\right)^2 - (-1)$$
    $$= \frac{3}{9} + \frac{1}{4} + 1$$
    $$= \frac{1}{3} + \frac{1}{4} + 1$$
    $$= \frac{4}{12} + \frac{3}{12} + \frac{12}{12}$$
    $$= \frac{19}{12}$$

8. Let θ be the angle of depression.

    $$\tan \theta = \frac{5.93}{12.02}$$
    $$\tan \theta \approx .4933$$
    $$\theta \approx 26.259°$$

    The angle of depression is about $26.3°$.

9. $33\frac{1}{3}$ revolutions per min = $\left(33\frac{1}{3}\right)(2\pi)$ radians per min

$$\left(33\frac{1}{3}\right)(2\pi) = \frac{100}{3}(2\pi)$$

$$= \frac{200\pi}{3} \text{ radians per min}$$

$v = r\omega$

$v = (6 \text{ in})\left(\frac{200\pi}{3} \text{ radians per min}\right)$

$v = 400\pi$ in per min

$v \approx 1260$ in per min

The needle is moving at 1260 in per min.

If ω is constant, as r decreases in $v = r\omega$, v also decreases. When the needle is less than 6 in from the center, it is moving slower.

10. $y = \sin 2x$

The amplitude is 1.
The period is $2\pi/2 = \pi$.
Graph a sine function with an amplitude of 1 and a period of π.

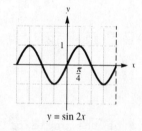

$y = \sin 2x$

11. $y = .5 \cos\left(4x - \frac{\pi}{2}\right)$

$y = .5 \cos 4\left(x - \frac{\pi}{8}\right)$

The amplitude is .5.
The period is $2\pi/4 = \pi/2$.

The phase shift is $\pi/8$ units to the right.
Graph a cosine function with a period of $\pi/2$ and an amplitude of .5, then translate the graph $\pi/8$ units to the right.

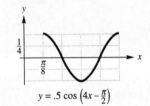

$y = .5 \cos\left(4x - \frac{\pi}{2}\right)$

12. $\dfrac{1}{\cot^2 \beta} + \tan \beta \cot \beta$

$= \tan^2 \beta + \tan \beta\left(\dfrac{1}{\tan \beta}\right)$

$= \tan^2 \beta + 1$

$= \sec^2 \beta$

13. $\cos s = -\dfrac{15}{17}$, s in quadrant II;

$\sin t = \dfrac{4}{5}$, t in quadrant I

$\sin^2 s = 1 - \cos^2 s$

$= 1 - \left(-\dfrac{15}{17}\right)^2$

$= \dfrac{64}{289}$

$\sin s = \pm\sqrt{\dfrac{64}{289}} = \pm\dfrac{8}{17}$

Since s is in quadrant II, $\sin s = \dfrac{8}{17}$.

$\cos^2 t = 1 - \sin^2 t$

$= 1 - \left(\dfrac{4}{5}\right)^2$

$= \dfrac{9}{25}$

$$\cos t = \pm\sqrt{\frac{9}{25}} = \pm\frac{3}{5}$$

Since t is in quadrant I, $\cos t = \frac{3}{5}$.

$\cos (s + t)$

$= \cos s \cos t - \sin s \sin t$

$= \left(-\frac{15}{17}\right)\left(\frac{3}{5}\right) - \left(\frac{8}{17}\right)\left(\frac{4}{5}\right)$

$= -\frac{45}{85} - \frac{32}{85}$

$= -\frac{77}{85}$

$\cos (s - t)$

$= \cos s \cos t + \sin s \sin t$

$= \left(-\frac{15}{17}\right)\left(\frac{3}{5}\right) + \left(\frac{8}{17}\right)\left(\frac{4}{5}\right)$

$= -\frac{45}{85} + \frac{32}{85}$

$= -\frac{13}{85}$

14. $\sin s = -\frac{8}{17}$, $\cos t = -\frac{8}{17}$, s and t in quadrant III

$\cos^2 s = 1 - \sin^2 s$

$= 1 - \left(-\frac{8}{17}\right)^2$

$= \frac{225}{289}$

$\cos s = \pm\sqrt{\frac{225}{289}} = \pm\frac{15}{17}$

$\cos s = -\frac{15}{17}$ since s is in quadrant III.

$\sin^2 t = 1 - \cos^2 t$

$= 1 - \left(-\frac{8}{17}\right)^2$

$= \frac{225}{289}$

$\sin t = \pm\sqrt{\frac{225}{289}} = \pm\frac{15}{17}$

$\sin t = -\frac{15}{17}$ since t is in quadrant III.

$\tan s = \frac{\sin s}{\cos s} = \frac{8}{15}$

$\tan t = \frac{\sin t}{\cos t} = \frac{15}{8}$

$\sin (s + t)$

$= \sin s \cos t + \cos s \sin t$

$= \left(-\frac{8}{17}\right)\left(-\frac{8}{17}\right) + \left(-\frac{15}{17}\right)\left(-\frac{15}{17}\right)$

$= \frac{64}{289} + \frac{225}{289}$

$= 1$

$\sin (s - t)$

$= \sin s \cos t - \cos s \sin t$

$= \left(-\frac{8}{17}\right)\left(-\frac{8}{17}\right) + \left(-\frac{15}{17}\right)\left(-\frac{15}{17}\right)$

$= \frac{64}{289} - \frac{225}{289}$

$= -\frac{161}{289}$

$\tan (s + t) = \frac{\tan s + \tan t}{1 - \tan s \tan t}$

$= \frac{\frac{8}{15} + \frac{15}{8}}{1 - \left(\frac{8}{15}\right)\left(\frac{15}{8}\right)}$

$= \frac{\frac{8}{15} + \frac{15}{8}}{0}$

$\tan (s + t)$ is undefined.

$\tan (s - t) = \frac{\tan s - \tan t}{1 + \tan s \tan t}$

$= \frac{\frac{8}{15} - \frac{15}{8}}{1 + \left(\frac{8}{15}\right)\left(\frac{15}{8}\right)}$

$= \frac{\frac{64}{120} - \frac{225}{120}}{2}$

$= \frac{-\frac{161}{120}}{2} = -\frac{161}{240}$

Since tan (s + t) is undefined,
s + t is a quandrantal angle.
Since sin (s – t) is negative and
tan (s – t) is negative, s – t is
in quadrant IV.

15. $\dfrac{\tan 20°}{1 - \tan^2 20°} = \dfrac{1}{2}\left(\dfrac{2\tan 20°}{1 - \tan^2 20°}\right)$

$\qquad\qquad = \dfrac{1}{2}\tan[2(20°)]$

$\qquad\qquad = \dfrac{1}{2}\tan 40°$

16. This is a quadrant I angle.

$\cos 75° = \cos\left(\dfrac{150°}{2}\right)$

$\qquad = \sqrt{\dfrac{1 + \cos 150°}{2}}$

$\qquad = \sqrt{\dfrac{1 + \left(-\dfrac{\sqrt{3}}{2}\right)}{2}}$

$\qquad = \sqrt{\dfrac{2 - \sqrt{3}}{4}}$

$\qquad = \dfrac{1}{2}\sqrt{2 - \sqrt{3}}$

17. $\cos \beta = \dfrac{5}{7}$, β in quadrant IV

Since β is in quadrant IV,

$$\dfrac{3\pi}{2} < \beta < 2\pi$$

and

$$\dfrac{3\pi}{4} < \dfrac{\beta}{2} < \pi.$$

Therefore, $\dfrac{\beta}{2}$ is in quadrant II;
cosine and tangent are negative, and
sine is positive.

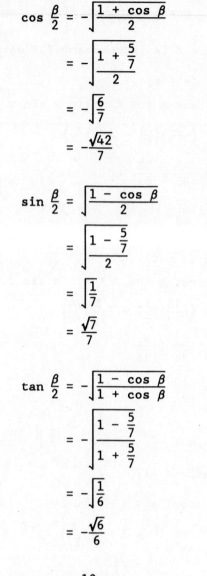

$\cos \dfrac{\beta}{2} = -\sqrt{\dfrac{1 + \cos \beta}{2}}$

$\qquad = -\sqrt{\dfrac{1 + \dfrac{5}{7}}{2}}$

$\qquad = -\sqrt{\dfrac{6}{7}}$

$\qquad = -\dfrac{\sqrt{42}}{7}$

$\sin \dfrac{\beta}{2} = \sqrt{\dfrac{1 - \cos \beta}{2}}$

$\qquad = \sqrt{\dfrac{1 - \dfrac{5}{7}}{2}}$

$\qquad = \sqrt{\dfrac{1}{7}}$

$\qquad = \dfrac{\sqrt{7}}{7}$

$\tan \dfrac{\beta}{2} = -\sqrt{\dfrac{1 - \cos \beta}{1 + \cos \beta}}$

$\qquad = -\sqrt{\dfrac{1 - \dfrac{5}{7}}{1 + \dfrac{5}{7}}}$

$\qquad = -\sqrt{\dfrac{1}{6}}$

$\qquad = -\dfrac{\sqrt{6}}{6}$

18. $\cos\left(\arcsin \dfrac{12}{13}\right)$

Let $\theta = \arcsin \dfrac{12}{13}$.   Then $\sin \theta = \dfrac{12}{13}$,
and θ is in quadrant I.

$\cos^2 \theta = 1 - \sin^2 \theta$

$\qquad = 1 - \left(\dfrac{12}{13}\right)^2$

$\qquad = \dfrac{169}{169} - \dfrac{144}{169}$

$\qquad = \dfrac{25}{169}$

$\cos \theta = \pm\sqrt{\dfrac{25}{169}} = \pm\dfrac{5}{13}$

$\cos \theta = \dfrac{5}{13}$ since θ is in quadrant I.

$\cos \left(\arcsin \dfrac{12}{13}\right) = \cos \theta = \dfrac{5}{13}$

19.
$$4 \sin^2 x \tan x = \tan x$$
$$4 \sin^2 x \tan x - \tan x = 0$$
$$\tan x (4 \sin^2 x - 1) = 0$$
$$\tan x = 0$$
$$x = 0, \pi$$

or

$$4 \sin^2 x - 1 = 0$$
$$\sin^2 x = \dfrac{1}{4}$$
$$\sin x = \pm\dfrac{1}{2}$$
$$x = \dfrac{\pi}{6}, \dfrac{5\pi}{6}, \dfrac{7\pi}{6}, \dfrac{11\pi}{6}$$

$$x = 0, \dfrac{\pi}{6}, \dfrac{5\pi}{6}, \pi, \dfrac{7\pi}{6}, \dfrac{11\pi}{6}$$

20.
$$2 \sin^2 2\theta + 5 \sin 2\theta = 2$$
$$2 \sin^2 2\theta + 5 \sin 2\theta + 2 = 0$$
$$(2 \sin 2\theta + 1)(\sin 2\theta + 2) = 0$$
$$2 \sin 2\theta + 1 = 0$$
$$\sin 2\theta = -\dfrac{1}{2}$$

or

$$\sin 2\theta + 2 = 0$$
$$\sin 2\theta = -2$$

The equation $\sin 2\theta = -2$ has no solution since $-1 \le \sin 2\theta \le 1$.

If $\sin 2\theta = -\dfrac{1}{2}$, then

$$2\theta = 210°, 330°, 570°, \text{ or } 690°$$
and $\quad \theta = 105°, 165°, 285°, \text{ or } 345°.$

21.
$$\arctan x = \arcsin \dfrac{7}{25}$$
$$\tan (\arctan x) = \tan \left(\arcsin \dfrac{7}{25}\right)$$
$$x = \tan \left(\arcsin \dfrac{7}{25}\right)$$

Let $\omega = \arcsin \dfrac{7}{25}$. Then $\sin \omega = \dfrac{7}{25}$, and ω is in quadrant I.

$$\cos^2 \omega = 1 - \sin^2 \omega$$
$$= 1 - \left(\dfrac{7}{25}\right)^2$$
$$= \dfrac{576}{625}$$
$$\cos \omega = \pm\sqrt{\dfrac{576}{625}} = \pm\dfrac{24}{25}$$

$\cos \omega = \dfrac{24}{25}$ since ω is in quadrant I.

$$\tan \omega = \dfrac{\sin \omega}{\cos \omega} = \dfrac{\dfrac{7}{25}}{\dfrac{24}{25}} = \dfrac{7}{24}$$

$$x = \tan \left(\arcsin \dfrac{7}{25}\right)$$
$$x = \tan \omega$$
$$x = \dfrac{7}{24}$$

22. In the triangle below, the distance from the ship to the lighthouse is a.

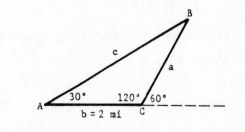

$B = 180° - (A + C) = 180° - 150°$
$$= 30°$$

Since $A = B$, the triangle is isosceles and $a = b$.

Therefore, the distance is 2 mi.

23. In the triangle below, x is the height of the clock tower.

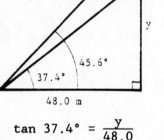

45.6°
37.4°
48.0 m

$\tan 37.4° = \dfrac{y}{48.0}$

$y = 48.0 \tan 37.4°$

$\tan 45.6° = \dfrac{x + y}{48.0}$

$x + y = 48.0 \tan 45.6°$

Substitute $48.0 \tan 37.4°$ for y in this equation.

$x + 48.0 \tan 37.4° = 48.0 \tan 45.6°$

$x = 48.0 \tan 45.6° - 48.0 \tan 37.4°$

$x = 48.0(\tan 45.6° - \tan 37.4°)$

$x \approx 12.317215$

The height of the clock tower is about 12.3 m.

24. Since $a = 9$, $b = 10$, and $c = 75°$, we have two sides and an included angle. Therefore, one triangle is possible.

25. $A = 30.2°$, $b = 20.3$ ft, $c = 13.5$ ft

$a^2 = b^2 + c^2 - 2bc \cos A$

$a^2 = (20.3)^2 + (13.5)^2$
$\qquad - 2(20.3)(13.5) \cos 30.2°$

$a^2 = 120.630981$

$a = 10.9832136$

$\dfrac{\sin B}{b} = \dfrac{\sin A}{a}$

$\sin B = \dfrac{b \sin A}{a}$

$\sin B = \dfrac{20.3 \sin 30.2°}{10.9832136}$

$\sin B = .92971923$

$B = 68.4°$

$C = 180° - (A + B)$

$\quad = 180° - (30.2° + 68.4°)$

$\quad = 180° - 98.6°$

$C = 81.4°$

26. Let x = distance from Town A to Town B.

Using the law of cosines,

$x^2 = (3428)^2 + (5631)^2$
$\qquad - 2(3428)(5631) \cos 43.33°$

$x^2 = 15,376,718.01$

$x = 3921.315852.$

The distance is about 3921 ft.

27. The sides of the triangle have the following lengths:

$$1.625 + 1.875 = 3.500,$$
$$1.875 + 1.250 = 3.125,$$
$$\text{and } 1.625 + 1.250 = 2.875.$$

Using the law of cosines,

$$(2.875)^2 = (3.500)^2 + (3.125)^2$$
$$- 2(3.500)(3.125) \cos \theta$$

$$\cos \theta = \frac{(3.500)^2 + (3.125)^2 - (2.875)^2}{2(3.500)(3.125)}$$

$$\cos \theta = .6285714286$$

$$\theta = 51.06°.$$

28. Let **u** be the horizontal component.

$$|\mathbf{u}| = |\mathbf{v}| \cos 25° = 14 \cos 25°$$
$$\approx 12.688309$$

The horizontal component has a magnitude of 13.

Let **w** be the vertical component.

$$|\mathbf{w}| = |\mathbf{v}| \sin 25° = 14 \sin 25°$$
$$\approx 5.9166557$$

The vertical component has a magnitude of 5.9.

29. Let **v** be the resultant force.

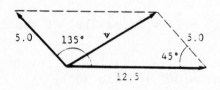

Apply the law of cosines.

$$|\mathbf{v}|^2 = (12.5)^2 + (5.0)^2$$
$$- 2(12.5)(5.0) \cos 45°$$
$$|\mathbf{v}|^2 = 92.861652$$
$$|\mathbf{v}| = 9.6364751$$

The magnitude of the resultant force is about 9.6 lb.

30. Let **v** be the resultant of the airspeed and windspeed vectors.

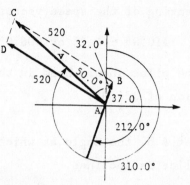

The groundspeed is $|\mathbf{v}|$. The direction in which the pilot should head will be the bearing for the airspeed vector.

Note in the figure that

$$50.0° + 32.0° = 82.0°.$$

Use the law of sines with triangle ABC.

$$\frac{\sin C}{37.0} = \frac{\sin 82.0°}{520}$$

$$\sin C = \frac{37.0 \sin 82.0°}{520}$$

$$\sin C = .0704613818$$

$$C = 4.0°$$

$$B = 180° - (A + C)$$
$$= 180° - (82.0° + 4.0°)$$
$$= 94.0°$$

Use the law of cosines to find $|\mathbf{v}|$.

$$|\mathbf{v}|^2 = (520)^2 + (37.0)^2$$
$$- 2(520)(37.0) \cos 94.0°$$
$$|\mathbf{v}|^2 = 274,453.2291$$
$$|\mathbf{v}| = 523.8828391$$

The groundspeed is 524 mph.

Since angle C in triangle ABC is 4.0°, angle CAD is 4.0° and the bearing of the speed vector is

    310.0° – 4.0° or 306.0°.

The pilot should head in the direction of 306.0°.

31. Let θ be the angle at which the slide is inclined.

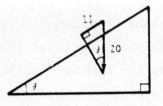

$$\sin \theta = \frac{11}{20}$$

$$\sin \theta = .55$$

$$\theta = 33.367013°$$

The slide is inclined at an angle of 33°.

32. $\sqrt{-\frac{16}{25}} = i\sqrt{\frac{16}{25}} = i\left(\frac{4}{5}\right) = \frac{4}{5}i$

33. $(2 + 3i) + (3 - 2i)$

$= (2 + 3) + (3 - 2)i$

$= 5 + i$

$(2 + 3i)(3 - 2i)$

$= 6 - 4i + 9i - 6i^2$

$= 6 + 5i - 6(-1)$

$= 6 + 5i + 6$

$= 12 + 5i$

$$\frac{2 + 3i}{3 - 2i} = \frac{(2 + 3i)}{(3 - 2i)} \cdot \frac{(3 + 2i)}{(3 + 2i)}$$

$$= \frac{6 + 4i + 9i + 6i^2}{9 - 4i^2}$$

$$= \frac{6 + 13i + 6(-1)}{9 - 4(-1)}$$

$$= \frac{13i}{13}$$

$$= i$$

34. $-\frac{5}{2}\sqrt{3} - \frac{5}{2}i$

$$r = \sqrt{\left(-\frac{5}{2}\sqrt{3}\right)^2 + \left(-\frac{5}{2}\right)^2}$$

$$= \sqrt{\frac{75}{4} + \frac{25}{4}}$$

$$= \sqrt{\frac{100}{4}}$$

$$= \sqrt{25}$$

$$r = 5$$

$$\cos \theta = \frac{-\frac{5}{2}\sqrt{3}}{5} \quad \text{and} \quad \sin \theta = \frac{-\frac{5}{2}}{5}$$

$$\cos \theta = -\frac{\sqrt{3}}{2} \quad \text{and} \quad \sin \theta = -\frac{1}{2}$$

Therefore, $\theta = 210°$ and

$$-\frac{5}{2}\sqrt{3} - \frac{5}{2}i = 5(\cos 210° + i \sin 210°).$$

35. $[6(\cos 30° + i \sin 30°)]$

$\cdot [7(\cos 150° + i \sin 150°)]$

$= (6)(7)[\cos (30° + 150°)$

$+ i \sin (30° + 150°)]$

$= 42(\cos 180° + i \sin 180°)$

$= 42(-1 + 0i)$

$= -42$

36. $\dfrac{8 \text{ cis } 85°}{2 \text{ cis } 55°} = \dfrac{8}{2} \text{ cis } (85° - 55°)$

$= 4 \text{ cis } 30°$

$= 4(\cos 30° + i \sin 30°)$

$= 4\left(\dfrac{\sqrt{3}}{2} + \dfrac{1}{2}i\right)$

$= 2\sqrt{3} + 2i$

37. $[2(\cos 30° + i \sin 30°)]^5$

$= 2^5[\cos (5 \cdot 30°) + i \sin (5 \cdot 30°)]$

$= 32(\cos 150° + i \sin 150°)$

$= 32(-\frac{\sqrt{3}}{2} + \frac{1}{2}i)$

$= -16\sqrt{3} + 16i$

38. $1 = 1(\cos 0° + i \sin 0°)$

Thus, $r = 1$ and $\theta = 0°$.

$\sqrt[6]{1} = 1$

$\alpha = \frac{0°}{6} + \frac{360° \cdot k}{6}$ for k = 0, 1, 2, 3, 4, 5

k = 0: $\alpha = 0°$

k = 1: $\alpha = 60°$

k = 2: $\alpha = 120°$

k = 3: $\alpha = 180°$

k = 4: $\alpha = 240°$

k = 5: $\alpha = 300°$

The six sixth roots of one are cis 0°, cis 60°, cis 120°, cis 180°, cis 240°, and cis 300°.

39. $r = 2 \sin 3\theta$

θ	0°	15°	30°	45°	60°
3θ	0°	45°	90°	135°	180°
$2 \sin 3\theta$	0	1.4	2	1.4	0

θ	75°	90°	105°	120°
3θ	225°	270°	315°	360°
$2 \sin 3\theta$	-1.4	-2	-1.4	0

θ	135°	150°	165°	180°
3θ	405°	450°	495°	540°
$2 \sin 3\theta$	1.4	2	1.4	0

The graph is a three-leaved rose.

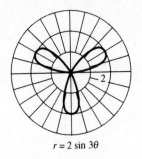

$r = 2 \sin 3\theta$

40.
$3x - 2y = 4$

$3r \cos \theta - 2r \sin \theta = 4$

$r(3 \cos \theta - 2 \sin \theta) = 4$

$r = \dfrac{4}{3 \cos \theta - 2 \sin \theta}$

41. $x = t + 4$, $y = t^3$, for t in [-1, 1]

t	-1	-.5	0	.5	1
x	3	3.5	4	4.5	5
y	-1	-.125	0	.125	1

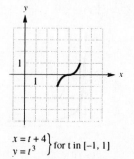

$\left.\begin{array}{l} x = t + 4 \\ y = t^3 \end{array}\right\}$ for t in [-1, 1]

If $x = t + 4$, then $t = x - 4$. Since $y = t^3$,

$$y = (x - 4)^3$$

for x in [-1 + 4, 1 + 4] or [3, 5].

42. $x = e^{2t}$, $y = e^{-t}$, for t in $(-\infty, \infty)$

Notice that both x and y are always positive.

$x = (e^t)^2$ and $y = \dfrac{1}{e^t}$

Since $y = \frac{1}{e^t}$, $e^t = \frac{1}{y}$ and $x = \left(\frac{1}{y}\right)^2$.

$$x = \frac{1}{y^2}$$

$$xy^2 = 1$$

$$y^2 = \frac{1}{x}$$

$$y = \pm\sqrt{\frac{1}{x}}$$

Since y is always positive,

$$y = \frac{1}{\sqrt{x}}.$$

t	-3	-1	-.5	0
$x = e^{2t}$	.002	.135	.368	1
$y = e^{-t}$	20.086	2.718	1.649	1

t	.5	1	3
$x = e^{2t}$	2.718	7.389	403.429
$y = e^{-t}$	.606	.368	.050

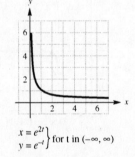

$\left.\begin{array}{l} x = e^{2t} \\ y = e^{-t} \end{array}\right\}$ for t in $(-\infty, \infty)$

43. $81^x = 3^{2-x}$

$(3^4)^x = 3^{2-x}$

$3^{4x} = 3^{2-x}$

$4x = 2 - x$

$5x = 2$

$x = \frac{2}{5}$

44.   Use $A = P\left(1 + \frac{r}{n}\right)^{nt}$ with A = 20,000,

r = .04, n = 4, and t = 2.

$$20,000 = P\left(1 + \frac{.04}{4}\right)^{4(2)}$$

$$P = \frac{20,000}{(1.01)^8}$$

$$P = 18,469.66445$$

The present value is \$18,469.66.

45.   Since $b^{\log_b x} = x$, $5^{\log_5 12} = 12$.

46. $\log_2 \frac{3\sqrt{2}}{4} = \log_2 (3\sqrt{2}) - \log_2 4$

$= \log_2 3 + \log_2 2^{1/2}$

$\qquad - \log_2 2^2$

$= \log_2 3 + \frac{1}{2} \log_2 2$

$\qquad - 2 \log_2 2$

$= \log_2 3 + \frac{1}{2}(1) - 2(1)$

$= \log_2 3 - \frac{3}{2}$

47. $\dfrac{\log_2 25}{\log_2 5} = \dfrac{\log_2 5^2}{\log_2 5}$

$= \dfrac{2 \log_2 5}{\log_2 5}$

$= 2$

48. $\log_3 8 = \dfrac{\ln 8}{\ln 3} = 1.8927893$

49. $\log (2 + 3x) - \log (4 + x) = \log 2$

$\log \left(\dfrac{2 + 3x}{4 + x}\right) = \log 2$

$\dfrac{2 + 3x}{4 + x} = 2$

$2 + 3x = 8 + 2x$

$x = 6$

50. $5e^{2x-3} = 10$

                        $e^{2x-3} = 2$

              $\ln e^{2x-3} = \ln 2$

       $(2x - 3) \ln e = \ln 2$

         $(2x - 3) \cdot 1 = \ln 2$

                           $2x = 3 + \ln 2$

                             $x = \frac{1}{2}(3 + \ln 2)$

or $x = .5(3 + \ln 2)$